SPECIAL RELATIVITY
from Einstein to Strings

The traditional undergraduate physics treatment of special relativity is too cursory to warrant a textbook. The graduate treatment of special relativity is deeper, but often fragmented between different courses such as general relativity and quantum field theory. For this reason physics students need one book that ties it all together. With this in mind, this book is written as a textbook for the self-learner whose physics background includes a minimum of one year of university physics with calculus. More advanced mathematical topics, such as group theory, are explained as they arise. The readership is expected to include high school and college physics educators seeking to improve and update their own understanding of special relativity in order that they may teach it better, science and engineering undergraduates who want to extend their cursory knowledge of relativity to greater depth, and physics graduate students looking for a simple unified treatment of material that usually appears in the graduate physics curriculum in a somewhat disconnected fashion.

The main difference between this book and existing books on special relativity is that it extends the topic list beyond the standard basic topics of spacetime geometry and physics, to include the more current and more advanced (but still accessible) topics of relativistic classical fields, causality, relativistic quantum mechanics, basic supersymmetry, and an introduction to the relativistic string. Another difference is that in most cases the dimension of space is allowed to be arbitrary.

PATRICIA M. SCHWARZ received her BA from San Francisco State University and her PhD in theoretical physics from the California Institute of Technology. Her research specialty is spacetime geometry in general relativity and string theory. She is an expert in multimedia and online education technology. She has created award-winning educational resources, and has hosted online physics courses with teachers and students from around the world.

JOHN SCHWARZ, the Harold Brown Professor of Theoretical Physics at the California Institute of Technology, is one of the founders of superstring theory. He co-authored a two-volume monograph *Superstring Theory* with Michael Green and Edward Witten in 1987. He is a MacArthur Fellow and a member of the National Academy of Sciences.

SPECIAL RELATIVITY

From Einstein to Strings

PATRICIA M. SCHWARZ AND JOHN H. SCHWARZ

Pasadena, California

CAMBRIDGE
UNIVERSITY PRESS

CAMBRIDGE
UNIVERSITY PRESS

University Printing House, Cambridge CB2 8BS, United Kingdom

One Liberty Plaza, 20th Floor, New York, NY 10006, USA

477 Williamstown Road, Port Melbourne, VIC 3207, Australia

314-321, 3rd Floor, Plot 3, Splendor Forum, Jasola District Centre, New Delhi - 110025, India

103 Penang Road, #05-06/07, Visioncrest Commercial, Singapore 238467

Cambridge University Press is part of the University of Cambridge.

It furthers the University's mission by disseminating knowledge in the pursuit of
education, learning and research at the highest international levels of excellence.

www.cambridge.org
Information on this title: www.cambridge.org/9781009197328

First published 2004
Reprinted 2005
First paperback edition 2022

A catalogue record for this publication is available from the British Library

Library of Congress Cataloging in Publication data
Schwarz, Patricia M. (Patricia Margaret), 1956–
Special relativity: from Einstein to strings / Patricia M. Schwarz and John H. Schwarz
p. cm.
Includes bibliographical references and index.
ISBN 0 521 81260 7
1. Special relativity (Physics) I. Schwarz, John H. II. Title.
QC173.65.S39 2004
530.11–dc22 2003055747

ISBN 978-1-009-19732-8 Paperback

To spacetime and everyone who has ever tried to understand it

Contents

Preface

Towards the end of the nineteenth century many physicists believed that all the fundamental laws that describe the physical Universe were known, and that all that remained to complete the understanding was an elaboration of details. The mind-boggling error of this viewpoint was laid bare within a few short years. Max Planck introduced the quantum in 1899 and Albert Einstein's breakthrough work on special relativity appeared in 1905. The ensuing relativity and quantum revolutions each led to surprising and unexpected concepts and phenomena that have profoundly altered our view of physical reality. The science and the history associated with each of these revolutions has been told many times before. But they are worth coming back to again and again with the added benefit of historical perspective. After all, they have changed the world scientifically, technologically, and philosophically. Perhaps due to the lesson from a century ago, very few people today are so foolish as to speak of an "end of science". In fact, revolutionary advances in theoretical physics are currently in progress, and we seem to be a long way from achieving a settled and final picture of physical reality.

As the title indicates, this book is about the special theory of relativity. This theory overthrew the classical view of space and time as distinct and absolute entities that provide the backdrop on which physical reality is superimposed. In special relativity space and time must be viewed together (as spacetime) to make sense of the constancy of the speed of light and the structure of Maxwell's electromagnetic theory. The basic consequences of special relativity can be described by simple algebraic formulas, but a deeper understanding requires a geometric description. This becomes absolutely crucial for the extension to include gravity.

This book is divided into two parts – entitled "Fundamentals" and "Advanced Topics." The first part gives a detailed explanation of special relativity. It starts with simple mathematics and intuitive explanations and gradually builds up more advanced mathematical tools and concepts. Ultimately, it becomes possible to . recast Maxwell's electromagnetic theory in terms of two simple equations

xi

($dF = 0$ and $d * F = *j$) that incorporate relativistic geometry in a simple and beautiful way. Each chapter in Part I of the book starts with a "hands-on exercise." These are intended to help the reader develop spatial awareness. They are not supposed to be scientific experiments, rather they are exercises to limber up the mind.

The second part of the book includes advanced topics that illustrate how relativity has impacted subsequent developments in theoretical physics up to and including modern work on superstring theory. Relativity and quantum mechanics each raised a host of new issues. Their merger led to many more. This is discussed in Chapter 7. One aspect of the structure of spacetime implied by special relativity is its symmetry. To describe this properly requires a branch of algebra called group theory. This is explored in Chapter 8. Chapter 9 raises the question of whether the symmetry of spacetime can be extended in a nontrivial way, and it describes the unique answer, which is supersymmetry. The last chapter gives a brief overview of modern theoretical physics starting with the well-established theories: general relativity and the standard model of elementary particles. It then discusses more speculative current research topics, especially supersymmetry and string theory, and concludes with a list of unsolved problems. These are topics that one would not ordinarily find in a book about special relativity. We hope the reader will enjoy finding them in a form that is more detailed than a popular book, but less technical than a textbook for a graduate-level course.

Part I

Fundamentals

1

From Pythagoras to spacetime geometry

Hands-on exercise:[1] measuring the lengths of lines

Physics is about describing the physical world. In physics courses we get used to doing this using mathematics, and sometimes it can seem as if the mathematics is the physics. But our goal is to learn about the physical world, and so sometimes we have to just put the math aside and let the physical world be our teacher. It is in this spirit that we begin this chapter with a hands-on exercise that requires measuring the physical world with your hands. To complete this exercise you will need the following supplies:

- Three cloth or paper measuring tapes, preferably from computer printouts of the file *measures.html*, available to download from the Cambridge website.
- Some Scotch tape.
- One large spherical object such as a large melon, a beach ball or a globe, with a diameter roughly between 15 and 20 cm.
- One flat table or desk.
- A pencil and some graph paper.

If you have printed out the page with the measuring tapes on them, cut them out with the edges of the paper aligned with the measuring edges of the printed tapes. Tape measures A and B should be taped together at a right angle to one another with the measuring edges facing one another. We will call this taped-together object the Side Measurer. The Side Measurer will be used to measure the lengths of the two sides of a right triangle, while tape measure C, which we will call the Hypotenuse Measurer, will be used to measure the length of the hypotenuse, in the common set of units inscribed on the three measures.

---

[1] Each chapter in Part I of the book starts with a "hands-on exercise." These are intended to help the reader develop spatial awareness. They are not supposed to be scientific experiments, rather they are exercises to limber up the mind. The reader is free to skip them, of course.

Go to your desk or table and tape the corner of the Side Measurer onto some convenient location on its surface. Now use the Hypotenuse Measurer to measure the distances between the locations on the Side Measurer marked by the numbers 1, 2, 3, 4, 5, 6. In other words, measure the distances from 1 to 1, 2 to 2 and so on. Make a table on your graph paper to record your measurements. Plot the results on the graph paper with the side lengths on the x axis and the hypotenuse lengths on the y axis.

Next untape the Side Measurer from the desk or table. Grab your large spherical object (henceforth referred to as the LSO) and tape the corner of the Side Measurer onto some convenient location on its surface, taking care to preserve the right angle where tape measures A and B are taped together. Now use the Hypotenuse Measurer to measure the same set of distances that you measured previously when the Side Measurer was taped to the table or desk. Write them down in a table as you did above, and then plot the data on the plot you made above.

Now on the same plot, draw the line $y = \sqrt{2}\,x$. Write down any impressions you have or conclusions you arrive at by looking at these data, and save them for later.

1.1 Pythagoras and the measurement of space

What does the previous hands-on exercise have to do with special relativity? Special relativity is a theory of spacetime geometry. Before we try to understand the geometry of spacetime, let's go back over what we've already learned about the geometry of space. In the exercise above we were exploring the applicability of the Pythagorean theorem on two different surfaces. The Pythagorean theorem states that:

Given a right triangle, the sum of the squares bounding the two legs of the triangle is equal to the square bounding the hypotenuse of the triangle.

In Pythagoras' time, there was only geometry – algebra was still 1300 years in the future. Pythagoras wasn't talking about the squares of the lengths of the sides as numbers. He proved his theorem by cutting up the squares on the legs and showing that the pieces could be reassembled into the square on the hypotenuse, so that the two squares truly were equal. But now that we have algebra, we can say that if the lengths of the two sides of a right triangle are denoted by A and B, then the length C of the hypotenuse of the right triangle is given by solving the equation

$$A^2 + B^2 = C^2 \tag{1.1}$$

for the value of C.

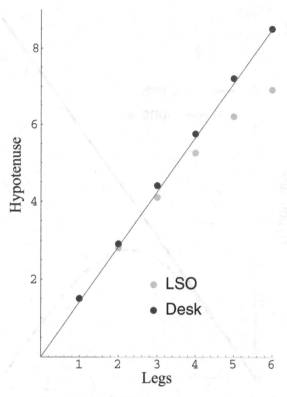

Fig. 1.1. Data from hands-on exercise.

As you should be able to see in the hands-on exercise, this formula works quite reliably when we're measuring right triangles on a desk or table but begins to fail when we measure right triangles on the LSO. One set of data from this exercise is plotted in Figure 1.1.

Now let's use math to explore this issue further. Consider a right isosceles triangle on a two-dimensional sphere of radius R with azimuthal angle θ and polar angle ϕ. A triangle in flat space is determined by three straight lines. The closest analog to a straight line on a sphere is a great circle. Let's make the legs of our right triangle extend from the north pole of the sphere along the great circles determined by $\phi = 0$ and $\phi = \pi/2$, beginning at $\theta = 0$ and terminating at $\theta = \theta_0$. The arc of a great circle of radius R subtending an angle θ_0 has arc length $R\theta_0$, so we can say that $A = B = R\theta_0$. The arc length of the great circle serving as the hypotenuse is given by

$$C = R \cos^{-1}(\cos^2 \theta_0). \tag{1.2}$$

We leave the derivation of this result as an exercise for the reader.

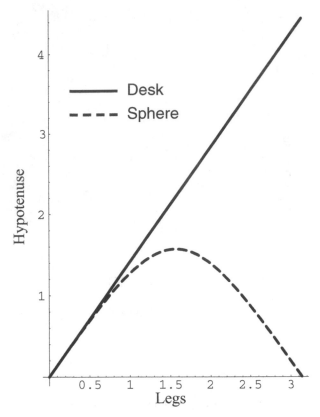

Fig. 1.2. Mathematical solution plotted for $R = 1$.

According to the Pythagorean formula, the hypotenuse should have a length

$$C = \sqrt{2}\, R\theta_0. \tag{1.3}$$

For small values of θ, where $C/R \ll 1$, the Pythagorean rule works fairly well, although not exactly, on the sphere, but eventually the formula fails. We can see how badly it fails in Figure 1.2. This is because the sphere is curved, and the Pythagorean formula only works on flat surfaces. The formula works approximately on the sphere when the distance being measured is small compared to the radius of curvature of the sphere. The mathematical way of saying this is that the sphere is *locally flat*.

But what does this have to do with Einstein's Special Theory of Relativity? In this book we're going to develop the concept of spacetime, but we're only going to study flat spacetime, because that's what special relativity is all about. Everything you will learn in this book will apply to flat spacetime in the same way that the Pythagorean formula applies to flat space. In the real world we experience the force of gravity, and gravity can only be consistently described in terms of a spacetime that is curved, not flat. But a curved spacetime can be approximated as being flat when the force of gravity is small, or, equivalently, when distance scales being measured are small compared to the radius of curvature of spacetime. So we can learn a lot about the Universe just by studying special relativity and flat spacetime, even though in the strictest sense there is no such thing as a completely flat space or spacetime – these geometries exist as mathematical idealizations, not in the material gravitating world.

Even though the Pythagorean formula is only approximately true, it is true enough at the distance scales accessible to Newtonian physics that all of classical physics depends on it. The mathematical and philosophical revolution that made this possible was the marriage of algebra and geometry in the Cartesian coordinate grid. In 1619 a young philosopher named René Descartes dreamt that an "Angel of Truth" came to him from God with the very Pythagorean message that mathematics was all that was needed to unlock all of the secrets of nature. One outcome of this insight was the description of space in terms of algebraic coordinates on an infinite rectangular grid. If space has two dimensions, the distances between any two points in this grid can be calculated by applying the Pythagorean rule, with the distance L_{12} between the two points P_1 and P_2 given by the length of the hypotenuse of the right triangle whose two legs are the differences Δx and Δy between the x and y coordinates of the two points as projected on the two orthogonal axes of the grid

$$L_{12}^2 = \Delta x^2 + \Delta y^2 = (x_1 - x_2)^2 + (y_1 - y_2)^2. \tag{1.4}$$

The world we know seems to have three space dimensions, but this is no problem because a rectangular coordinate system can be defined just as easily in any number of dimensions. In D space dimensions we can describe each point P at which an object could be located or an event could take place by a position vector $\vec{r}$ representing a collection of D coordinates $(x^1, x^2, \ldots, x^D)$ in a D-dimensional rectangular grid. The distance r_{12} between two points $\vec{r}_1$ and $\vec{r}_2$ is given by the Pythagorean formula generalized to D dimensions

$$|r_{12}|^2 = \sum_{i=1}^{D} (x_1^i - x_2^i)^2. \tag{1.5}$$

Any position vector $\vec{r}$ in this D-dimensional space can be written in terms of the D coordinate components x^i in a basis of D orthonormal vectors $\hat{e}_i$ as

$$\vec{r} = \sum_{i=1}^{D} x^i \hat{e}_i. \tag{1.6}$$

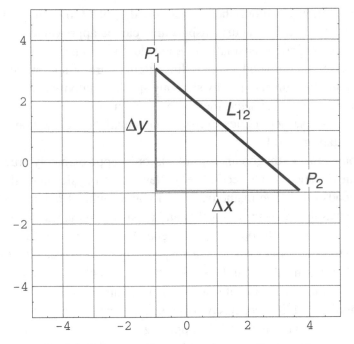

Fig. 1.3. Space as a flat rectangular coordinate grid.

A set of orthonormal basis vectors has the inner product

$$\hat{e}_i \cdot \hat{e}_j = \delta_{ij}, \tag{1.7}$$

where δ_{ij} is the Kronecker delta symbol given by the relation

$$\delta_{ij} = \begin{cases} 1 & i = j \\ 0 & i \neq j. \end{cases} \tag{1.8}$$

Any other vector $\vec{V}$ in this space can then be represented in this orthonormal basis as

$$\vec{V} = \sum_{i=1}^{D} V^i \hat{e}_i, \quad V^i = \vec{V} \cdot \hat{e}_i, \tag{1.9}$$

where the set of D numbers V^i are said to be the components of the vector $\vec{V}$ in this specific basis. Note that the same vector can be represented in more than one basis. This is an extremely important thing to remember, and we will return to it again and again in this book, in greater and greater detail, because this is one of the basic mathematical ideas behind the principle of relativity, both special and general.

1.2 The differential version, in D dimensions

The Pythagorean formula gives us the length of a straight line between two points in a D-dimensional flat space. In order to calculate the length of a line that isn't straight, we can approximate the line as being made up of an infinite number of tiny straight lines with infinitesimal length dl, each of which satisfies an infinitesimal version of the Pythagorean rule. In two space dimensions we write this as

$$dl^2 = dx^2 + dy^2 \tag{1.10}$$

and in D dimensions it becomes

$$dl^2 = \sum_{i=1}^{D} (dx^i)^2. \tag{1.11}$$

If we use the Kronecker delta function δ_{ij} as defined in Eq. (1.8), and adopt the convention of summing over repeated indices, this expression can be rewritten as

$$dl^2 = \delta_{ij} \, dx^i dx^j. \tag{1.12}$$

In differential geometry this object is called the Euclidean metric in rectangular coordinates. Euclidean space is another name for flat space. A metric is another name for an infinitesimal line element. Using the methods of differential geometry, the curvature of a given space can be calculated from the first and second derivatives of metric functions $g_{ij}(x)$, which replace the constant δ_{ij} if the space has nonzero curvature. If we were to calculate the curvature of the D-dimensional space whose metric is Eq. (1.12), we would find that it is exactly zero, because all of the derivatives of the components of δ_{ij} are zero. But that's a subject for another book.

Now that we have an infinitesimal line element, we can integrate it to find the lengths of lines that are not straight but curved, using the differential version of the Pythagorean formula, also known as the Euclidean metric in rectangular coordinates. If we have some curve C between points P_1 and P_2 then the length ΔL of the curve is given by

$$\Delta L = \int_{P_1}^{P_2} dl. \tag{1.13}$$

A curve in D-dimensional Euclidean space can be described as a subspace of the D-dimensional space where the D coordinates x^i are given by single-valued functions of some parameter t, in which case the length of a curve from $P_1 = x(t_1)$ to $P_2 = x(t_2)$ can be written

$$\Delta L = \int_{t_1}^{t_2} \sqrt{\delta_{ij} \, \dot{x}^i \dot{x}^j} \, dt, \quad \dot{x}^i = \frac{dx^i}{dt}. \tag{1.14}$$

For example, we can calculate the circumference of a circle of radius R in two-dimensional Euclidean space described by $\{x^1 = R\cos t, \; x^2 = R\sin t\}$. In this case,

$$\delta_{ij}\dot{x}^i\dot{x}^j = R^2(\sin^2 t + \cos^2 t) = R^2 \tag{1.15}$$

and

$$\Delta L = \int_0^{2\pi}\sqrt{\delta_{ij}\,\dot{x}^i\dot{x}^j}\,dt = \int_0^{2\pi} R\,dt = 2\pi R. \tag{1.16}$$

Since it's guaranteed that $\delta_{ij}\dot{x}^i\dot{x}^j \geq 0$, formula (1.14) for the length of a curve is always positive as long as the curve itself is well-behaved. This won't continue to be the case when we graduate from space to spacetime.

1.3 Rotations preserve the Euclidean metric

At first glance, a description of space as a rectangular coordinate grid seems like turning the Universe into a giant prison ruled by straight lines that point in fixed directions and tell us how we have to describe everything around us in their terms. We know that, in the real world, we possess free will and can turn ourselves around and look at any object from a different angle, a different point of view. We see that the object looks different from that point of view, but we know it is not a different object, but just the same object seen from a different angle.

Luckily for us, the same wisdom emerges from the mathematics of Euclidean space. We don't have to stick with one rigid coordinate system – we can turn the whole coordinate grid around to see the same object from a different angle. This can be done in any number of dimensions, but for the sake of brevity we will stick with $D = 2$ with the traditional choice $x^1 = x, x^2 = y$.

A general linear transformation from coordinates (x, y) to coordinates $(\tilde{x}, \tilde{y})$ can be written in matrix form as

$$\begin{pmatrix} \tilde{x} \\ \tilde{y} \end{pmatrix} = \begin{pmatrix} a_{11} & a_{12} \\ a_{21} & a_{22} \end{pmatrix}\begin{pmatrix} x \\ y \end{pmatrix} + \begin{pmatrix} b_1 \\ b_2 \end{pmatrix}. \tag{1.17}$$

The constants b_1 and b_2 represent a shift in the origin of the coordinate system. Taking the differential of this expression automatically gets rid of b_1 and b_2, and this reflects the freedom with which we can set the origin of the coordinate system anywhere in the space without changing the metric. This freedom is called translation invariance, and ultimately, as we shall show in a later chapter, it leads to conservation of momentum for objects moving in this space.

If we require that the metric remain unchanged under the rest of the transformation, so that

$$d\tilde{x}^2 + d\tilde{y}^2 = dx^2 + dy^2, \tag{1.18}$$

then it must be true that

$$a_{11}^2 + a_{12}^2 = a_{21}^2 + a_{22}^2 = 1, \qquad a_{11}a_{12} = -a_{21}a_{22}. \tag{1.19}$$

We have three equations for four variables, so instead of a unique solution, we get a continuous one-parameter family of solutions that can be written in terms of an angular parameter θ as

$$a_{11} = a_{22} = \cos\theta, \qquad a_{12} = -a_{21} = \sin\theta. \tag{1.20}$$

This describes a rotation by an angle θ. We also get a second family of solutions

$$a_{11} = -a_{22} = \cos\theta, \qquad a_{12} = a_{21} = \sin\theta, \tag{1.21}$$

which represents a reflection about an axis characterized by θ. Transformations like this will be discussed in more detail in Chapter 8.

Expanding out the matrix multiplication for the solution in (1.20), our coordinate transformation becomes

$$\tilde{x} = x\cos\theta + y\sin\theta$$
$$\tilde{y} = -x\sin\theta + y\cos\theta. \tag{1.22}$$

This transformation is a rotation of the rectangular coordinate system by an angle θ about the origin at $x = y = 0$. To see this, look at the $\tilde{x}$ and $\tilde{y}$ axes, represented by the line $\tilde{y} = 0$ and $\tilde{x} = 0$, respectively. In the (x, y) coordinate system, they become

$$\tilde{y} = 0 \rightarrow y = x\tan\theta$$
$$\tilde{x} = 0 \rightarrow x = -y\tan\theta, \tag{1.23}$$

and if we plot this we can see that the $\tilde{x}$ and $\tilde{y}$ axes are both rotated counterclockwise by an angle θ compared to the x and y axes.

As the angle from which we view an object changes, we don't expect the object itself to change. Suppose we have a position vector $\vec{r}$ pointing to some object in two-dimensional Euclidean space. In the (x, y) coordinate system, $\vec{r}$ can be written

$$\vec{r} = x\hat{e}_x + y\hat{e}_y. \tag{1.24}$$

If we require that the position vector itself remain unchanged as we rotate the coordinate system in which the vector is described, so that

$$\vec{r} = x\hat{e}_x + y\hat{e}_y = \tilde{x}\hat{e}_{\tilde{x}} + \tilde{y}\hat{e}_{\tilde{y}}, \tag{1.25}$$

then the transformation of the coordinate components of the vector must cancel the transformation of the unit basis vectors. Therefore the transformation rule for the orthonormal basis vectors must be

$$\hat{e}_{\tilde{x}} = \hat{e}_x \cos\theta + \hat{e}_y \sin\theta$$
$$\hat{e}_{\tilde{y}} = -\hat{e}_x \sin\theta + \hat{e}_y \cos\theta. \tag{1.26}$$

The rotation transformation can be written in matrix form as

$$R(\theta) = \begin{pmatrix} \cos\theta & \sin\theta \\ -\sin\theta & \cos\theta \end{pmatrix}. \tag{1.27}$$

The inverse transformation is a rotation in the opposite direction

$$R(\theta)^{-1} = R(-\theta) = \begin{pmatrix} \cos\theta & -\sin\theta \\ \sin\theta & \cos\theta \end{pmatrix}. \tag{1.28}$$

The matrix $R(\theta)$ satisfies the conditions $\det R = 1$ and $R^T I R = I$, where I is the identity matrix

$$I = \begin{pmatrix} 1 & 0 \\ 0 & 1 \end{pmatrix}. \tag{1.29}$$

The first condition classifies R as a special, as opposed to a general, linear transformation. The second condition classifies R as an orthogonal matrix. The full name of the group of linear transformations represented by $R(\theta)$ is the special orthogonal group in two space dimensions, or $SO(2)$[2] for short. Transformations by $R(\theta)$ take place around a circle, which can be thought of as a one-dimensional sphere, known as S^1 for short.

A transformation matrix can have the properties of being special and orthogonal in any number of dimensions, so rotational invariance is easily generalized to D space dimensions, although the matrices get more complicated as D increases. As one would expect, since we have $SO(2)$ for $D = 2$, we have $SO(D)$ for arbitrary D. A rotation in D space dimensions takes place on a $(D - 1)$-dimensional sphere, called S^{D-1} for short. We will examine rotational invariance in D space dimensions in greater detail in a later chapter. For now, everything we want to accomplish in this chapter can be achieved using the simplest case of $D = 2$.

1.4 Infinitesimal rotations

So far this seems like elementary stuff. Why are we looking at rotations in space, when our goal in this book is to learn about spacetime? We will see shortly that the

---

[2] The group becomes known as $O(2)$ if we include reflections in addition to rotations. For a reflection, the determinant is -1 instead of $+1$.

mathematics of rotational invariance of Euclidean space has a very close analog in the relativistic invariance of spacetime. To build the case for this, and build our first glimpse of spacetime, we need to study infinitesimal rotations, that is, rotations for which θ is close to zero.

A very small rotation with $\theta \sim 0$ can be written:

$$\tilde{x} \simeq x + \theta y + O(\theta^2)$$
$$\tilde{y} \simeq y - \theta x + O(\theta^2) \tag{1.30}$$

In this infinitesimal limit, the rotation matrix $R(\theta)$ can be written in terms of a matrix $\mathbf{r}$

$$R(\theta) \simeq I + \theta \mathbf{r} + \cdots, \qquad \mathbf{r} = \begin{pmatrix} 0 & 1 \\ -1 & 0 \end{pmatrix}, \tag{1.31}$$

which we will call the generator of the rotation transformation. Notice that $\mathbf{r}$ is an antisymmetric matrix, that is $\mathbf{r}^T = -\mathbf{r}$, where the matrix transpose is defined by $(M^T)_{ij} = M_{ji}$.

Written in this manner, an infinitesimal rotation looks like the first two terms in the expansion of an exponential

$$e^{\alpha x} \sim 1 + \alpha x + \cdots \tag{1.32}$$

This is no coincidence. A non-infinitesimal rotation $R(\theta)$ can be obtained from the exponential of the generator matrix $\mathbf{r}$

$$R(\theta) = \exp{(\theta \mathbf{r})} = \begin{pmatrix} \cos\theta & \sin\theta \\ -\sin\theta & \cos\theta \end{pmatrix}, \tag{1.33}$$

as you will be asked to prove in an exercise at the end of this chapter.

Now suppose we want to consider unifying space and time into a two-dimensional spacetime, hopefully something as simple and symmetric as Euclidean space. What would be the analog of rotational invariance in that case? The most obvious difference between space and time is that we can't turn ourselves around to face backwards in time like we can in space. So a rotation transformation presents a problem. Is there a transformation that is like a rotation but without the periodicity that conflicts with the knowledge that we can't rotate our personal coordinate frames to face backwards in time?

An antisymmetric generator matrix will always lead to a rotation when we take the exponential to get the full transformation. Suppose the generator matrix is not antisymmetric but instead symmetric, so that the matrix is equal to its own transpose? Such a generator matrix yields a new transformation that looks almost like a rotation, but is not. Let's call the new generator matrix ℓ and the new transformation parameter ξ. The new infinitesimal transformation is

$$L(\xi) \simeq I + \xi\ell + \cdots, \qquad \ell = \begin{pmatrix} 0 & -1 \\ -1 & 0 \end{pmatrix}. \tag{1.34}$$

Because of the sign difference in the generator ℓ, the exponential of the generator gives the unbounded functions $\cosh\xi$ and $\sinh\xi$ instead of the periodic functions $\cos\theta$ and $\sin\theta$

$$L(\xi) = \exp(\xi\ell) = \begin{pmatrix} \cosh\xi & -\sinh\xi \\ -\sinh\xi & \cosh\xi \end{pmatrix}. \tag{1.35}$$

The action of $L(\xi)$ on a set of rectangular coordinates axes is not to rotate them but to skew them like scissors. The θ in $R(\theta)$ lives on the interval $(0, 2\pi)$, but the parameter ξ in $L(\xi)$ can vary between $(-\infty, \infty)$. In the limit $\xi \to \pm\infty$, the transformation degenerates and the axes collapse together into a single line, as you will be asked to show in an exercise.

As with the rotation transformation $R(\theta)$, the inverse transformation of $L(\xi)$ is a transformation in the opposite direction

$$L(\xi)^{-1} = L(-\xi) = \begin{pmatrix} \cosh\xi & \sinh\xi \\ \sinh\xi & \cosh\xi \end{pmatrix}. \tag{1.36}$$

This new transformation satisfies the special condition $\det L = 1$. However, the orthogonality condition $R^T I R = I$ is amended to

$$L^T \eta L = \eta, \qquad \eta = \begin{pmatrix} -1 & 0 \\ 0 & 1 \end{pmatrix}. \tag{1.37}$$

Because of the minus sign, this type of transformation is called a special orthogonal transformation in $(1, 1)$ dimensions, or $SO(1, 1)$ for short. It's going to turn out that this $(1, 1)$ refers to one space and one time dimension. In order to develop that idea further, it's time to bring time into the discussion.

1.5 Could a line element include time?

The journey towards the description of physical space by an infinite Cartesian coordinate grid was a rough one, because enormous spiritual and moral significance was given to the organization of physical space by European culture in the Middle Ages. At one time, even the assertion that empty space existed was considered heresy. By contrast, the question of time was not as controversial. It seemed obvious from common experience that the passage of time was something absolute that was universally experienced by all observers and objects simultaneously in the same way. This didn't conflict with the story of the Creation told in the Bible, so the concept of absolute time did not present a challenge to devout Christians such as Isaac Newton, who held it to be an obvious and unquestionable truth.

In this old picture of space and time, space and time are inherently separate, and both absolute. An object at a location marked in absolute space by the position vector $\vec{r}$ moves along a path in absolute space that can be written as a function of time such that

$$\vec{r} = \vec{r}(t) = x^i(t)\hat{e}_i, \tag{1.38}$$

where the implied sum over the repeated index i runs over all of the dimensions in the space. For now, let's restrict space to one dimension, so we're only dealing with one function of time $x(t)$.

In Newtonian physics in one space dimension, in the absence of any forces, the motion of an object is determined by the solution to the differential equation

$$\frac{d^2x(t)}{dt^2} = 0. \tag{1.39}$$

This equation is invariant under the transformation

$$\tilde{t} = t$$
$$\tilde{x} = x - vt, \tag{1.40}$$

where v is the velocity of an observer in the $\tilde{x}$ coordinate system relative to an observer in the x coordinate system.

This invariance principle was first proposed by Galileo based on general physical and philosophical arguments, long before Newton's equation existed. Galileo argued that when comparing a moving ship with dry land, the natural laws governing the motion of an object on the ship should not depend on the motion of the ship relative to the dry land, as long as that motion was smooth motion at a constant velocity.

This sounds suspiciously like what we learned about Euclidean space and rotations – an object should not change as we rotate the coordinate system used to describe it. But now we're not talking about rotating space into space, we're talking about space and time, and instead of a rotation, which involves a dimensionless angle, we have motion, which involves the dimensionful quantity of velocity.

Let's consider the possibility that Galilean invariance could be an infinitesimal version of some full spacetime invariance principle, a version valid only for velocities close to zero. Suppose we use as our candidate spacetime transformation the $SO(1, 1)$ transformation developed in the previous section.

At first it would appear that Galilean invariance is not consistent with a small rotation by $L(\xi)$. However, the units by which we measure space and time are not the same. Time is measured in units of time such as seconds or years, and space is measured in units of length such as feet or meters. If we want to compare a Galilean transformation to a rotation by $L(\xi)$, we should scale the coordinate t by some dimensionful constant c with units of length/time, so that $\tau = ct$ has the dimension of length

$$\tau = ct, \qquad [\tau] = L \rightarrow [c] = L/T. \tag{1.41}$$

If we took the transformation $L(\xi)$ literally as a kind of rotation in two-dimensional spacetime with time coordinate τ and space coordinate x, then this rotation would change coordinate components (τ, x) to $(\tilde{\tau}, \tilde{x})$ by

$$\tilde{\tau} = \tau \cosh \xi - x \sinh \xi$$
$$\tilde{x} = -\tau \sinh \xi + x \cosh \xi. \tag{1.42}$$

The Galilean transformation contains a parameter v with units of velocity. Using v and c, we can define a dimensionless parameter $\beta \equiv v/c$. If we assume that $\xi \simeq \beta$ then an infinitesimal rotation by $L(\xi)$ for small ξ looks like

$$\tilde{\tau} \simeq \tau - \beta x$$
$$\tilde{x} \simeq x - \beta \tau. \tag{1.43}$$

So far, this looks nothing like a Galilean transformation. However, maybe one of the terms is smaller than the others and can be neglected. If we rewrite the above equations back in terms of t and c then we get:

$$\tilde{t} \simeq t - \frac{vx}{c^2} \simeq t, \qquad c \rightarrow \infty \tag{1.44}$$
$$\tilde{x} \simeq x - vt. \tag{1.45}$$

The infinitesimal limit $\beta \rightarrow 0$ corresponds to the limit $c \rightarrow \infty$. In this limit the second term in the first equation can be safely neglected because it's much smaller than the others. So our postulated spacetime rotation $L(\xi)$ for very small values of $\xi \sim \beta$ does appear to be consistent with a Galilean transformation.

So here comes the big question: What line element is left invariant by the spacetime rotation $L(\xi)$? What is the spacetime analog of the differential version of the Pythagorean theorem? One can show, as you will in an exercise, that the line element

$$ds^2 = -d\tau^2 + dx^2 \tag{1.46}$$

is invariant under the coordinate transformation given by $L(\xi)$ in (1.42) such that

$$-d\tau^2 + dx^2 = -d\tilde{\tau}^2 + d\tilde{x}^2. \tag{1.47}$$

The metric (1.46) gives us at last the analog of the Pythagorean rule for space-time. If curvature comes from derivatives of the metric, then this strange metric with a minus sign must be flat. This metric for flat spacetime is called the Minkowski metric, and flat spacetime is also known as Minkowski spacetime. Note that unlike the Euclidean metric, the Minkowski metric is not positive-definite. This is extremely important and we will explore the implications of this fact in greater detail in later chapters.

What happens for higher dimensions? A flat spacetime with D space dimensions and one time dimension has a line element

$$ds^2 = -d\tau^2 + dl^2, \qquad dl^2 = \delta_{ij}dx^i dx^j. \qquad (1.48)$$

The Latin indices (i, j) refer to directions in space, and by convention take the values $(1, 2, \ldots, D)$. When dealing with spacetime, it has become the convention to use a set of Greek indices (μ, ν) and appoint the 0th direction as being the time direction so that $dx^0 = d\tau$, in which case the Minkowski metric can be written

$$ds^2 = \eta_{\mu\nu}dx^\mu dx^\nu, \qquad (1.49)$$

where

$$\eta_{00} = -1,$$
$$\eta_{0i} = \eta_{i0} = 0,$$
$$\eta_{ij} = \delta_{ij}. \qquad (1.50)$$

In two spacetime dimensions the metric is invariant under an $SO(1, 1)$ transformation of the coordinates. As one might suspect, in $d = D + 1$ spacetime dimensions, the transformation is called $SO(1, D)$ or $SO(D, 1)$. In either case, this indicates that it pertains to D space dimensions and one time dimension. (We never consider more than one time dimension!)

1.6 The Lorentz transformation

The transformation $L(\xi)$ is known as the Lorentz transformation, the invariance principle is called Lorentz invariance and the transformation group $SO(1, D)$ is known as – no surprise here – the Lorentz group, which we'll study in greater detail in a later chapter. But we don't yet know the relationship between the Lorentz transformation parameter ξ and the velocity parameter $\beta \equiv v/c$ that gives the relative velocity between the two coordinate systems in question.

Suppose we have an observer who is at rest in the (τ, x) coordinate frame in a flat spacetime with metric (1.46). The $(\tilde{\tau}, \tilde{y})$ coordinate frame is moving at velocity β relative to the (τ, x) frame. (Note: Even though β is a dimensionless parameter proportional to the velocity, for the sake of brevity we shall refer to it as the velocity.) Therefore an observer measuring space and time in the $(\tilde{\tau}, \tilde{y})$ coordinate system sees the observer in the (τ, x) frame not as being at rest, but moving with velocity $-\beta$ so that

$$\frac{d\tilde{x}}{d\tilde{\tau}} = -\beta. \tag{1.51}$$

Since the observer in question is at rest in the (τ, x) coordinate frame, for her/him $dx = 0$ and therefore

$$ds^2 = -d\tau^2. \tag{1.52}$$

But according to the $(\tilde{\tau}, \tilde{y})$ coordinate system,

$$ds^2 = -d\tilde{\tau}^2 + d\tilde{x}^2 = -d\tilde{\tau}^2 + \beta^2 d\tilde{\tau}^2. \tag{1.53}$$

Because the two coordinate systems (τ, x) and $(\tilde{\tau}, \tilde{y})$ differ only by a Lorentz transformation, and because we're in flat spacetime, where the metric is Lorentz-invariant, it must be true that

$$d\tilde{\tau} = \frac{d\tau}{\sqrt{1 - \beta^2}}. \tag{1.54}$$

Taking the differential of the Lorentz transformation relating the two frames in Eq. (1.42) gives

$$d\tilde{\tau} = d\tau \cosh\xi - dx \sinh\xi$$
$$d\tilde{x} = dx \cosh\xi - d\tau \sinh\xi, \tag{1.55}$$

so the Lorentz transformation parameter ξ is related to the velocity β by

$$\cosh\xi = \gamma, \qquad \sinh\xi = \gamma\beta, \qquad \gamma \equiv \frac{1}{\sqrt{1 - \beta^2}}. \tag{1.56}$$

The Lorentz transformation $L(\xi)$, rewritten in terms of γ and β, becomes

$$L(\beta) = \begin{pmatrix} \gamma & -\gamma\beta \\ -\gamma\beta & \gamma \end{pmatrix}. \tag{1.57}$$

Something very strange and interesting has happened. At first it seemed from the infinitesimal transformation (1.43) that we would end up with $\xi = \beta$, and the velocity β would live in the interval $(-\infty, \infty)$. However, the Lorentz transformation is only real and finite for

$$-1 < \beta < 1, \tag{1.58}$$

which means that

$$-c < v < c. \tag{1.59}$$

So the velocity c, at first introduced only to create a dimensionally balanced coordinate transformation, ends up being the maximum allowed velocity in the spacetime.

It's probably no secret that this maximum velocity c imposed by the geometry of flat spacetime is the speed of light. But that association is something physical that can't be proven by geometry alone. To show that c is the speed of light, we need to appeal to the physics of light, which is the subject of the next chapter.

Exercises

1.1 Verify that the rotation $R(\theta)$ leaves the Euclidean line element dl^2 invariant.

1.2 Let's look at a two-sphere of radius R whose center is at the origin of three-dimensional flat Euclidean space, with coordinates related by

$$(x, y, z) = (R \sin\theta \cos\phi, R \sin\theta \sin\phi, R \cos\theta). \tag{E1.1}$$

Consider the three great circles passing through the north pole $\vec{x}_0 = (0, 0, R)$ and the points $\vec{x}_1 = (R \sin\theta_0, 0, R \cos\theta_0)$ and $\vec{x}_2 = (0, R \sin\theta_0, R \cos\theta_0)$ on the sphere. These three circles define the right triangle on the sphere described at the beginning of the chapter. Recall that the Euclidean dot product of two vectors $\vec{x}_i \cdot \vec{x}_j = |x_i||x_j| \cos\theta_{ij}$, where θ_{ij} is the angle between the two vectors in the two-dimensional plane they determine.

  (a) Use this result to verify that the arc length of the hypotenuse of this right triangle is given by Eq. (1.2).

  (b) Expand Eq. (1.2) for small θ_0 to check whether the Pythagorean rule is obeyed in that limit.

1.3 Using the transformation rule for basis vectors, compute the components of the vector $\vec{V}$ in the $(\tilde{x}, \tilde{y})$ coordinate system of Eq. (1.17).

1.4 On a sheet of graph paper, represent the (x, y) coordinate system as a two-dimensional rectangular grid, with y on the vertical axis and x on the horizontal axis. Using the transformation $R(\theta)$ to relate the two coordinate systems $(\tilde{x}, \tilde{y})$ and (x, y), plot the two lines $\tilde{x} = 0$ and $\tilde{y} = 0$ in the (x, y) coordinate system for the values $\theta = \pi/4, \pi/2, 3\pi/4$ and π. Then draw another coordinate grid with $(\tilde{x}, \tilde{y})$ on the axes, and plot the two lines $x = 0$ and $y = 0$ in the $(\tilde{x}, \tilde{y})$ coordinate system for the same values of θ.

1.5 On a sheet of graph paper, represent the (τ, x) coordinate system as a two-dimensional rectangular grid, as you did above for the (x, y) coordinate system, but with τ replacing y on the vertical axis, and x on the horizontal axis. Using the transformation $L(\xi)$ to relate the two coordinate systems (τ, x) and $(\tilde{\tau}, \tilde{x})$, plot the two lines $\tilde{\tau} = 0$ and $\tilde{x} = 0$ in the (τ, x) coordinate system for $\xi = 1/3, 1/2, 1$ and 2. Then draw another coordinate grid with $(\tilde{\tau}, \tilde{x})$ on the axes, and plot the two lines $\tau = 0$ and $x = 0$ in the $(\tilde{\tau}, \tilde{x})$ coordinate system for the same values of ξ. Using (1.56), calculate β and γ for $\xi = 1/3, 1/2, 1$ and 2.

1.6   Given the 2×2 antisymmetric matrix

$$A = \begin{pmatrix} 0 & 1 \\ -1 & 0 \end{pmatrix}, \qquad \text{(E1.2)}$$

compute the first four terms in the Taylor expansion of the exponential $e^{\theta A}$ around $\theta = 0$ and derive a general formula for the elements of $e^{\theta A}$ as an infinite sum of powers of θ.

1.7   Given the 2×2 symmetric matrix

$$S = \begin{pmatrix} 0 & -1 \\ -1 & 0 \end{pmatrix}, \qquad \text{(E1.3)}$$

compute the first four terms in the Taylor expansion of the exponential $e^{\xi S}$ around $\xi = 0$ and derive a general formula for the elements of $e^{\xi S}$ as an infinite sum of powers of ξ.

1.8   Multiply two rotation matrices $R(\theta_1)$ and $R(\theta_2)$. Is the result a third rotation matrix $R(\theta_3)$? If so, what is the angle θ_3 of the resulting transformation in terms of θ_1 and θ_2?

1.9   Multiply two Lorentz transformation matrices $L(\xi_1)$ and $L(\xi_2)$. Is the result another Lorentz transformation matrix $L(\xi_3)$? If so, what is the transformation parameter ξ_3 of the resulting matrix in terms of ξ_1 and ξ_2?

2

Light surprises everyone

Hands-on exercise: wave and particle properties

The purpose of this exercise is for you to observe some basic wave and particle properties. To complete this exercise you will need the following:

- Tub of water, or access to a quiet pond, lake or swimming pool.
- Things to float on the surface of the water.
- Pen or pencil and some drawing paper.
- Small projectile such as a stone.

Disturb the middle of the tub just until you are able to make a visible wave on the surface. Watch how the wave propagates. Wait until the surface of the water returns to being flat and make another wave. Keep doing this as many times as necessary to be able to draw what you see on the paper and answer the following questions:

- Does the wave have a definite location at any one moment in time?
- Does the wave have a definite direction as it propagates?
- Approximately how far does the wave travel in 1 s?
- Describe the motion of the water in which the wave moves.

Throw your small projectile in the air at various angles, letting it drop down (not in the water). Keep doing this as many times as necessary to be able to draw what you see and answer the following questions:

- Does the object have a definite location at any one moment in time?
- Does the object have a definite direction as it propagates?
- Approximately how far does the object travel in 1 s?
- Describe the motion of the air in which the object moves.
- Is there any difference between the vertical and horizontal motion of the object?

2.1 Conflicting ideas about space and light

What we call known physics today, what we learn in school or teach ourselves from books, at one time was the unknown. It was what people did not understand, and sought to understand through exhausting and often frustrating intellectual labor. In the process of moving from the unknown to the known, a lot of wrong ideas can come up that seem very right at the time. The story of the classical understanding of space and light from Aristotle to Einstein is a story in which almost everyone involved was both right and wrong at the same time.

Aristotle started with an idea that seems right enough – *Nature abhors a vacuum* – and used it to argue that empty space could not exist, period. Every last tiny space in the Universe was filled with a universal substance, which later came to be called the *ether*. According to Aristotle, the space taken up by a material object was the surface area, not the volume, of the object. Using Aristotle's logic, the amount of space taken up by a round ball of radius R would be $4\pi R^2$ rather than $4\pi R^3/3$. Aristotle was such a powerful figure in Western culture that it took until the fifteenth century for his argument against spatial volume to be refuted. But even so, Aristotle's argument that empty space could not exist formed the root of the wrong understandings of both space and light that troubled classical physics until Einstein showed up with a brilliant idea that put the controversies to rest, at least until quantum theory showed up.

Even though Descartes' work on analytic geometry laid the mathematical foundation for the Newtonian description of physical space as an empty and absolute backdrop for the actions of matter, Descartes shared Aristotle's abhorrence of the void. Descartes believed that a type of material substance called the *plenum* must fill the entire universe, down to every nook and cranny, and that vortices swirling in this fluid were what moved the planets in their orbits. Descartes believed that light was an instantaneous disturbance in the plenum between the observer and the observed. He believed so strongly that light propagates instantaneously that he swore that if this were ever proved false, he would confess to knowing absolutely nothing.

Newton learned geometry by reading Descartes, but he inserted into physics his own belief that space was both empty and absolute. To Newton, a devout Christian, to question the absoluteness of space was to question the absoluteness of God – not merely an intellectual error, but an actual sin against God. In Newton's Universal Law of Gravitation, the force between two gravitating objects varies inversely as the square of the distance between them, and the distance between gravitating objects is treated as an empty space devoid of any intervening substance. Newton's powerful and concise theory was a huge success in explaining all of known astronomy at the time, but Newton's many critics rightly complained that the Universal

Law of Gravitation provides no mechanical means for transmitting the gravitational force between bodies, other than the literal hand of God – a conclusion that Newton was not unsatisfied with himself.

Field theory had not been invented yet. This was still the age of mechanics. Things happened because one thing pushed or pulled on another. Pushing or pulling is not something that anyone envisioned could be done across empty space. If there is no material substance filling the space between the planets, then how would one planet sense the introduction or removal of any other planet?

The lack of any causal mechanism of force transmission in Newton's theory of gravity led many of his contemporaries to label his theory as nonsense. One such person was Dutch physicist Christian Huygens, who, like Newton, learned his vocation by reading Descartes. Huygens, however, was appointed by fate to be the undoing of his own master. It was Huygens who made the first numerical estimate of the speed of light and proved Descartes wrong about the very thing Descartes was the most certain he was right.

In 1667 Galileo had tried to measure the speed of light using lanterns and mountain tops but he never had a chance, because light travels too fast for the time interval in question to have been measured by any existing timekeeping device. The speed of light is 3×10^8 m/s. At such an enormous speed, light only needs 8 min. to cross the 1.5×10^{11} m from the Sun to the Earth. If we can only measure time to within a few seconds, then in order to measure the speed of light, we have to observe light propagation over the distance scale of the solar system. And this is how the first successful measurement of the speed of light was made in 1676.

Danish astronomer Ole Roemer, who spent 10 years making careful observations of the orbital periods of Jupiter's moon Io, was quite surprised when the period he observed seemed to fluctuate with the distance between Jupiter and Earth, with the period being longer when Jupiter and Earth were moving farther apart. In 1676 he announced that this discrepancy could only come from the time it took light to travel from Io to the Earth. Two years later Huygens provided a numerical estimate for this speed of 144 000 miles/s. Huygens had proved that his hero Descartes was wrong about light. Luckily for Descartes, he didn't live long enough to have to fulfill his promise to confess to knowing absolutely nothing.

Huygens took the finite speed of light as evidence for his wave theory of light. In his treatise on optics *Le traité de la lumière*, he put forward his model of wave propagation that physics students now learn as Huygens' Principle: given a particular wave front, each point on that wave front acts as the source point for a spherical secondary wave that advances the wave front in time. Huygens' Principle is illustrated in Figure 2.1.

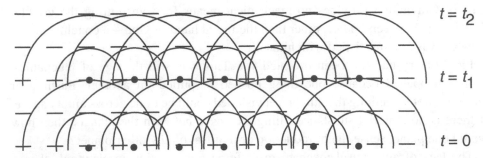

Fig. 2.1. According to Huygens' Principle, each point on a wave front acts as a source point for a spherical secondary wave that determines the wave front at some later time. The dashed lines represent a wave front advancing in time t.

Huygens' optics of wave fronts made little impact in his own time, for he was living in the age of Isaac Newton, Superstar. Consistent with his belief in absolute empty space, Newton envisioned light as a swarm of particles he called "corpuscles" moving through empty space, each corpuscle moving at a different speed depending on the color of the light it represented. Newton had a corpuscular explanation for refraction that was wrong, but the error was not experimentally measurable at the time. Snell's Law of Refraction[1] states that if a light ray is incident on the interface between two transparent media, the angle θ_1 made by the incident ray with the normal to the plane of the interface and the angle θ_2 made by the refracted ray are related through the formula

$$\frac{\sin \theta_1}{\sin \theta_2} = \frac{v_1}{v_2},\tag{2.1}$$

where v_1 and v_2 are the speeds of light in the two transparent media. Snell's law is illustrated in Figure 2.2.

Newton wrongly argued that the ray angle would be smaller in the medium where the speed of light was the largest. The debate over Huygens' wave theory and Newton's particle theory could have been settled over refraction alone, except for one problem: the best known value for the speed of light in air was still off by 25 percent, so they had no hope of being able to measure the difference between the speed of light in air and its speed in water. When this was done in 1850 by Foucault, Newton's theory was conclusively ruled out and Huygens was vindicated.

Despite the fact that he was wrong, Newton made enormous contributions to optics. In 1669 he built the first reflecting telescope, which used a curved mirror instead of a lens, and revolutionized astronomy. Newton ground the mirror himself. When Newton published his book *Opticks* in 1704, it created a scientific and

---

[1] Also known as Descartes' Law.

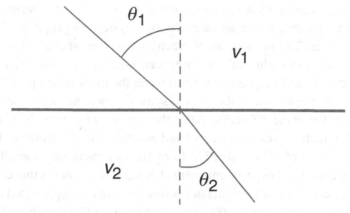

Fig. 2.2. Snell's Law of Refraction describes the bending of a ray of light when it travels from a medium in which the speed of light is v_1 into a medium in which the speed is $v_2 \neq v_1$.

a popular sensation all over England and Europe. Voltaire published a popularization of Newton's work, and discussions of Newtonian optics were all the rage at amateur science and philosophy clubs throughout the educated upper and middle class.

No matter how popular it became and how it revolutionized astronomy, Newtonian optics could not explain the phenomenon of diffraction, where light appears to bend around the edges of objects. Diffraction was first observed in 1665 in Italy by Father Francesco Grimaldi, but both Newton and Huygens regarded it as irrelevant to the wave vs. particle debate. Diffraction could not be ignored forever, however, and eventually wave optics had to be brought back into the picture.

In 1746 in his book *Nova theoria lucis et colorum* (*New Theory of Light and Colour*), Swiss mathematician Leonhard Euler advanced the notion that light consists of wavelike vibrations in the ether. Euler argued that light propagates in the ether, just as sound propagates in the air. Unfortunately, Euler's theory was ultimately no better than Newton's in explaining diffraction. A very bright eye doctor named Thomas Young refined Euler's wave theory to make it consistent with Huygens' Principle. Young and Augustin Fresnel finally proved through their understanding and demonstrations of interference and diffraction that light definitely had wavelike properties that could not be explained by tiny corpuscles flying through space.

2.2 Maxwell's transverse undulations

A real understanding of the nature of light required an understanding of electromagnetism, and that took 100 years to happen, if we start counting from Newton.

In Newton's time, electricity was only known about through the experience of electrostatic shock, and magnetism was a useful but mysterious property of a mineral used in compasses. The unification of electricity and magnetism was the greatest accomplishment of eighteenth- and nineteenth-century physicists. The unified theory of electricity and magnetism is what made the modern age possible.

Electricity and magnetism, in the form of static forces of attraction, were known, and named, by the ancient Greeks. After the success of Newton's $1/R^2$ law of gravitational attraction, scientists wondered whether the electrostatic force also obeyed such a law. From 1771 to 1773 Lord Henry Cavendish, a wealthy aristocratic recluse, conducted experiments with charged metal spheres that confirmed a $1/R^2$ law of electrostatic attraction, but his results were not appreciated by the scientific community until almost 100 years later because Cavendish suffered from paranoia and could not be persuaded to publish.[2] What could have been called Cavendish's Law is now known to us as Coulomb's Law, named after Charles-Augustin de Coulomb, a French military engineer whose brilliance in both the theoretical and practical aspects of applied mechanics enabled him to perfect a torsion balance sensitive enough to detect the $1/R^2$ dependence on the force between two charged spheres. Coulomb published his work from 1785 to 1791, just a dozen years after the work done by Cavendish, but a whole century before Cavendish's work on electrostatics would be known to the science community.

The first inkling that there might be a unified theory of electricity and magnetism arose in 1819, when Hans Christian Oersted, a Danish philosopher and physicist, discovered during the course of a class demonstration that an electric current moving through a wire caused the deflection of a magnetic needle of a compass. His experiment inspired André-Marie Ampère in France to measure the forces between two parallel electric currents. Ampère found that the forces were magnetic and not electric in nature. He also discovered that when electric current is passed through a wire wrapped around a coil, the coil as a whole exerts a magnetic force as if it were a bar magnet.[3] In 1826 Ampère published his work on electrodynamics *Memoir on the Mathematical Theory of Electrodynamic Phenomena, Uniquely Deduced from Experience*, where he argued that electricity and magnetism were related, and that electric and magnetic forces could be added up around a circuit.

After it was accepted that electricity could produce magnetism, the next obvious question was: Could magnetism produce electricity? In 1831 London bookbinder and self-taught experimental physicist Michael Faraday showed that a changing magnetic flux through a coil of wire could induce an electric current in a nearby

---

[2] Cavendish's discovery did not see print until 1879, when James Clerk Maxwell published his research notes under the title *The Electrical Researches of the Honorable Henry Cavendish*.
[3] This device is called a solenoid.

coil of wire. Faraday discovered electromagnetic field theory by mapping out electric and magnetic lines of force in drawings.

Faraday was a brilliant experimenter and thinker but lacked the mathematical background necessary to turn his discoveries into the foundation for a unified theory of electromagnetism. The mathematician of the next generation who was destined to do this job was James Clerk Maxwell, born the same year that Faraday made his most important discovery. Maxwell's work on electromagnetism began with his study of Faraday's drawings in 1856, and culminated in 1873 with the publication of *A Treatise on Electricity and Magnetism*, wherein Faraday's description of electric and magnetic lines of force, and the laws and relations of the previous 100 years of experiment, could be reduced to solutions of a set of linear partial differential equations that physics students everywhere now learn as the Maxwell equations. Maxwell's unified theory of electricity and magnetism led to a surprising conclusion about the nature of light that revolutionized science and eventually, human society.

The Maxwell equations outside of any sources of currents or charge, also known as the vacuum Maxwell equations, can be written in rationalized mksA units as

$$\vec{\nabla} \cdot \vec{E} = 0$$
$$\vec{\nabla} \cdot \vec{B} = 0$$
$$\vec{\nabla} \times \vec{E} + \frac{\partial \vec{B}}{\partial t} = 0$$
$$\vec{\nabla} \times \vec{B} - \mu_0 \epsilon_0 \frac{\partial \vec{E}}{\partial t} = 0, \tag{2.2}$$

where the constant ϵ_0 represents the permittivity of free space, and the constant μ_0 represents the magnetic permeability of free space. Using the vector calculus relation

$$\vec{\nabla} \times (\vec{\nabla} \times \vec{V}) = \vec{\nabla}(\vec{\nabla} \cdot \vec{V}) - \nabla^2 \vec{V}, \tag{2.3}$$

the above four equations can be reduced to wave equations for $\vec{E}$ and $\vec{B}$

$$\nabla^2 \vec{E} - \frac{1}{c^2} \frac{\partial^2 \vec{E}}{\partial t^2} = 0$$
$$\nabla^2 \vec{B} - \frac{1}{c^2} \frac{\partial^2 \vec{B}}{\partial t^2} = 0 \tag{2.4}$$

with the speed of the wave given by $c = (\mu_0 \epsilon_0)^{-1/2}$.

In 1862 Maxwell made the first link between the speed of light and the speed of propagation of an electromagnetic field in terms of the two vacuum constants ϵ_0 and μ_0, concluding, in his now-famous words:

We can scarcely avoid the conclusion that light consists in the transverse undulations of the same medium which is the cause of electric and magnetic phenomena.

Maxwell's reference to a medium for light transmission shows that he was not a believer in empty space, but a proponent of the ether. However, the ether that Maxwell and other physicists of his day believed in was not the plenum of Descartes but existed solely to transmit electromagnetic effects through space. Maxwell failed to see in his own equations that no such medium was necessary in order for electromagnetic waves to exist. But it was a huge leap of faith for scientists of the nineteenth century to imagine waves without a medium to do the waving, and most, if not all, were unable to make it.

Maxwell's theory that light consists of electromagnetic waves was confirmed in 1887 when Heinrich Hertz was able to generate radio waves from one loop of wire and receive them with another. Hertz showed that the speed of his generated waves matched the measured speed of light, and that his generated waves acted like light waves when it came to optical behavior such as reflection, refraction, interference and diffraction.

The unified mathematical theory of electromagnetism was confirmed and a powerful new means of transmitting information across large distances was born at the same time. Both the mathematics and the technology unleashed by the new unified theory brought further changes in physics and in society that were not even hinted at in 1887. Maxwell's set of differential equations contains a piece of information so deep that nobody even understood to look for it until Einstein appeared on the scene.

2.3 Galilean relativity and the ether

After Maxwell's electromagnetic equations of light propagation were confirmed by Hertz, the next item on the agenda was to find evidence for the medium in which light propagated. This is where the concept of relativity entered the picture. Galileo first postulated that the laws of physics should be the same for all observers moving at constant relative velocity to one another. In such a system, it is meaningless to say that any one observer is at rest, because a state of absolute rest or motion cannot be detected using the laws of physics. For example, consider some skater A tossing a tennis ball up and down while riding on roller skates at a constant velocity V relative to some other skater B. (Assume that both skate in a straight line and don't speed up or slow down during the tossing, so that the acceleration is always zero.)

In the coordinate frame $\tilde{S}$ attached to skater A, the ball is going up and down along what we will call the $\tilde{y}$ axis with zero velocity component along the

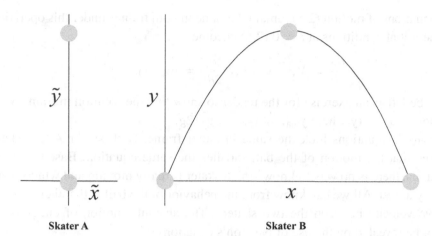

Skater A   Skater B

Fig. 2.3. In the rest frame of skater A, the tennis ball has no initial velocity in the $\tilde{x}$ direction and so only travels straight up and down. In the rest frame of skater B, the tennis ball has an initial velocity in the x direction and so skater B sees the tennis ball travel in a parabolic trajectory.

corresponding $\tilde{x}$ axis. In this frame the Newtonian equations of motion are

$$m\,\frac{d^2\tilde{x}}{d\tilde{t}^2} = 0, \qquad m\,\frac{d^2\tilde{y}}{d\tilde{t}^2} = -mg, \tag{2.5}$$

where m is the mass of the ball, and $g = 9.8 \text{ m/s}^2$ is the gravitational acceleration near the Earth's surface. The initial conditions at $\tilde{t} = 0$ are

$$\frac{d\tilde{x}}{d\tilde{t}} = 0, \qquad \tilde{x} = 0, \qquad \frac{d\tilde{y}}{d\tilde{t}} = v_0, \qquad \tilde{y} = 0, \tag{2.6}$$

where v_0 is the initial upward velocity of the ball after leaving skater A's hand at $\tilde{t} = 0$.

This set of equations is solved by the ball going straight up until it reaches a height of $\tilde{y}_{\max} = v_0^2/2g$, when it stops and then falls back into skater A's hand.

But what happens to the ball in the frame S of skater B who sees skater A skate by with velocity V along the x axis? The Galilean transformation between frames S and $\tilde{S}$ can be written as

$$\tilde{t} = t$$
$$\tilde{x} = x - Vt$$
$$\tilde{y} = y, \tag{2.7}$$

or, equivalently,

$$t = \tilde{t}$$
$$x = \tilde{x} + V\tilde{t}$$
$$y = \tilde{y}. \tag{2.8}$$

The equations of motion (2.5) remain the same in both frames under this operation, but the initial conditions at $t = 0$ (2.6) become

$$\frac{dx}{dt} = V, \quad x = 0, \quad \frac{dy}{dt} = v_0, \quad y = 0. \tag{2.9}$$

It will be left as an exercise for the reader to show that the solution in frame S is a parabolic trajectory where $y_{max} = \tilde{y}_{max} = v_0^2/2g$.

Newton's equations look the same in either frame. Both skater A and skater B agree that the motion of the ball satisfies the same equation. Based on these equations, there is no way to know which skater is really moving and which skater is really at rest. All we can know from the behavior of the ball is that there is some relative velocity between the two skaters. The absolute motion of either skater cannot be revealed by the use of Newton's equations.

This is the principle of Galilean relativity. The solutions may look different in different frames, but the equations they solve look the same. One cannot determine which frame is picked out as special by the laws of physics, because Newton's law looks the same in both frames, as long as there is no acceleration in the direction of motion. That is, as long as we are dealing with what are commonly called *inertial frames*.

This equivalence of inertial frames does not apply, however, when waves are propagating in some medium (such as the putative ether), because a wave equation is not invariant under a Galilean transformation. The rest frame of the medium in which the waves propagate picks out a special frame for the equations of motion.

Suppose we have a wave equation in two space dimensions in some coordinate frame $\tilde{S}$

$$\tilde{\nabla}^2 \psi - \frac{1}{c^2}\frac{\partial^2 \psi}{\partial \tilde{t}^2} = \frac{\partial^2 \psi}{\partial \tilde{x}^2} + \frac{\partial^2 \psi}{\partial \tilde{y}^2} - \frac{1}{c^2}\frac{\partial^2 \psi}{\partial \tilde{t}^2} = 0. \tag{2.10}$$

This equation is isotropic in space. If we rotate the $(\tilde{x}, \tilde{y})$ coordinates by some rotation matrix $R(\theta)$, Eq. (2.10) will remain the same.[4] We can write solutions to this equation that are plane waves traveling in a particular direction, but any such solution can be rotated at some angle θ in the $(\tilde{x}, \tilde{y})$ plane and still be a solution to the above equation. But now suppose that coordinates in frame $\tilde{S}$ are related to coordinates in frame S by the Galilean transformation (2.7). This equation, written in terms of coordinates in frame S, takes the completely different form

$$\frac{\partial^2 \psi}{\partial x^2}\left(1 - \frac{V^2}{c^2}\right) + \frac{\partial^2 \psi}{\partial y^2} - \frac{2V}{c^2}\frac{\partial^2 \psi}{\partial x \partial t} - \frac{1}{c^2}\frac{\partial^2 \psi}{\partial t^2} = 0. \tag{2.11}$$

---

[4] This equation assumes that the medium in which the waves propagate is an isotropic medium, with the same elastic properties in all directions.

This equation is not isotropic. Rotating the coordinates (x, y) by $R(\theta)$ will not leave this equation unchanged. What was an isotropic wave equation to observers in frame $\tilde{S}$, the rest frame of the medium, is not an isotropic equation to observers in frame S. The direction of motion in the transformation has spoiled the symmetry that made all directions equal in (2.10).

Think of a small boat traveling at speed V across a large lake into which someone has just thrown a big rock. (Imagine that this boat is so light that it doesn't make its own waves that interfere with the waves from the rock.) From the point of view of an observer at rest relative to the lake, the waves spread out at the same speed in all directions from the point where the rock entered the lake. But to an observer in the boat, the speed of the waves around the boat depends on the angle relative to the direction of the boat. An observer in the boat can use the waves around the boat to detect whether the boat is moving relative to the lake. Waves traveling in the same direction as the boat will appear to be the slowest and waves traveling in the opposite direction from the boat will appear to be the fastest. Only if the boat is at rest relative to the lake will the wave speeds in all directions be equal.

2.4 The Michelson–Morley experiment

If the propagation of light waves requires the presence of ether, just as the propagation of water waves requires the presence of water, then it should be possible to detect the passage of the Earth through this ether by measuring the speed of light waves in different directions relative to the direction of the Earth's travel through the ether. That was the goal of the Michelson–Morley experiment – to find the ether that everyone believed must be there, by measuring the speed of light in different directions as the Earth travels around the Sun.

Before we discuss the experiment itself, let's consider the problem of two swimmers in a river of width L where the current travels at speed V relative to the river bank, as shown in Figure 2.4. Each swimmer swims with a speed c relative to the rest frame of the water. Swimmer A swims across the river to a place on the bank directly opposite the starting point and back, while swimmer B swims a distance L down the bank and then swims back to the starting point. If we call the frame in which the water is at rest $\tilde{S}$, and the frame in which the river bank is at rest S, then coordinates in the two frames are related by

$$\tilde{t} = t$$
$$\tilde{x} = x + Vt$$
$$\tilde{y} = y. \tag{2.12}$$

The sign of V is positive because we are assuming that the river is traveling in the $-x$ direction.

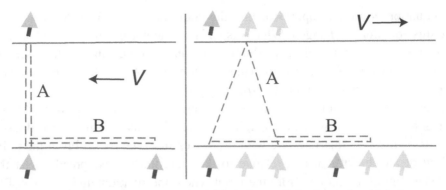

Fig. 2.4. The paths of swimmers A and B in frame S, the frame in which the trees on the river bank are at rest and the river water is moving, are shown on the left. The paths of the two swimmers in frame $\tilde{S}$, the frame in which the river water is at rest and the trees on the river bank are moving, are shown on the right.

In the water frame $\tilde{S}$ both swimmers travel with a speed that we will call c, but in the river bank frame S their speeds are not the same. The time derivative of the inverse of (2.12) gives

$$v_x = v_{\tilde{x}} - V$$
$$v_y = v_{\tilde{y}}. \tag{2.13}$$

In frame S coordinates, swimmer B swims a distance L up the river at speed $v_x = c - V$ and then swims back down the river a distance L at speed $v_x = c + V$. (Note that we must assume that $c > V$, otherwise $v_x \leq 0$ when swimmer B starts out, and he will be carried down the river by the current instead of swimming up the river against the current.) The total time elapsed is

$$\Delta t_B = \frac{L}{c + V} + \frac{L}{c - V} = \frac{2Lc}{c^2 - V^2}. \tag{2.14}$$

To get the velocity of swimmer A, we need to think a little harder. According to coordinates in the river bank frame S, swimmer A travels across the river in the y direction with some velocity $\pm v_y$, plus on the way out and minus on the way back. But according to the frame in which the water is at rest, swimmer A travels a diagonal path across the water, with total speed $c = \sqrt{v_{\tilde{x}}^2 + v_{\tilde{y}}^2}$. The condition $v_x = 0$ tells us that $v_{\tilde{x}} = V$, therefore $v_y = \pm\sqrt{c^2 - V^2}$. The total time elapsed by swimmer A going across the river and back is then

$$\Delta t_A = \frac{2L}{\sqrt{c^2 - V^2}}. \tag{2.15}$$

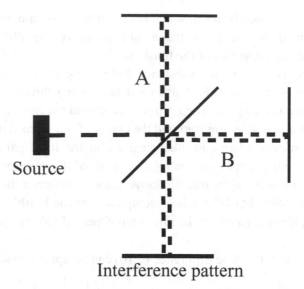

Fig. 2.5. The Michelson interferometer used in the Michelson–Morley experiment has a half-silvered mirror that splits a beam of monochromatic light from a source into two beams. One beam travels along arm A to a mirror and back, the other travels along arm B to a mirror and back, and then the two beams meet and re-combine to produce an interference pattern. This diagram shows the paths of the light beams in the rest frame of the apparatus.

The difference between the times of swimmers A and B is

$$\Delta t_B - \Delta t_A = \frac{2L}{c}\left(\frac{1}{1 - V^2/c^2} - \frac{1}{\sqrt{1 - V^2/c^2}}\right). \qquad (2.16)$$

For finite L and c, this quantity vanishes if and only if the velocity V of the river relative to the river bank is zero.

This is basically how the Michelson–Morley experiment was set up to detect the presence of the ether, using two light beams instead of two swimmers, and a set of mirrors and a detector instead of a river bank, shown in Figure 2.5. A light beam was sent through a half-silvered mirror, which split the beam into two beams, traveling orthogonal paths of equal length through the ether, according to the rest frame of the instrument. Mirrors at the ends of both arms of the instrument reflected the light beams back to a detector where their travel times could be compared.

Comparing the two times is where the problems begin. In the case of two swimmers traveling at speeds of the order of meters per second, swimming in a river flowing at a speed of the same relative order of magnitude, the kind of stopwatch available at the end of the nineteenth century was a perfectly adequate timekeeping device for the time resolution needed to measure V. But light travels too fast for its travel time to be measured by a stopwatch and, unlike our swimmers, the two light beams can't be made to wear labels that would distinguish them from one another when they arrive close together at the finish line.

If the Universe is filled with a substance that propagates light waves as water propagates water waves, then the Earth must be moving through that substance with some relative velocity V_E. The Earth orbits around the Sun at about 30 km/s, our solar system as a whole orbits around the center of our galaxy, the Milky Way, and our galaxy moves relative to other galaxies in the local galactic cluster. If there is an ether, whatever its motion relative to all of these astronomical group-ings, there must be some measurable relative velocity between the Earth and the ether, and that velocity should be at least comparable to the Earth's orbital velocity around the Sun, if not greater. So the minimum expected value of V_E/c should be about 10^{-4}.

In the limit $V \ll c$, the time difference (2.16) can be approximated by

$$\Delta t_{AB} = \Delta t_B - \Delta t_A \sim \frac{L}{c}\left(\frac{V}{c}\right)^2. \tag{2.17}$$

If $L \sim 1\,\mathrm{m}$ and $V_E/c \sim 10^{-4}$, then the time resolution necessary to measure the minimum expected value of V_E is $\Delta t \sim 3 \times 10^{-17}$ s, a feat of time measurement beyond any existing watch or clock of the era.

Luckily the wave nature of light offered physicist Albert Michelson a way to let the light beams themselves achieve the needed resolution, through the phe-nomenon of wave interference. Wave interference happens when two or more waves of the same frequency combine with different phases, making a pattern of light and dark fringes where the waves add together or cancel each other out, respectively. The Michelson interferometer, as it is now known, uses a beam of monochromatic light split through a half-silvered mirror to two orthogonal mirrors at distances L_A and L_B, then reflected back to recombine with a phase difference (when $L_A - L_B \neq 0$) that causes a set of interference fringes to be visible. When the phase difference between the light beams changes (for example by moving one of the mirrors so that $L_A - L_B$ changes) the fringes in the interference pattern will shift. If $L_A - L_B$ changes by a half wavelength of the light being used, the interference pattern will shift by one fringe. Visible light has a wavelength of the order of about 5000 Å, or 5×10^{-7} m, so a Michelson interferometer can be used to measure distances very precisely.

In the Michelson–Morley experiment, both arms of the interferometer had the same length. The source of the phase difference in the recombined beams was the time difference (2.17). A fringe shift should have occurred when the arms of the interferometer were rotated by $90°$. Such a rotation flips the sign of Δt_{AB} and so the total phase difference shifts by $2\Delta t_{AB}$. If the period of the light is $T = \lambda/c$

then the interference pattern should then shift by ΔN fringes, where

$$\Delta N = \frac{2\Delta t_{AB}}{T} \sim \frac{2L}{\lambda}\left(\frac{V}{c}\right)^2. \tag{2.18}$$

When Michelson did his first experiment in 1881, $L = 1.2\,\text{m}$ and $\lambda \sim 5 \times 10^{-7}\,\text{m}$, so the expected minimum value of ΔN was about $1/20$. This small expected value was not observed, prompting Michelson and Morley to do a more sensitive experiment in 1887, with $L = 11\,\text{m}$, achieved through multiple reflections. The interference pattern should have shifted by one half of a fringe, but again the pattern didn't appear to shift at all. Michelson repeated the experiment as the Earth revolved around the Sun, expecting to get a measurable value for V_E at some point, but his best efforts resulted in failure, at least in his own mind. In the minds of physicists today, he succeeded brilliantly, because after many further attempts using more sophisticated apparatus, Michelson's null result is now recognized to be the right result, and the ether has been banished from physics.

Philosophical support for the ether was so strong within physics that the null result of Michelson and Morley provoked some of the best physicists of the day into feats of intellectual gymnastics to try to explain it away as some complicated effect of the ether, rather than accepting it as proof that the ether does not exist. Some of these ether excuses were just plain wrong, such as the ether drag theory, where the null ether velocity came about because a layer of ether was being dragged around the Earth.

One of the wrong arguments trying to save the ether turned out to be on the right path for eliminating the ether at last. One way to cancel out the effect of the ether would be for one of the arms of the interferometer to change length to compensate the change in phase from the ether. In 1892 George Fitzgerald and Hendrik Lorentz posited that an object of length L_0 moving through the ether would be contracted in length so that

$$L = L_0\sqrt{1 - \frac{V^2}{c^2}}. \tag{2.19}$$

If the arm of the interferometer parallel to the direction of V_E was shortened by just this amount, that would compensate for the extra travel time of the light beam and produce a null result. This phenomenon relied on some interaction between the ether and the charged particles in the interferometer arm. It was never a physically compelling idea, but it took until 1932 to be experimentally ruled out. As we will see below, the idea that a length contraction of some kind is taking place was a good guess. The length contraction that we will learn about later in this chapter, named after Lorentz, who first worked out the mathematical transformation that now bears his name, comes about not because of properties of electrons in solids, but because of the relativity of time and space that Einstein informed the physics world in about 1905.

2.5 Einstein ponders electromagnetism and relativity

Albert Einstein had been fascinated by electromagnetism since he was young. One problem that really vexed him, however, was the inconsistency between the principle of Galilean relativity and the electromagnetic theory of Maxwell. Einstein was obsessed by the fact that a simple Galilean transformation has the power to render the Maxwell equations inconsistent. A Galilean transformation can be made between frames of arbitrary relative velocity, including the velocity of light. So in principle, if not in practice, it should be possible for an observer to ride along with a light wave, at the same speed. What would such an observer see?

Let's consider a plane electromagnetic wave in three space dimensions (the minimum number of dimensions where such a wave is possible). Let's orient the coordinate system in frame S so that the wave is moving in the $+x$ direction at velocity c. The Maxwell equations (2.2) are solved by

$$\vec{E} = \hat{E} \, e^{i(kx - \omega t)}$$
$$\vec{B} = \hat{B} \, e^{i(kx - \omega t)}$$
$$\hat{x} \cdot \hat{E} = \hat{x} \cdot \hat{B} = 0$$
$$\hat{x} \times \hat{E} = c\hat{B}$$
$$\hat{x} \times \hat{B} = -\hat{E}/c, \tag{2.20}$$

where $\omega/k = c$, and $\hat{E}$ and $\hat{B}$ are independent of x and t. An observer moving relative to the origin of frame S with velocity V would see the wave as

$$\vec{E} = \hat{E} \, e^{ik(\tilde{x} - (c - V)t)}, \qquad \vec{B} = \hat{B} \, e^{ik(\tilde{x} - (c - V)t)}. \tag{2.21}$$

If $V = c$, the time dependence cancels out, as one would expect for an observer traveling in the rest frame of a plane wave, and the plane wave becomes static.

According to the Maxwell equations (2.2) a static plane electromagnetic wave is not possible. A plane electromagnetic wave propagates by oscillating. The oscillation in time of the electric field in the wave produces the magnetic field, and the oscillation in time of the magnetic field in turn produces the electric field. The oscillations of each field act as the source for the other field in a radiative solution. If there were such a thing as a rest frame for a plane electromagnetic wave, it would be a frame in which the Maxwell equations should predict that the wave does not exist. Therefore the rest frame of an electromagnetic wave either must not exist, or this rest frame must define a special frame where everything that had been learned about electromagnetism since Cavendish and Coulomb no longer applies.

This inconsistency bothered Einstein a great deal. He could not ignore it, and so he decided that either Galilean relativity or the Maxwell equations had to go. Given the amount of fascination and respect Einstein had for the theory of electromagnetism, it is no surprise that the edifice he chose to attack with his intellect turned out to be the principle of Galilean relativity. But he also had great respect for the idea that there should be no special frame of reference in the laws of physics. Einstein reasoned that there must be some other principle of relativity, one respected by the theory of electromagnetism, to take the place of Galilean relativity.

Einstein was able to figure out what that new principle of relativity should be, and that is what this book is about.

2.6 Einstein's two postulates

Einstein resolved his desperate intellectual quandary, and moved physics into a new age, with the two simple but powerful postulates that he published in his famous 1905 paper *On the Electrodynamics of Moving Bodies*:

(i) All physical laws valid in one frame of reference are equally valid in any other frame of reference moving uniformly relative to the first.
(ii) The speed of light (in a vacuum) is the same in all inertial frames of reference, regardless of the motion of the light source.

The two postulates together resolved the paradoxes caused by the violation of Galilean relativity of the Maxwell equations. If the speed of light is the same in all frames of reference moving uniformly relative to one another, then one can never find a frame in which the speed of light is zero, and so one can never see a frozen electromagnetic wave.

Implications of new theory

If the speed of light is independent of the motion of the source, then time and space are relative, not absolute. The two postulates don't say that explicitly. At first reading it may not seem very radical to say that the speed of light is independent of the motion of the source. But what are the full implications of such a statement? The speed of light tells us how much space light can cross in a given amount of time. If the source is in motion, then the source is itself traversing space as a function of time. The only way for the speed of light to remain the same in all inertial frames is if space and time are themselves relative, so that the change in space and time can somehow make up for the motion of the source of light in a frame in which the source is moving. This relativity of space and time is why

Einstein's theory is called *special relativity*. (General relativity includes gravity and curved, rather than flat, spacetime.)

   Newton would have been happy that special relativity dispenses with the Cartesian plenum, but one could imagine Newton being displeased by being forced to consider that space and time are not absolute but dependent on the observer. As a few "thought experiments" will show, the relativity of time and space is an inescapable conclusion of Einstein's postulates.

Relativity of simultaneity

The first major implication of special relativity is that simultaneity is not an absolute property of time, but a relative one. That is to say, two events that happen at the same time (but not the same place) for one set of observers *happen at different times* according to any other observers moving at some uniform velocity relative to the first set.

   This is a direct outcome of the second postulate. To prove this, let's consider a thought experiment with a subway car moving at velocity v past a platform. If we label the frame of reference of the observers on the subway car as $\tilde{S}$ and label the frame of reference of the observers on the platform S, then the subway frame $\tilde{S}$ is moving at velocity v relative to the platform frame S.

   The conductor is standing in the middle of the subway car with a laser that fires two pulses at the same time in opposite directions, shown in Figure 2.6. At $\tilde{t} = 0$, she fires the laser so that pulse 1 is aimed at a detector at the rear of the

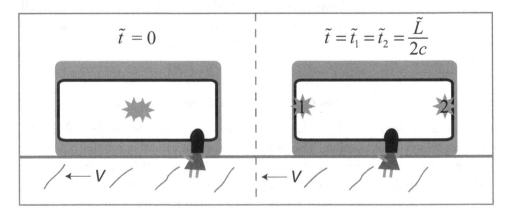

Fig. 2.6. In the rest frame of the subway car, the car stays still while the platform whizzes by at velocity V in the $-\tilde{x}$ direction. The distance pulse 1 travels to hit the rear end of the car is equal to the distance pulse 2 travels to hit the front end of the car. If both light pulses travel at the same speed c, then pulse 1 hits the rear end of the car at the same exact time that pulse 2 hits the front of the car. Events 1 and 2 are simultaneous in the rest frame of the subway car.

subway car (relative to the platform) and pulse 2 is aimed at a detector in the front of the subway car. At time $\tilde{t} = \tilde{t}_1 = \Delta\tilde{x}_1/c$, pulse 1 hits the rear of the car, and at time $\tilde{t} = \tilde{t}_2 = \Delta\tilde{x}_2/c$, pulse 2 hits the front of the car. If the conductor fires the two pulses from the exact midpoint between the two detectors, so that $\Delta\tilde{x}_1 = \Delta\tilde{x}_2 \equiv \tilde{L}/2$, then the two light pulses have to travel equal distances to get to their respective detectors and hence $\tilde{t}_1 = \tilde{t}_2$.

According to observers standing on the platform, the subway car is moving at velocity v relative to the platform, shown in Figure 2.7. By time t_1, the car has moved forward by a distance vt_1. Let's say that L is the length of the car according to observers on the platform.[5] In the platform frame, the distance pulse 1 travels to hit the detector is not $L/2$ but $L/2 - vt_1$. In Galilean relativity the speed of the light pulse 1 is not c but $c - v$. For t_1 this adds up to

$$(c - v)t_1 = \frac{L}{2} - vt_1 \rightarrow t_1 = \frac{L}{2c}. \tag{2.22}$$

Likewise, according to platform observers, the speed of pulse 2 is not c but $c + v$. Because the car is moving in this frame, the distance pulse 2 must cross is not $L/2$ but $L/2 - vt_1$. This gives

$$(c + v)t_2 = \frac{L}{2} + vt_2 \rightarrow t_2 = \frac{L}{2c}. \tag{2.23}$$

According to Galilean relativity, $t_1 = t_2$ for the observers standing on the platform, and the events are simultaneous in both frames.

This is not what happens if the speed of light is c for both the observers riding on the subway car and those standing on the platform. In that case for pulse 1 we get

$$ct_1 = \frac{L}{2} - vt_1 \rightarrow t_1 = \frac{L}{2(c + v)}. \tag{2.24}$$

For pulse 2 the answer is

$$ct_2 = \frac{L}{2} + vt_2 \rightarrow t_2 = \frac{L}{2(c - v)}. \tag{2.25}$$

The time difference between the two events according to the platform observers is

$$\Delta t = t_2 - t_1 = \frac{vL}{c^2(1 - v^2/c^2)} = \frac{\gamma^2 \beta}{c} L, \tag{2.26}$$

---

[5] In Galilean relativity, it is true that $\tilde{L} = L$, but this turns out to be a wrong assumption in special relativity, as will be shown later.

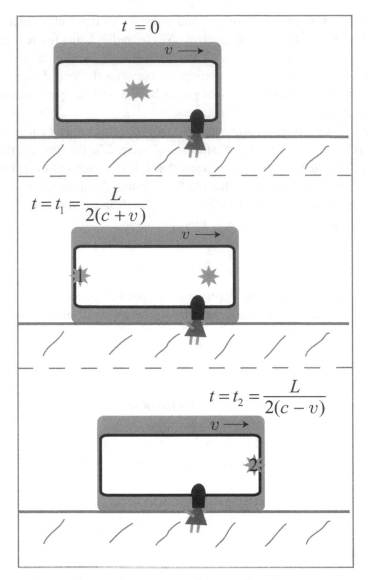

Fig. 2.7. In the rest frame of the subway platform, the subway car keeps moving in the $+x$ direction after the two light pulses are fired, and so the distance pulse 1 needs to travel to hit the rear of the car is shorter than the distance pulse 2 needs to travel to hit the front of the car. If both pulses 1 and 2 travel at the same speed c, then pulse 1 hits the rear of the car before pulse 2 has time to hit the front of the car. Events 1 and 2 happen at different times in the rest frame of the subway platform, even though they happen at the same time in the rest frame of the subway car. This *relativity of simultaneity* is mandated by the invariance of the speed of light.

where

$$\gamma \equiv \frac{1}{\sqrt{1-\beta^2}}, \qquad \beta \equiv \frac{v}{c}. \tag{2.27}$$

Clearly, $v \neq 0 \rightarrow t_1 \neq t_2$. If the speed of light is the same in both frames, then these two events that happen at the same time according to observers riding in the car do not happen at the same time according to observers standing on the platform watching the car go by.

So the second postulate of Einstein is not as simple as it sounds. If the speed of light is in fact the same in all inertial frames, then the passage of time cannot possibly be the same in all inertial frames.

Time is not absolute, but relative. A bedrock certainty of the Newtonian era was reduced to shifting sand by Einstein's work.

Time dilation

If two observers moving at uniform velocity relative to one another cannot agree on whether two events happen at the same time or not, what else are they bound to disagree on? What about the length of time that passes between two events? Does Einstein's second postulate cause disagreement here?

The classic thought experiment for testing this possibility is to have the subway conductor in the above example fire just one pulse of light, but in a direction perpendicular to the motion of the subway car, at a mirror located a distance $\Delta \tilde{y}$ from the source, shown in Figure 2.8. In the subway frame $\tilde{S}$, the total time for the light pulse to hit the mirror and return to the conductor is $\Delta \tilde{t} = 2\Delta \tilde{y}/c$.

In the platform rest frame S, between the time the pulse leaves the conductor's laser gun, hits the mirror, and returns, the subway car moves a distance $\Delta x = v\Delta t$. The light pulse according to observers standing on the platform travels a path whose total length is

$$d = 2\sqrt{\Delta y^2 + \left(\frac{\Delta x}{2}\right)^2} = 2\sqrt{\Delta y^2 + \left(\frac{v\Delta t}{2}\right)^2}. \tag{2.28}$$

If the speed of light is the same in both frames, $d = c\Delta t$, where Δt is the time interval for the laser pulse's trip to the mirror and back, according to observers on the platform. Solving for Δt gives

$$\Delta t = \frac{2\Delta y}{\sqrt{c^2 - v^2}} = \frac{2\Delta y}{c\sqrt{1 - v^2/c^2}}. \tag{2.29}$$

Because the motion of the subway car is entirely in the x direction, it seems valid to assume that $\Delta \tilde{y} = \Delta y$. If this is true, we can relate the time interval $\Delta \tilde{t}$

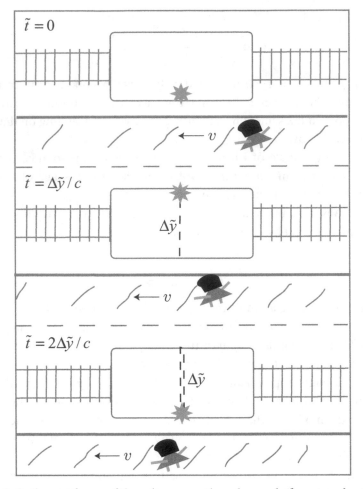

Fig. 2.8. In the rest frame of the subway car, the subway platform travels at speed v in the $-\tilde{x}$ direction, while the light pulse travels in the $\tilde{y}$ direction. The pulse travels a distance $\Delta\tilde{y}$ to the mirror on the opposite side of the car, and another $\Delta\tilde{y}$ again back to the starting point. The light pulse travels at speed c both ways, so the total travel time is $\Delta\tilde{t} = 2\Delta\tilde{y}/c$.

that has passed for observers on the subway car to the time interval Δt that has passed for observers standing on the platform through the equation

$$\Delta t = \gamma\,\Delta\tilde{t}, \tag{2.30}$$

where γ is given by (2.27). Since $\gamma \geq 1$, the time interval measured by observers for whom the subway car is moving is greater than the time interval measured by observers riding on the car. This is called **relativistic time dilation**, shown in Figure 2.9.

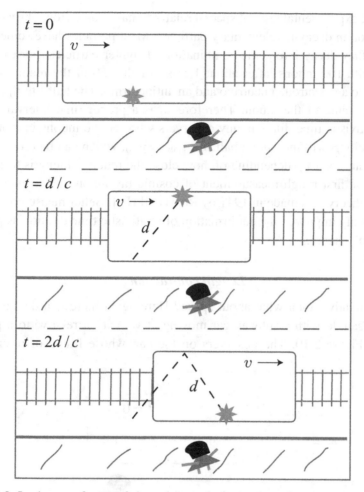

Fig. 2.9. In the rest frame of the subway platform, the subway car is moving at speed v in the $+x$ direction. The light pulse travels a diagonal path in the x and y directions, for a total distance $d = 2\sqrt{\Delta y^2 + (\Delta x/2)^2}$, with $\Delta x = v\Delta t$. If the light pulse travels at speed c both ways, then the total travel time is $\Delta t = d/c$. Solving for Δt gives $\Delta t = 2\Delta y/c\sqrt{1 - v^2/c^2}$. Since $\Delta y = \Delta\tilde{y}$, and $1/\sqrt{1 - v^2/c^2} > 1$, we can conclude that $\Delta t > \Delta\tilde{t}$. The total time Δt elapsed according to observers in the rest frame S of the subway platform is greater than the total time $\Delta\tilde{t}$ elapsed according to observers in the rest frame $\tilde{S}$ of the platform. This phenomenon is called *relativistic time dilation*.

The subway car is at rest according to observers riding inside it, so in that frame, the light pulse returns to the same location in space it left from. The time interval between two events that occur at the same location in space is called the **proper time** between those two events. The proper time between two events is the shortest possible time interval between those events. This will be proven in the next chapter.

The best experimental tests of special relativity have come from observations of time dilation in decaying elementary particles. Most particles have some half life τ_0 before they decay into some combination of lighter particles. For example, if we start with 100 muons, after about 1.5 µs, roughly 50 of the muons will have decayed into an electron, a neutrino and an antineutrino. The half life τ_0 is defined in the rest frame of the muon. Therefore τ_0 is a proper time interval. According to relativistic time dilation, any observers who see the muons in motion with some velocity β will measure a half life $\tau_{\text{lab}} = \gamma \tau_0$, which can be orders of magnitude greater than τ_0, depending on how close the muon velocity is to the speed of light. The first rough measurement of cosmic ray muon half lives to confirm Einstein's theory was made in 1941 by Rossi and Hall. Better measurements since then have only improved the confirmation of relativistic time dilation as predicted by Einstein's theory.

Length contraction

If time is relative, then what about space? Here again there is a thought experiment one can do with a subway car moving at velocity v relative to a platform, shown in Figure 2.10. The observers on the car, whose frame of reference we

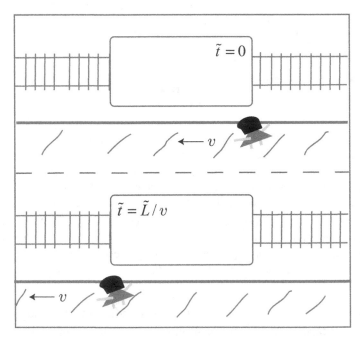

Fig. 2.10. In the rest frame of the subway car, the subway platform travels at speed v in the $-\tilde{x}$ direction. If an observer at the front end of the car sees the passenger standing on the platform go by at $\tilde{t} = 0$, then an observer at the rear end of the car will see that passenger go past at $\tilde{t} = \tilde{L}/v$, where $\tilde{L}$ is the length of the car measured in its rest frame $\tilde{S}$.

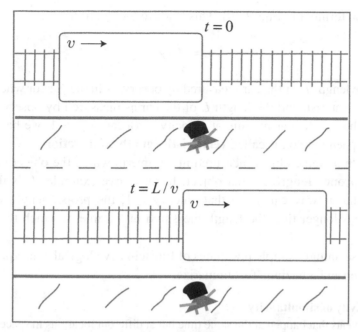

Fig. 2.11. In the rest frame of the subway platform, the subway car is moving at speed v in the $+x$ direction. The passenger waiting on the platform sees the front end of the car go by at $t = 0$ and sees the rear end go by at $t = L/v$, where L is the length of the subway car as measured in frame S where it is moving. Since the waiting passenger measures both times at the same location in space in her/his own frame S, the time interval between the measurements is a proper time interval. Using this fact to relate the passenger's time interval to the time interval measured by the observers on the subway car, we obtain the relation $L = \tilde{L}/\gamma$, which implies that $L < \tilde{L}$, since $\gamma > 1$ for $v \neq 0$. This phenomenon is called *relativistic length contraction.*

have chosen to call $\tilde{S}$, measure the length of their car to be $\tilde{L}$. They verify this by watching the front end of the car and then the rear end of the car pass the center of the platform, and measuring the amount of time $\Delta\tilde{t}$ that passes between those two events. According to the observers on the car, the car is at rest and the subway platform is moving at a velocity $-v$ relative to the car. Therefore $\tilde{L} = v\Delta\tilde{t}$.

To an observer standing at the center of the platform when the subway car goes by, the car has a length L that can be calculated by measuring the time that passes between the time when the front end of the car reaches the center of the platform and the time when the rear end of the car reaches the center of the platform, shown in Figure 2.11. Therefore according to frame S it is true that $L = v\Delta t$.

The interval Δt is a proper time interval for the observer standing on the platform, because the two events whose times are being measured both happen at the center of the platform. Therefore the time intervals $\Delta\tilde{t}$ and Δt are related to the time dilation formula in Eq. (2.30). This leads to the relation

$$L = \tilde{L}/\gamma \qquad\qquad (2.31)$$

between the length $\tilde{L}$ of the car measured by observers in the car for whom the car appears to be at rest, and the length L of the car as measured by observers on the platform who see the car moving at velocity v. Because $\gamma \geq 1$, we have $L \leq \tilde{L}$, and so this phenomenon is called **relativistic length contraction**.

The length of some object measured in a frame in which the object is at rest is called the **proper length** of that object. In the above example, $\tilde{L}$ is the proper length of the subway car. Note that since $\gamma \geq 1$, the proper length of an object is always longer than the length measured in a frame in which the object is moving.

The two seemingly simple postulates of Einstein have logical consequences that are very deep and amazing. To summarize:

- **Relativity of simultaneity**
  Two events that happen at the same time but at different locations in space according to one set of observers do not happen at the same time to any other observers moving at constant velocity relative to the first set.
- **Time dilation**
  The time interval between two events that happen at the same location according to one set of observers is always shorter than the time interval between those same two events as measured by observers moving at constant velocity relative to the first set.
- **Length contraction**
  The length of an object as measured by observers in a frame in which the object is at rest is always greater than the length of that same object as measured by observers for whom the object is in motion.

2.7 From light waves to spacetime geometry

In Chapter 1 we entertained the possibility that there could be a distance function on a unified space and time that would be analogous to the Pythagorean rule in space. The Pythagorean rule on its face tells us how to calculate the length ΔL of the hypotenuse of a right triangle, given the lengths ΔX and ΔY of the two legs of the triangle, by the formula

$$\Delta L^2 = \Delta X^2 + \Delta Y^2. \qquad\qquad (2.32)$$

But if we can describe space everywhere with a rectangular coordinate system, then the distance between any two points in space can be calculated using this rule. In differential form in rectangular coordinates (x, y) it becomes

$$dl^2 = dx^2 + dy^2, \tag{2.33}$$

and is called the Euclidean metric in two dimensions.

This metric is left invariant if we rotate the coordinate system by some angle θ

$$\begin{pmatrix} \tilde{x} \\ \tilde{y} \end{pmatrix} = R(\theta) \begin{pmatrix} x \\ y \end{pmatrix}, \tag{2.34}$$

where

$$R(\theta) = \begin{pmatrix} \cos\theta & \sin\theta \\ -\sin\theta & \cos\theta \end{pmatrix}. \tag{2.35}$$

Invariance means

$$dl^2 = dx^2 + dy^2 = d\tilde{x}^2 + d\tilde{y}^2. \tag{2.36}$$

The distance between two points in space doesn't change when we rotate the co-ordinate system in which the two points are described. The distance between two points is invariant under rotations.

We would want a metric for a unified flat spacetime to have some analogous invariance principle, but the rotation $R(\theta)$ is not the right one because, in measurable material reality, we don't seem to be able to rotate ourselves backward in time the way we can rotate ourselves backwards in space. Our search for a non-periodic version of a rotation led us to the matrix $L(\xi)$, given by

$$L(\xi) = \begin{pmatrix} \cosh\xi & -\sinh\xi \\ -\sinh\xi & \cosh\xi \end{pmatrix}. \tag{2.37}$$

This matrix acts on spacetime coordinates (τ, x), with $\tau = ct$, as

$$\begin{pmatrix} \tilde{\tau} \\ \tilde{x} \end{pmatrix} = L(\xi) \begin{pmatrix} \tau \\ x \end{pmatrix}. \tag{2.38}$$

We compared $L(\xi)$ with a Galilean transformation at velocity v for small values of the parameter ξ and found that we could rewrite the transformation in terms of a dimensionless parameter β, where $\beta = v/c$, with c having units of velocity. Using the relations $\beta = \tanh\xi$ and $\gamma = 1/\sqrt{1 - \beta^2} = \cosh\xi$, the matrix $L(\xi)$ can be rewritten as

$$L(\beta) = \begin{pmatrix} \gamma & -\beta\gamma \\ -\beta\gamma & \gamma \end{pmatrix}. \tag{2.39}$$

The analog of a Pythagorean rule that is preserved by this coordinate transformation is

$$ds^2 = -d\tau^2 + dx^2, \tag{2.40}$$

and invariance in this case means

$$ds^2 = -d\tau^2 + dx^2 = -d\tilde{\tau}^2 + d\tilde{x}^2. \tag{2.41}$$

We showed that c served as the limiting velocity of this transformation, and we promised that in this chapter we would show you that c is the speed of light.

In this chapter we traced scientific thinking about the nature of space and light through philosophical and experimental stages of understanding from Aristotle and Descartes to Cavendish and Coulomb to Maxwell, Michelson and Morley. At the end of the journey, light was understood to consist of oscillations of an electromagnetic field, one that apparently propagated without any measurable medium. A young physicist named Albert Einstein became obsessed with the fact that according to the Maxwell equations, it should not be possible to transform to a frame in which a light wave is at rest, but the principle of Galilean relativity does not forbid such a transformation. Einstein resolved that conflict through his famous two postulates, whose radical consequences for space and time we examined in the preceding section.

Now it's time to connect these two ideas together. How does the spacetime Pythagorean rule that we played with in Chapter 1 connect to the two postulates of Einstein introduced in the previous section?

The three radical consequences of Einstein's postulates for our understanding of space and time are the relativity of simultaneity, time dilation and length contraction. These same phenomena arise out of simple geometrical arguments when we employ the spacetime metric and its invariance under a Lorentz transformation between two inertial frames.

- **Relativity of simultaneity**
  Two events that happen at the same time but at different locations in space according to one set of observers do not happen at the same time to any other observers moving at constant velocity relative to the first set.

  For any two events which in some coordinate frame $\tilde{S}$ satisfy $\Delta\tilde{\tau} = 0$ and $\Delta\tilde{x} \neq 0$, it will be true in any other frame S moving at constant velocity relative to $\tilde{S}$ that $\Delta\tau \neq 0$.

  The invariant interval tells us that

  $$-\Delta\tau^2 + \Delta x^2 = -\Delta\tilde{\tau}^2 + \Delta\tilde{x}^2, \tag{2.42}$$

  where Δx^2 is shorthand for $(\Delta x)^2$, etc. If $\Delta\tilde{\tau} = 0$ and $\Delta\tilde{x} \neq 0$ then we have

  $$\Delta\tau^2 = \Delta x^2 - \Delta\tilde{x}^2. \tag{2.43}$$

This quantity is only zero if $\Delta x^2 = \Delta \tilde{x}^2$, but the only value of β in a Lorentz transformation for which this could be true is $\beta = 0$. So two events that happen at the same time in $\tilde{S}$ will not happen at the same time according to an observer in some frame S moving with velocity $\beta \neq 0$ relative to $\tilde{S}$.

The Lorentz transformation between S and $\tilde{S}$ is

$$\Delta \tau = \gamma \, \Delta \tilde{\tau} + \gamma \beta \, \Delta \tilde{x}$$
$$\Delta x = \gamma \, \Delta \tilde{x} + \gamma \beta \, \Delta \tilde{\tau}. \tag{2.44}$$

For $\Delta \tilde{\tau} = 0$,

$$\Delta \tau = \gamma \beta \, \Delta \tilde{x}. \tag{2.45}$$

The space interval $\Delta \tilde{x}$ is equal to the proper length $\tilde{L}$ in the subway car illustration. It will be left as an exercise to show that this relation is equivalent to (2.26).

- **Time dilation**

The time interval between two events that happen at the same location according to one set of observers is always shorter than the time interval between those same two events as measured by observers moving at constant velocity to the first set.

The invariant interval tells us that

$$-\Delta \tau^2 + \Delta x^2 = -\Delta \tilde{\tau}^2 + \Delta \tilde{x}^2. \tag{2.46}$$

If $\Delta \tilde{x} = 0$ and $\Delta \tilde{\tau} \neq 0$ then we have

$$\Delta \tau^2 = \Delta x^2 + \Delta \tilde{\tau}^2. \tag{2.47}$$

Since $\Delta x^2 > 0$, it must be true that $\Delta \tau > \Delta \tilde{\tau}$.

When this condition is imposed on (2.44), this gives

$$\Delta \tau = \gamma \, \Delta \tilde{\tau}, \tag{2.48}$$

which agrees with (2.30).

- **Length contraction**

The length of an object as measured by observers in a frame in which the object is at rest is always greater than the length of that same object as measured by observers for whom the object is in motion.

The proper length of an object is the interval in space $|\Delta \tilde{x}|$ between the endpoints of the object, as measured in some coordinate frame $\tilde{S}$ in which the object is at rest. But how do we measure the length of a moving object? The only reasonable way to do this is to measure both ends of the object at the same time, otherwise the measured length will end up including the distance that the object moved in the time between the measurements. Notice that this problem does not occur in the rest frame of the object, so the endpoints of an object at rest can be compared at different times and still give the proper length as their difference.

Therefore we are seeking to calculate $|\Delta x|$ for $\Delta \tau = 0$. When this condition is applied to the invariant interval (2.42) we get

$$\Delta x^2 = \Delta \tilde{x}^2 - \Delta \tilde{\tau}^2, \tag{2.49}$$

and this guarantees that $|\Delta x| < |\Delta \tilde{x}|$ for all $\beta \neq 0$. So an object's length, as measured in a frame in which it is moving, is less than its length as measured in the frame in which it is at rest.

The Lorentz transformation between the two frames gives

$$0 = \gamma \, \Delta \tilde{\tau} + \gamma \beta \, \tilde{L}$$
$$L = \gamma \, \tilde{L} + \gamma \beta \, \Delta \tilde{\tau}. \tag{2.50}$$

Eliminating $\Delta \tilde{\tau}$ between the two equations gives the relativistic length contraction formula in (2.31), in this case derived independently from time dilation.

Einstein's second postulate is that the speed of light is the same for all observers. In Galilean relativity, velocities add and subtract linearly so that if observer 1 travels at velocity v_1 with respect to observer 2, who travels at velocity v_2 in the frame of some other observer 3, the velocity of observer 1 in frame 3 is simply

$$v_3 = v_1 + v_2. \tag{2.51}$$

The Lorentz transformation preserves the speed of light through the velocity transformation rule

$$v_3 = \frac{v_1 + v_2}{1 + (v_1 v_2 / c^2)}, \tag{2.52}$$

or, using $\beta = v/c$,

$$\beta_3 = \frac{\beta_1 + \beta_2}{1 + \beta_1 \beta_2}. \tag{2.53}$$

Derivation of this formula will be left to the reader as an exercise.

Notice that if β_1 or $\beta_2 = 1$ then $\beta_3 = 1$ as well. There is no way to add velocities in special relativity to go faster than the speed of light. The Lorentz symmetry of flat spacetime preserves the speed of light in all frames. The speed of light is the same even according to an observer traveling at the speed of light. So light has no rest frame. One can never find a frame in spacetime in which the speed of light is zero.

Greek geometry originated with Pythagoras and his rule for measuring space. Einstein realized that electromagnetism was only consistent if one could not define a rest frame for an electromagnetic wave, which could only be true if the speed of light is the same for all observers. If the speed of light is the same for all observers and if all physical laws must appear the same for all inertial observers, then space and time can be treated as a unified geometrical object, called spacetime. And so now we have a geometry of spacetime, with a spacetime version of the Pythagorean rule, which in four spacetime dimensions can be written

$$ds^2 = -d\tau^2 + dx^2 + dy^2 + dz^2. \tag{2.54}$$

This object is known as the Minkowski metric, or the metric of flat spacetime. This metric is left invariant by a Lorentz transformation. This Lorentz invariance of the spacetime metric is the geometric expression of the principle of relativity proposed by Einstein's first postulate. The study of special relativity is equivalent to the study of the geometry of flat spacetime. In Chapter 3 we will examine flat spacetime geometry in four dimensions (three space plus one time) in detail.

Exercises

2.1 Find the shortest and longest distances between Jupiter and the Earth, and calculate the difference between the times it takes light to travel those two distances.

2.2 Check whether Newton's equation (2.5) looks the same in frame S as it does in frame $\tilde{S}$. Find the exact solution of this equation in frame S for the trajectory $(x(t), y(t))$ of the tennis ball for the initial conditions (2.9). What happens in the limit $V \to 0$? Does this match the solution of Newton's equation in frame $\tilde{S}$?

2.3 Find the solution to the tennis ball problem if skater A is traveling in the $\tilde{x}$ direction with constant acceleration a relative to skater B. Does the principle of Galilean relativity apply here? Why, or why not?

2.4 Suppose skaters A and B from the tennis ball problem are caught in the rain. Skater B claims that the rain is falling straight downward in frame S in the y direction with velocity V_d. Does skater A agree that the rain is falling straight downward in frame $\tilde{S}$? If not, then at what angle with the vertical axis does the rain appear to be falling in frame $\tilde{S}$? If the two skaters disagree about the angle at which the rain is falling, is there some way for them to determine which of them is right? (Assume that the raindrops have reached the constant terminal velocity V_d from air resistance and hence are no longer accelerating due to gravity.)

2.5 Suppose, instead of rain in the exercise above, we have light arriving on Earth from a distant star, making $V_d = c$, and instead of skater A, we have the planet Earth moving at orbital velocity $V_E = 30$ km/s. At what angle does the light from this star appear to hit the Earth? What happens to that angle 6 months later? What considerations have been left out of the statement of this exercise, and how are they resolved? This phenomenon is called *stellar aberration*.

2.6   According to Newton's law of gravity, the magnitude of the gravitational force between two objects of mass m_1 and m_2 separated by distance r_{12} is

$$F_{12} = \frac{Gm_1m_2}{r_{12}^2}, \tag{E2.1}$$

where G is Newton's gravitational constant and

$$r_{12} = \sqrt{(x_1 - x_2)^2 + (y_1 - y_2)^2 + (z_1 - z_2)^2}. \tag{E2.2}$$

Is this force law consistent with the principle of Galilean relativity? Why, or why not?

2.7   Represent the Galilean transformation

$$\tilde{t} = t$$
$$\tilde{x} = x - Vt$$
$$\tilde{y} = y, \tag{E2.3}$$

as a matrix equation $\tilde{X} = GX$, where $\tilde{X}$ and X are column vectors and G is a matrix. Do the same with the Galilean transformation

$$\tilde{t} = t$$
$$\tilde{x} = x$$
$$\tilde{y} = y - Ut, \tag{E2.4}$$

but call the resulting matrix H. Compute the matrix products G^2, H^2, HG, GH and $HG - GH$. Does each resulting matrix represent a Galilean transformation between two inertial frames? Find the direction and velocity of the resulting transformation in each case where this is true.

2.8   Using the Galilean transformation (E2.3), compute the differentials $(d\tilde{t}, d\tilde{x})$ in terms of the differentials (dt, dx).

2.9   Using the Galilean transformation (E2.3), compute the partial derivatives $(\partial f/\partial \tilde{t}, \partial f/\partial \tilde{x}, \partial f/\partial \tilde{y})$ in terms of the partial derivatives $(\partial f/\partial t, \partial f/\partial x, \partial f/\partial y)$. Use the chain rule for partial derivatives of some function $f(x)$

$$\frac{\partial f}{\partial \tilde{x}^\mu} = \frac{\partial f}{\partial x^\lambda} \frac{\partial x^\lambda}{\partial \tilde{x}^\mu}, \tag{E2.5}$$

where $x^\mu = (t, x, y)$ are coordinates in frame S, $\tilde{x}^\mu = (\tilde{t}, \tilde{x}, \tilde{y})$ are coordinates in frame $\tilde{S}$, and repeated indices are summed over all values.

2.10   Using the above result, show that the Galilean transformation of Eq. (2.10) gives Eq. (2.11).

2.11 Find the plane wave solution of Eq. (2.11) and check whether the dispersion relation (the relationship between the frequency and wavelength of the wave) is given by $\omega(k) = ck$. •

2.12 Using the rotation transformation

$$\tilde{t} = t$$
$$\tilde{x} = x\cos\theta + y\sin\theta$$
$$\tilde{y} = -x\sin\theta + y\cos\theta, \qquad (E2.6)$$

compute the partial derivatives $(\partial f/\partial\tilde{t}, \partial f/\partial\tilde{x}, \partial f/\partial\tilde{y})$ in terms of the partial derivatives $(\partial f/\partial t, \partial f/\partial x, \partial f/\partial y)$, and check whether this transformation preserves the form of Eq. (2.10).

2.13 Using the Lorentz transformation

$$\tilde{\tau} = \gamma\tau - \gamma\beta x$$
$$\tilde{x} = -\gamma\beta\tau + \gamma x$$
$$\tilde{y} = y, \qquad (E2.7)$$

with $\tau = ct$, $\beta = v/c$ and $\gamma = (1 - \beta^2)^{-1/2}$, compute the partial derivatives $(\partial f/\partial\tilde{\tau}, \partial f/\partial\tilde{x}, \partial f/\partial\tilde{y})$ in terms of the partial derivatives $(\partial f/\partial t, \partial f/\partial x, \partial f/\partial y)$, and check whether this transformation preserves the form of Eq. (2.10).

2.14 In the relativity of simultaneity example, if $L = 1$ m, at what fraction of the speed of light does the subway car have to be moving relative to the platform in order for observers on the platform to measure $\Delta t = 1$ s between the arrival times of the two light flashes?

2.15 In the simultaneity problem, given the relation (2.44), and bearing in mind that for this problem $\Delta\tilde{x}$ is the proper length of the subway car, derive the relation (2.26).

2.16 Suppose there is some type of particle that has a lifetime of 1 s in its own rest frame. How fast is a beam of these particles traveling according to an observer who measures the particle lifetime to be 10 s?

2.17 Verify that eliminating $\Delta\tilde{\tau}$ from (2.50) yields (2.31).

2.18 Suppose someone on rollerblades is skating on a subway car, being watched by a security guard in the car, and both of them are being watched by a passenger on the subway platform. The skater's frame has coordinates (τ_s, x_s), the subway car frame has coordinates (τ_c, x_c) and the platform frame has coordinates (τ_p, x_p). According to the security guard in the car frame, the skater travels in the $+x_c$ direction at velocity β_s, and according to the passenger on the platform, the subway car travels in the $+x_p$ direction at velocity β_c.

Light surprises everyone

(a) After a time interval $\Delta\tau_s$ in the skater frame, the skater trips and falls down. According to the security guard in the car, how far does the skater travel before falling down? Over what time interval $\Delta\tau_c$ in the subway car frame does this occur?

(b) How far does the skater travel before falling down according to the passenger on the platform? After what period of time $\Delta\tau_p$ does this happen according to the passenger?

(c) According to the passenger on the platform, how fast was the skater going before falling down?

(d) What happens if both the skater and the subway car are traveling at the speed of light?

The relation you should end up with is the velocity addition rule for special relativity.

2.19   Consider the product of two Lorentz transformations

$$L(\xi_1) = \begin{pmatrix} \cosh\xi_1 & -\sinh\xi_1 \\ -\sinh\xi_1 & \cosh\xi_1 \end{pmatrix} \tag{E2.8}$$

and

$$L(\xi_2) = \begin{pmatrix} \cosh\xi_2 & -\sinh\xi_2 \\ -\sinh\xi_2 & \cosh\xi_2 \end{pmatrix}. \tag{E2.9}$$

Show that $L(\xi_1)L(\xi_2) = L(\xi_1 + \xi_2)$, and then use the relation $\beta = \tanh\xi$ and $\gamma = \cosh\xi$ to derive the velocity addition rule for special relativity.

2.20   Suppose that in the velocity addition rule (2.53) we keep β_1 fixed and vary β_2. What is the maximum value of β_3 that is possible?

2.21   What problem arises when we try to take seriously the notion that the spacetime interval

$$ds^2 = -d\tau^2 + dx^2 \tag{E2.10}$$

measures the distance between two spacetime events? Can this problem be resolved, and if so, how?

3

Elements of spacetime geometry

Hands-on exercise: manifolds and coordinate patches

To complete this exercise you will need the following supplies:

- Two rulers or other rigid length-measuring sticks.
- Large flat surface like the top of a table or desk, similar to that used in the Chapter 1 hands-on exercise.
- Large spherical object (LSO), similar to that used in the Chapter 1 hands-on exercise.
- A device for measuring time.

Build a model of E^2 (two-dimensional Euclidean space) with two rulers taped together at right angles to make rectangular coordinate axes. Place the coordinate axes on the surface of the table or desk that you used in the Chapter 1 hands-on exercise so that the corners are touching the surface. Move them around to see what it means for the surface to be both locally and globally like E^2. Could you accurately measure the distance between any two points on this surface using only these straight axes, if they were long enough and couldn't bend?

Place the axes on the surface of the LSO that you used in the Chapter 1 hands-on exercise so that they are tangent to the surface. Move them around from point to point and ponder what it means for the surface to be locally but not globally like E^2. Could you accurately measure the distance between any two points on this surface using only these straight axes, if they were long enough and couldn't bend?

How could you add a time axis to your model of space to make a model of spacetime? With what combination of objects or devices could you measure the distance between two events in this model spacetime?

3.1 Space and spacetime

We learned some surprising things about space and time in Chapter 2. Space and time have failed to be absolute structures that can always be distinguished from one another. According to everything that we've learned about the propagation of light, space and time must transform into one another, and so rather than existing in space that evolves in time, with space and time distinct phenomena, we exist in spacetime, with space and time mixing with one another, time able to turn into space and space able to turn into time through the Lorentz transformation, which in two spacetime dimensions looks like

$$\tilde{\tau} = \gamma\tau - \gamma\beta x$$
$$\tilde{x} = -\gamma\beta\tau + \gamma x. \tag{3.1}$$

It's more than just a cliché to say that this relativity of the phenomena of space and time has enormous implications. In this chapter, however, we will ignore the enormous implications and delve into the nuts and bolts of spacetime, to look into spacetime geometry from the dry exacting perspective of mathematics, to see what makes spacetime tick, so to speak.

Coordinates in space

Before we get to spacetime, let's first clarify what we mean by space. Let's start from the abstract concept of a manifold.

To make a manifold $\mathcal{M}$, we start with a well-behaved set of points, that is, a set of points fulfilling two requirements regarding how the points in $\mathcal{M}$ can be divided into open subsets or neighborhoods. First, near any point p in $\mathcal{M}$, we should be able to find some open subset of points $U(p)$ that is near p. Second, any two points p and q in $\mathcal{M}$ can be put into different open subsets $U(p)$ and $U(q)$ of $\mathcal{M}$ that don't intersect.

For any point p in $\mathcal{M}$, points in an open neighborhood $U(p)$ near p should be describable by D real numbers $(x^1, x^2, \ldots, x^D)$ representing the values of D coordinates in a Euclidean space $\mathbf{E}^D$ in D dimensions. These coordinates define a distance Δl_{AB} between any two points p_A and p_B given by

$$\Delta l_{AB} = \sqrt{(x_A^1 - x_B^1)^2 + (x_A^2 - x_B^2)^2 + \cdots + (x_A^D - x_B^D)^2} \tag{3.2}$$

in this local neighborhood of p. However, this distance is a property of the particular choice of coordinates and might not be the distance function that we want on the manifold. Later, we will discuss adding another mathematical structure (namely, a metric) that gives an intrinsic meaning to distances independent of the choice of local coordinates.

Near some other point q, points can be described by D real numbers $(w^1, w^2, \ldots, w^D)$, with the distance Δl_{AB} between any two points q_A and q_B

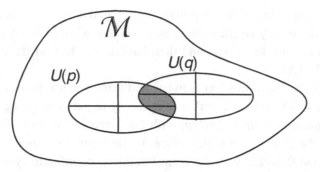

Fig. 3.1. A manifold $\mathcal{M}$ with two locally Euclidean coordinate patches $U(p)$ and $U(q)$. The intersection $U(p) \cap U(q)$ is the shaded area. The local coordinates in the two neighborhoods are related by a coordinate transformation.

the same as (3.2) but with $x \to w$. In the intersection $U(p) \cap U(q)$ of these two neighborhoods, points can be described by either local coordinate system, with a smooth coordinate transformation relating the two coordinate systems given by transition functions

$$w^i = w^i(x^1, \dots, x^D), \qquad i = 1, \dots, D. \tag{3.3}$$

This mapping from one set of local coordinates to another is one-to-one if the Jacobian

$$J = \det M, \qquad M^i_j = \frac{\partial w^i}{\partial x^j} \tag{3.4}$$

is everywhere nonzero. If $\mathcal{M}$ can be covered by a union of such coordinates patches, with the overlapping local patches of $\mathbf{E}^D$ related to one another as above, then $\mathcal{M}$ is a manifold of dimension D.[1]

The surface of the LSO in the exercise above is an example of a manifold, if we idealize it so that it is smooth at all distances scales, even the subatomic scale. Any point p on the surface of the LSO can serve as the origin of a local $\mathbf{E}^2$ coordinate patch, let's call it $U(p)$, with local coordinates (x, y) measured on the axes you made with the two rulers, with their origins located at p. Any nearby point P on the LSO could be described in terms of coordinates (x_P, y_P) on the two axes located at p. But if P is too far from p, the curvature of the LSO becomes important, and the location of P can't be described by coordinates on the axes located at p.

The way out of this problem is to define another local coordinate patch $U(q)$ at some other point q that is close to P, and then define transition functions that tell us

---

[1] The name *differentiable manifold* is also used. The transition functions define a *differentiable structure* on the manifold.

how the two coordinate patches at p and q are related in the regions $U(p) \cap U(q)$ on the LSO where they overlap. We can cover the whole LSO with coordinate patches and transition functions, and then have a way to find the coordinates of any point on the LSO.

On the LSO, the local set of axes twists and turns when it is moved from point to point on the LSO, and there isn't a simple way to patch them together and still have a rectangular coordinate system with one fixed set of axes. This is because the surface of the LSO is a curved surface. In differential geometry, the curvature can be computed from the way the local Euclidean coordinate system twists and turns from one patch to another. When there is no curvature, so that the local Euclidean coordinate system doesn't twist or turn from one patch to another, the whole manifold can be covered globally by a single Euclidean coordinate patch, and then we say that this manifold is flat.[2]

On the top of a desk or table, any point P can be described in terms of coordinates along the axes located at some point p, if we imagine the axes extended in either direction as far as necessary. We really only need one coordinate patch of $\mathbf{E}^2$ to specify the coordinates of any point q on the desk. This is because the desk is flat.

A flat space can be covered by Euclidean coordinates, but it's still flat when described in terms of any other coordinates. For example, flat space in three dimensions can be described in rectangular coordinates (x, y, z), or in spherical coordinates (r, θ, ϕ), related by the coordinate transformation

$$
\begin{aligned}
x &= r \sin\theta \, \cos\phi \\
y &= r \sin\theta \, \sin\phi \\
z &= r \cos\theta,
\end{aligned}
\tag{3.5}
$$

with $0 \leq \theta \leq \pi$ and $0 \leq \phi < 2\pi$. In these coordinates, the metric

$$
dl^2 = dx^2 + dy^2 + dz^2
\tag{3.6}
$$

becomes

$$
dl^2 = dr^2 + r^2 d\theta^2 + r^2 \sin^2\theta d\phi^2,
\tag{3.7}
$$

but the space is still flat – it's just $\mathbf{E}^3$ in spherical coordinates. We can always invert the coordinate transformation (3.5) to cover the space with rectangular coordinates again, and that's what makes the manifold flat.

---

[2] Note that this definition of flatness assumes that the manifold is infinite, or is finite with a boundary. There are flat manifolds that cannot be covered globally by one set of Euclidean coordinates because of topological issues such as periodic boundary conditions. A torus T^D in D dimensions is one such case.

Coordinates in spacetime

What happens if we add a time dimension to the flat space described above, and get flat spacetime? All of the definitions above can be recycled if we just substitute Minkowski space $\mathbf{M}^d$ for Euclidean space $\mathbf{E}^D$. The local coordinate patches have d spacetime coordinates $(x^0, x^1, x^2, \ldots, x^D)$, where $x^0 \equiv \tau = ct$, and any two points p_A and p_B in the local neighborhood $U(p)$ are separated by the Minkowski interval Δs_{AB}, given by

$$\Delta s_{AB}^2 = -(\tau_A - \tau_B)^2 + (x_A^1 - x_B^1)^2 + \cdots + (x_A^D - x_B^D)^2. \tag{3.8}$$

A flat spacetime is a spacetime manifold where one patch of Minkowski coordinates can cover the entire manifold (with a caveat regarding topological issues similar to that raised in the case of flat space). We will call this manifold $\mathbf{M}^d$, for Minkowski spacetime in d dimensions.

An important property of the Minkowski interval in (3.8) is the fact that it is the same in all inertial frames. In other words, it is unchanged when one makes a change of coordinates that corresponds to a Lorentz transformation. In saying this we are using the name Lorentz transformation in its most general sense, which includes both the possibility of a Lorentz boost by an arbitrary velocity (of magnitude less than c) and a spatial rotation, as well as an arbitrary combination of the two. In the special case of one space and one time dimension, there are no rotations and the only Lorentz transformations are boosts. When there are two or more spatial dimensions, both types of transformations are possible. The set of all such transformations forms the Lorentz group, which will be discussed in Chapter 8.

A point in spacetime has a location in space and a location in time, in other words, a point in spacetime represents an event, something that happens at a particular place at a particular time. Points p_A and p_B refer to events that we shall call E_A and E_B. The quantity Δs_{AB}^2 cannot be the square of the distance between two events, because it can be negative. But if it's not the distance between events, then what does this interval signify? It tells us about the *causal structure* of the spacetime – which event can be the cause of another event, and which cannot.

Normally, without taking special relativity into account, we would just use time to decide the question of causality, assuming that the time between two events is absolute and the same for all observers. If event E_A occurs at time t_A and event E_B occurs at time t_B, if $t_A < t_B$, then event E_A could be the cause of, or have influence on, event E_B, but if $t_B < t_A$, then the causal relationship is reversed. But this view doesn't take into account the speed of light, which is finite and the same for all observers. When we take into account the speed of light, we lose absolute time

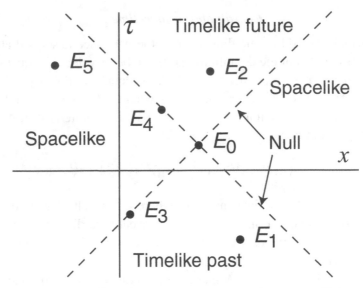

Fig. 3.2. The sign of Δs^2 divides the spacetime around any event E_0 into regions of timelike, null and spacelike separation. Events E_3 and E_4 take place on the null light cone $\mathcal{L}$ of E_0, shown above by dashed lines. Event E_2 takes place at a timelike interval to the future of E_0, and event E_1 takes place at a timelike interval to the past of E_0. Event E_5 happens at a spacelike interval from E_0, too soon for light to travel from E_0 to the location where E_5 happens. Event E_0 can only cause or influence events to the future of E_0 with a timelike or null separation from E_0. The principle of causality is encoded in the geometry of spacetime in a Lorentz invariant manner.

and learn that time is relative, that two inertial observers will not agree when two events occur at the same time.

So what happens to causality when the passage of time is relative? The answer is encoded into the spacetime geometry by the sign of the Minkowski interval, which is unchanged by a Lorentz transformation, and hence the same for all inertial observers. There are three possibilities for the sign of Δs_{AB}^2, and these three possibilities divide up the spacetime around each point into the three Lorentz invariant regions described below:

(i) $\Delta s_{AB}^2 > 0$: Events E_A and E_B are separated by a *spacelike* interval. There exists a Lorentz boost at some velocity β to a frame where events E_A and E_B happen at the same time, but there exists no Lorentz transformation to a frame where the two events happen at the same place.

(ii) $\Delta s_{AB}^2 = 0$: Events E_A and E_B are separated by a *lightlike* or *null* interval. This is the path a beam of light or a massless particle would take to get from event E_A to event E_B. There exists no Lorentz transformation at any velocity β to a frame where events E_A and E_B happen at the same time or the same place.

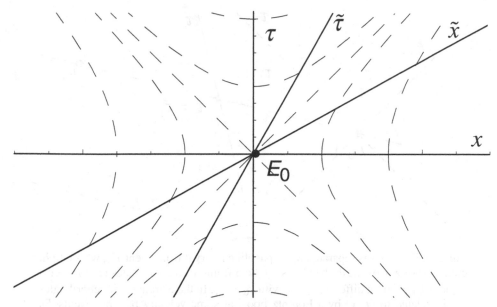

Fig. 3.3. In this spacetime diagram, the coordinates $(\tilde{\tau}, \tilde{x})$ are related to the coordinates (τ, x) by a Lorentz boost. The dashed lines represent curves of constant Δs^2 from event $E_0 = (\tau_0, x_0)$. These curves are hyperbolas, which degenerate to straight lines through the origin in the limit $\Delta s^2 \to 0$. All of the dashed lines on this diagram represent Lorentz invariant submanifolds of Minkowski spacetime. The invariant submanifold with $\Delta s^2 = 0$ is the light cone of event E_0.

(iii) $\Delta s_{AB}^2 < 0$: Events E_A and E_B are separated by a *timelike* interval. There exists a Lorentz transformation at some velocity β to a frame where events E_A and E_B happen at the same place, but there exists no Lorentz transformation to a frame where the two events happen at the same time.

The set of spacetime events that satisfy $\Delta s^2 = 0$ divides the timelike region of the spacetime of any event E_0 from the spacelike region of that event. In d spacetime dimensions, this set forms a submanifold of $\mathbf{M}^d$ whose coordinates satisfy the condition

$$(\tau - \tau_0)^2 = (x^1 - x_0^1)^2 + (x^2 - x_0^2)^2 + \cdots + (x^D - x_0^D)^2. \qquad (3.9)$$

This is called a *null hypersurface* of $\mathbf{M}^d$. Slices of constant τ on this null hypersurface are $(D-1)$ spheres representing the fronts of light waves either leaving from (for $\tau > \tau_0$) or converging at (for $\tau < \tau_0$) the event E_0. The total hypersurface $\mathcal{L}$ is a Lorentz-invariant manifold called the *light cone* or *null cone* of event E_0. The half of $\mathcal{L}$ with $\tau < \tau_0$, denoted $\mathcal{L}^-$, is called the past of event E_0 and the half of $\mathcal{N}$ with $\tau > \tau_0$, denoted $\mathcal{L}^+$, is called the future light cone of event E_0. Even though the time coordinate τ is itself relative, the light cone $\mathcal{L}$ of any event in $\mathbf{M}^d$

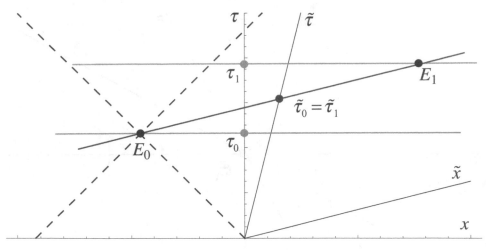

Fig. 3.4. Event E_1 is separated by a spacelike interval from event E_0, whose light cone $\mathcal{L}$ is shown by the dashed lines. In the S frame with coordinates (τ, x), events E_0 and E_1 occur at different times, with $\tau_0 < \tau_1$. In the $\tilde{S}$ frame, with coordinates $(\tilde{\tau}, \tilde{x})$ related to (τ, x) by a Lorentz boost at some velocity $\beta = \beta_0$, events E_0 and E_1 occur at the same time $\tilde{\tau}_0 = \tilde{\tau}_1$. If $\beta > \beta_0$, then E_1 happens *before* E_0. The time ordering of two events separated by a spacelike interval depends on the motion of the observer.

is invariant under a Lorentz transformation. We will show below that the light cone of an event serves as a boundary between the past and future of that event that is the same for all inertial observers.

The set of events in $\mathbf{M}^d$ with a timelike separation from an event E_0 satisfy the condition

$$\lambda^2 = (\tau - \tau_0)^2 - |\vec{x} - \vec{x}_0|^2, \tag{3.10}$$

where $\lambda^2 > 0$. Each possible value of λ defines a Lorentz-invariant submanifold of $\mathbf{M}^d$ with $(d - 1)$ dimensions. For $d = 2$, we get submanifolds of $\mathbf{M}^2$ that are the hyperbolas that cross the τ axis at $\tau = \lambda$, shown as dashed lines in the timelike regions of Figure 3.3. For $d > 2$, the submanifolds are hyperboloids of revolution in the $(d - 1)$ space coordinates. Since λ is a continuous parameter, this means the timelike region of any event can be seen as being filled by an infinite number of these surfaces. Rearranging the equation to

$$(\tau - \tau_0)^2 = \lambda^2 + |\vec{x} - \vec{x}_0|^2 \tag{3.11}$$

shows that each value of λ^2 corresponds to a minimum possible value for $(\tau - \tau_0)^2$, which occurs when $|\vec{x} - \vec{x}_0|^2 = 0$. So the parameter λ is the proper time of an observer at rest with respect to the (τ, x) coordinate system.

Since each value of λ^2 corresponds to a minimum of $(\tau - \tau_0)^2$, then $\tau - \tau_0$ can never pass through zero as long as λ doesn't pass through zero. In the limit $\lambda \to 0$, the Lorentz-invariant hyperboloid (3.10) degenerates to the light cone (3.9). Passing through the light cone takes us out of the timelike region of E_0 to the spacelike region. The past and the future timelike regions of an event E_0 are sets with zero intersection. The past light cone of E_0 is the Lorentz-invariant boundary of the timelike past of E_0, and the future light cone is the Lorentz-invariant boundary of the timelike future of E_0. Even though time and space mix in special relativity, every observer agrees about the boundaries of the timelike past and future of every event.

In the spacelike region of some event E_0, however, the picture is different. Events within the spacelike region of E_0 satisfy the condition

$$\rho^2 = |\vec{x} - \vec{x}_0|^2 - (\tau - \tau_0)^2, \tag{3.12}$$

where $\rho^2 > 0$. As with λ in the timelike region, every value of ρ defines a different Lorentz-invariant submanifold of $\mathbf{M}^d$ in the spacelike region of E_0. For $d = 2$, we get submanifolds of $\mathbf{M}^2$ that are the hyperbolas that cross the x axis at $x = \rho$, shown as dashed lines in the spacelike regions of Figure 3.3.

In the spacelike region of E_0, $\tau - \tau_0$ can pass through zero for $\vec{x} \neq \vec{x}_0$, for any value of $\rho \neq 0$. This is opposite from the situation in the timelike region. In the spacelike region of E_0, it makes no sense to say whether any event is to the future or the past of event E_0. There is always a Lorentz boost from a frame S where $\tau > \tau_0$ to some other frame $\tilde{S}$ where $\tilde{\tau} \leq \tilde{\tau}_0$, and vice versa.

Notice that for any event in the spacelike region of E_0, $|\vec{x} - \vec{x}_0|^2 \geq \rho^2$, with equality only for $\tau = \tau_0$. This means that $|\rho|$ measures the proper distance between the event and E_0. It also means that it is not possible in the spacelike region of E_0 to find a Lorentz boost to a frame where $|\vec{x} - \vec{x}_0|^2$ vanishes. In the timelike region of E_0, it is always possible to find a frame where a given event in the region takes place at $\vec{x}_0$. The spacelike and timelike regions of an event are in a sense dual to one another. Differences in location can be transformed away in the timelike region of an event, and differences in time can be transformed away in the spacelike region of an event.

The pole in the barn

The difference between spacetime and space can be appreciated by revisiting the difference between a rotation in space and a Lorentz boost in spacetime. A rotation of the rectangular coordinates in $\mathbf{E}^D$ is like a rotation of a rigid object. The axes all turn together in the same direction as much as needed. A Lorentz boost in some particular direction, for example the x^1 direction, of the coordinates in $\mathbf{M}^d$ looks like

$$\tilde{\tau} = \gamma\tau - \gamma\beta x^1$$
$$\tilde{x}^1 = -\gamma\beta\tau + \gamma x^1$$
$$\tilde{x}^i = x^i, \quad 2 \leq i \leq D. \tag{3.13}$$

When drawn in the (τ, x^1) plane, the $\tilde{\tau}$ axis makes an angle of $\Delta\phi = \pi/2 - 2\tan^{-1}\beta$ with the $\tilde{x}^1$ axis, so the angle between the axes goes to zero for $\beta \to 1$. The Lorentz boost squeezes the time and space coordinate axes into one another, as shown in Figure 3.3. That's not how a rigid body transforms. The Lorentz boost (3.13) could be said to shear the spacetime in the (τ, x^1) plane, which tells us that spacetime behaves more like an elastic medium than a rigid one.

There is a thought experiment that exemplifies this difference, in a situation that appears paradoxical according to the reasoning that we learn in rigid time and space, but which is not paradoxical at all once we understand special relativity. This experiment features a pole being moved through a barn. The pole and the barn both have proper length L_0. The pole is being carried on a rocket moving at velocity β through the barn, which has doors on the front and rear. Before the pole enters the barn, the front door is open but the rear door is closed. After the rear end of the pole passes the front door, the front door closes. When the front end of the pole is about to hit the rear door, the rear door opens.

According to observers in the rest frame S of the barn, the pole is Lorentz-contracted from its proper length L_0 to length $\Delta L_P = L_0/\gamma$, where $\gamma = 1/\sqrt{1 - \beta^2}$ as usual. The distance between the two barn doors is $\Delta L_B = L_0$. Since $\Delta L_P < \Delta L_B$, **the pole easily fits inside the barn with both doors closed**.

When we look at this same sequence of events in the rest frame $\tilde{S}$ of the pole, however, a problem arises. According to observers in the pole frame, the pole and the rocket are at rest, and the barn comes rushing at them with velocity $-\beta$. The barn is Lorentz-contracted to length $\Delta\tilde{L}_B = L_0/\gamma$, while the pole has length $\Delta\tilde{L}_P = L_0$. According to observers in this frame, $\Delta\tilde{L}_P > \Delta\tilde{L}_B$, in other words, the pole is longer than the barn, so **the pole cannot possibly fit inside the barn with both doors closed**.

This appears to violate both common sense and the symmetry that is supposed to be inherent in the principle of relativity. There shouldn't be one frame where the pole can be trapped inside the barn, and another frame where it crashes through the doors.

The key to this mystery is that it is spacetime, not space or time individually, that does the stretching and contracting. The length of each object is measured in

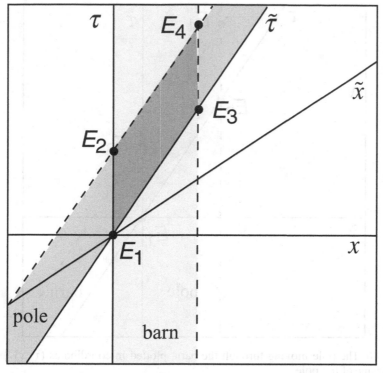

Fig. 3.5. The pole moving through the barn, plotted in coordinates (τ, x) in the rest frame of the barn.

the rest frame of the object, by comparing both ends of the object at the same time. But in special relativity, simultaneity is relative, and time ordering can be relative, if there is a spacelike separation between two events.

The sequence of events under examination is shown in Figure 3.5 in the rest frame of the barn, and in Figure 3.6 in the rest frame of the pole and rocket. In Figure 3.5, the coordinate axes (τ, x) represent the rest frame of the barn, and the coordinate axes $(\tilde{\tau}, \tilde{x})$ represent the rest frame of the pole, moving at velocity β in the $+x$ direction. The sequence of events according to the time τ in the barn rest frame is:

E1: The front end of the pole enters through the front door of the barn.
E2: The rear end of the pole enters through the front door of the barn.
E3: The front end of the pole leaves through the rear door of the barn.
E4: The rear end of the pole leaves through the rear door of the barn.

Between times τ_2 and τ_3, the pole, of length L_0/γ, is completely inside the barn, which in this frame has length L_0.

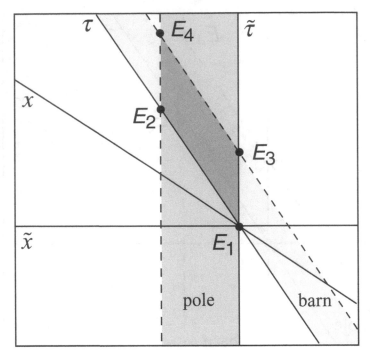

Fig. 3.6. The pole moving through the barn, plotted in coordinates $(\tilde{\tau}, \tilde{x})$ in the rest frame of the pole.

In Figure 3.6, the coordinate axes $(\tilde{\tau}, \tilde{x})$ represent the rest frame of the pole, and the coordinate axes (τ, x) represent the rest frame of the barn, moving at velocity $-\beta$ in the $+x$ direction. The sequence of events according to the time $\tilde{\tau}$ in the pole rest frame is:

E1: The front door of the barn passes the front end of the pole.
E3: The rear door of the barn passes the front end of the pole.
E2: The front door of the barn passes the rear end of the pole.
E4: The rear door of the barn passes the rear end of the pole.

In the pole frame, the sequence of events E_2 and E_3 is the opposite from what they were in the barn frame. This is possible because these two events – the rear end of the pole entering the front door of the barn, and the front end of the pole leaving the rear end of the barn – occur at a spacelike separation. (Proof of this fact will be left to the reader as an exercise.)

The pole can't be contained within the barn between times $\tilde{\tau}_2$ and $\tilde{\tau}_3$, because $\tilde{\tau}_2 > \tilde{\tau}_3$. According to observers for whom the pole is at rest and the barn is moving, the front end of the pole is already out of the rear door of the barn *before* the rear end of the pole has entered through the front door of the barn. According to the sequence of events as measured according to clocks in the rest frame of the pole, the pole is never completely contained within the barn at any time. And that is perfectly consistent with the pole, with length L_0, being longer than the barn, which in this frame has length L_0/γ.

If you feel reassured by this, then don't be! How do we normally define any object that exists in nature, such as a pole or a barn? We normally define an object to exist in space at distinct moments in time. A pole is supposed to have extent in space, not extent in time. But in special relativity, objects have extent in space and time. That's what spacetime means. The measurement of space is connected to the measurement of time. An object exists in spacetime, and the measurement of its length depends on a measurement of both space and time.

In Figures 3.5 and 3.6, the pole and barn are represented by the areas they sweep out as they move in time. These areas are called the *world sheets* of the pole and barn, for the one-dimensional representation of the pole and barn we're using here. In real life, the pole and barn sweep out world volumes in spacetime.

The pole at any moment in time is represented on the diagram by a slice of the world sheet of the pole at that time. But the time slices will be at different angles, depending on the angle of the time axis of the observer relative to the time axis in the rest frame of the pole. A slice of the pole world sheet at some time τ in the rest frame of the pole reveals a pole that has length L_0. A slice of the world sheet of the pole at some time $\tilde{\tau}$ in the barn rest frame reveals a pole that has length L_0/γ.

So what do we mean by a rigid object such as a pole or a barn? Is an object that we see in space just a particular time slice of the world volume of that object in spacetime? Is there really such a thing as a rigid object in relativity at all? We will ponder this question again in later chapters.

3.2 Vectors on a manifold

In introductory physics courses, students learn to describe an object by its location in space as a function in time, using a time-dependent vector representing the displacement from the origin of the coordinate system, which in Euclidean coordinates looks like

$$\vec{x}(t) = x(t)\hat{e}_x + y(t)\hat{e}_y + z(t)\hat{e}_z. \tag{3.14}$$

The basis vectors $(\hat{e}_x, \hat{e}_y, \hat{e}_z)$ in the set are mutually orthogonal and have unit magnitude, everywhere in space, for all values of the parameter t. If we have N such objects, each with mass m_i and position vector $\vec{x}_i(t)$, then we can talk about the collective motion of the whole ensemble by looking at the trajectory of the center of mass of the whole ensemble

$$\vec{X}_{\text{cm}}(t) = \sum_{i=1}^{N} \frac{m_i}{M} \, \vec{x}_i(t), \qquad M = \sum_{i=1}^{N} m_i. \qquad (3.15)$$

In introductory physics it is taken for granted that all of the above mathematical operations make sense. We can define vectors by the displacements in a coordinate basis, the basis vectors are the same everywhere, we can add and subtract vectors, multiply them by numbers, and check whether they are orthogonal, anywhere in this space with no problem. In other words, we assume that we are living in a vector space where all of these operations can be defined.

Newtonian physics makes sense because the physical space we are employing as a model for nature has the structure of a vector space, namely $\mathbf{E}^3$, Euclidean space in three dimensions. On a general manifold, a vector space can only be defined at each point in the manifold. In flat space and spacetime, it's possible to get away with ignoring this fact, but in this section, we will not ignore it, and we will show how vectors in spacetime are properly defined on a general manifold, before we make use of the convenient fact that the manifold we're dealing with is flat.

Properties of a vector space

Assume that $\mathbf{v}$, $\mathbf{u}$ and $\mathbf{w}$ are vectors belonging to some vector space that we will label $\mathbf{V}$, and that a and b are real numbers that we will call scalars. There are three operations defined in a vector space: addition, scalar multiplication, and vector multiplication through the inner product. Each operation comes with a set of axioms that, when taken together, guarantee that we can do the things that were done to define a vector like $\vec{X}_{\text{cm}}$ in (3.15).

The set of assumptions that we make about addition is:

(i) The sum of two vectors is a vector: $\mathbf{v} + \mathbf{u} \in \mathbf{V}$.
(ii) Addition is commutative: $\mathbf{v} + \mathbf{u} = \mathbf{u} + \mathbf{v}$.
(iii) Addition is associative: $(\mathbf{v} + \mathbf{u}) + \mathbf{w} = \mathbf{v} + (\mathbf{u} + \mathbf{w})$.
(iv) The vector $\mathbf{0} \in \mathbf{V}$ is the identity element under addition: $\mathbf{0} + \mathbf{v} = \mathbf{v}$.
(v) Every vector $\mathbf{v} \in \mathbf{V}$ has an inverse $-\mathbf{v}$ under addition: $\mathbf{v} + (-\mathbf{v}) = \mathbf{0}$.

The set of assumptions that we make about scalar multiplication is:

(i) A vector multiplied by a scalar is a vector: $a\mathbf{v} \in \mathbf{V}$.
(ii) Scalar multiplication is associative: $a(b\mathbf{v}) = (ab)\mathbf{v}$.
(iii) The scalar 1 is the identity element under scalar multiplication: $1\mathbf{v} = \mathbf{v}$.
(iv) Scalar multiplication is distributive over vectors: $a(\mathbf{v} + \mathbf{u}) = a\mathbf{v} + a\mathbf{u}$.
(v) Scalar multiplication is distributive over scalars: $(a + b)\mathbf{v} = a\mathbf{v} + b\mathbf{v}$.

Any vector in a vector space can be written as a linear combination of other vectors in the space. If we can find a set of vectors in the vector space such that any other vectors in the space can be written as a linear combination of vectors in that set, that set of vectors forms a *basis* for the vector space. If there are D such linearly independent vectors, then we say the vector space has dimension D. If we denote the set of D basis vectors by $\{\hat{e}_i\}$, where $1 \leq i \leq D$, then any vector $\mathbf{v} \in \mathbf{V}$ can be written as

$$\mathbf{v} = v^i \hat{e}_i, \tag{3.16}$$

where $\{v^i\}$ are called the components of $\mathbf{v}$ in that particular basis, and we are employing, as usual, the summation convention where pairs of upper and lower indices are summed over all values.

For example, $(\hat{e}_x, \hat{e}_y, \hat{e}_z)$ are the usual basis vectors for $\mathbf{E}^3$. Any vector in Euclidean space in three dimensions can be written as a linear combination of those three vectors.

The tangent space

Euclidean space in D dimensions has the structure of a vector space already. On a generic manifold that is not flat, or is flat but topologically nontrivial, there is no global vector space where vectors can be defined. There is only a local vector space defined from point to point. This local vector space is called the *tangent space* to the manifold. The tangent space to a manifold $\mathcal{M}$ at point p is commonly written as $T_p(\mathcal{M})$. But how are the vectors in this space defined?

In basic calculus, we learn that the derivative of a function $f(x)$ is the tangent to the curve $y = f(x)$ at point x, and this tangent line acts like a vector, because it points in some direction in the (x, y) plane. What isn't usually taught in basic calculus is that the set of derivatives operating on a function can be equivalent to a set of vectors in a vector space, and that such a set can be used to define the tangent space $T_p(\mathcal{M})$ for any $\mathcal{M}$. Let's look at the space of derivative operators $\mathbf{X} = a^i (\partial/\partial x^i)$, where the a^i are constants. $\mathbf{X}$ operates on some function $f(x)$, where x refers to all of the coordinates $\{x^i\}$ in the local patch of $\mathbf{E}^D$ in $\mathcal{M}$. The object

$$\mathbf{X}[f] = a^i \frac{\partial f}{\partial x^i} \tag{3.17}$$

represents the action of $\mathbf{X}$ on f.

If we have two such operators, say

$$\mathbf{X} = a^i \frac{\partial}{\partial x^i}, \qquad \mathbf{Y} = b^i \frac{\partial}{\partial x^i}, \tag{3.18}$$

then their sum is

$$\mathbf{X} + \mathbf{Y} = \mathbf{W}, \qquad \mathbf{W}[f] = c^i \frac{\partial f}{\partial x^i}, \tag{3.19}$$

where $c^i = a^i + b^i$. Addition and scalar multiplication of these operators is associative, commutative and distributive, one can define a $\mathbf{0}$ operator and an additive inverse. All of the properties of a vector space noted above apply to these derivative operators as well. The set of D operators $\{\partial/\partial x^i\}$ form a basis for this vector space, so the vector space has dimension D, the same as the manifold.

But what are the components $\{a^i\}$ that make $\mathbf{X}$ a vector in $T_p(\mathcal{M})$? Consider some curve γ in $\mathcal{M}$, that is, a mapping from an interval of some real parameter that we will call λ, to the manifold $\mathcal{M}$. This curve can be represented in the local coordinates of the neighborhood $U(p)$ as $x^i = x^i(\lambda)$, which passes through point p at $\lambda = 0$. The function $f(x)$ evaluated on this curve becomes $f(x(\lambda))$. If we associate the tangent to this curve evaluated at $\lambda = 0$ with the vector $\mathbf{X}$ through

$$\frac{df(x(\lambda))}{d\lambda}\bigg|_{\lambda=0} = \frac{\partial f}{\partial x^i} \frac{dx^i}{d\lambda}\bigg|_{\lambda=0} = \mathbf{X}[f], \tag{3.20}$$

then the components in question are

$$a^i = \frac{dx^i}{d\lambda}\bigg|_{\lambda=0}. \tag{3.21}$$

Partial derivatives commute with one another, that is,

$$\frac{\partial}{\partial x_i}\left(\frac{\partial f}{\partial x_j}\right) - \frac{\partial}{\partial x_j}\left(\frac{\partial f}{\partial x_i}\right) = 0. \tag{3.22}$$

When all of the basis vectors in a tangent space commute with one another, the basis is called a *coordinate basis* for $T_p(\mathcal{M})$. For example, the Euclidean basis vectors $(\hat{e}_x, \hat{e}_y, \hat{e}_z)$, used in physics when we use the coordinates (x, y, z) to describe space, correspond to the basis $(\partial_x, \partial_y, \partial_z)$. These derivatives all commute with one another, so this basis is a coordinate basis for $T_p(\mathbf{E}^3)$. As an example of a set of basis vectors that do not commute, consider the orthonormal basis

$$(\hat{e}_r, \hat{e}_\theta, \hat{e}_\phi) = \left(\frac{\partial}{\partial r}, \frac{1}{r}\frac{\partial}{\partial \theta}, \frac{1}{r\sin\theta}\frac{\partial}{\partial \phi}\right) \tag{3.23}$$

used with spherical coordinates (r, θ, ϕ) in $\mathbf{E}^3$. Commuting the first two of these, we get

$$\frac{\partial}{\partial r}\left(\frac{1}{r}\frac{\partial f}{\partial \theta}\right) - \frac{1}{r}\frac{\partial}{\partial \theta}\left(\frac{\partial f}{\partial r}\right) = -\frac{1}{r^2}\frac{\partial f}{\partial \theta} \neq 0. \tag{3.24}$$

The basis vectors (3.23) do not commute, so they form what is called a *non-coordinate basis* for $T_p(\mathbf{E}^3)$.

In flat spacetime in four dimensions, with coordinates (τ, x, y, z), the coordinate basis for $T_p(\mathbf{M}^4)$ is $(\partial_\tau, \partial_x, \partial_y, \partial_z)$. But it's not the act of adding an extra coordinate and calling it time that makes spacetime different from space. The difference between geometry in space and in spacetime has to do with the metric. To understand the metric, we also need to look at the other geometrical objects that can be defined on a manifold in addition to vectors.

The inner product and the metric

In Newtonian physics we normally employ the inner product as an operation between two vectors that yields a real number: $\mathbf{v} \cdot \mathbf{u} \in \mathbf{R}$. This is not strictly correct. In orthodox differential geometry, the inner product $\langle \mathbf{v}, \bar{\omega} \rangle \in \mathbf{R}$ takes as arguments a vector $\mathbf{v}$ in the tangent space $T_p(\mathcal{M})$, and an element $\bar{\omega}$ of a dual space to the tangent space called the cotangent space, or $T_p^*(\mathcal{M})$.

A basis $\{\bar{e}^i\}$ for $T_p^*(\mathcal{M})$ that is dual to a basis $\{\hat{e}_i\}$ for $T_p(\mathcal{M})$ can be constructed from the relation

$$\langle \hat{e}_j, \bar{e}^i \rangle = \delta_j^i. \tag{3.25}$$

If we're using a coordinate basis for $T_p(\mathcal{M})$, then the basis vectors $\{\hat{e}_i\}$ are the derivatives $\{\partial_i\}$ with respect to the local coordinates $\{x^i\}$. The corresponding dual basis for the cotangent space consists of differentials $\{dx^i\}$ of the local coordinates, which we can see from the relation

$$\left\langle \frac{\partial}{\partial x^j}, dx^i \right\rangle = \frac{\partial x^i}{\partial x^j} = \delta_j^i. \tag{3.26}$$

Here we are using dx^i as a basis one form in the cotangent space. We also use the notation dx^i in this book to represent an infinitesimal change in the coordinate x^i, so the reader should be mindful of the context in which the object dx^i is being employed.

The geometrical objects that live in the cotangent space are called *forms*, more specifically, *one forms*. Given a basis $\{dx^i\}$ of one forms, one can expand any one form $\bar{\omega} \in T_p^*(\mathcal{M})$ in components as

$$\bar{\omega} = \omega_i \, dx^i. \tag{3.27}$$

Combining (3.27) with (3.26) gives us the inner product of a vector $\mathbf{v} = v^i \partial_i$ and a form $\bar{\omega}$ in terms of their components

$$\langle \mathbf{v}, \bar{\omega} \rangle = v^i \omega_i. \tag{3.28}$$

We want a geometrical object that maps two vectors in $T_p(\mathcal{M})$ into a real number. Such an object is called the *metric tensor*. A tensor $\mathbf{T}$ is a generalization of a vector that we will discuss in greater detail later. The metric tensor $\mathbf{g}$ takes two vectors as arguments, and produces a real number. So it is a map from two copies of the tangent space $\mathbf{g} : T_p(\mathcal{M}) \otimes T_p(\mathcal{M}) \mapsto \mathbf{R}$. In a coordinate basis, using (3.26) and (3.27) we can write

$$\mathbf{g} = g_{ij}\, dx^i \otimes dx^j, \tag{3.29}$$

where here dx^i means the one form and not the infinitesimal line element. The components of the metric tensor are

$$g_{ij} = \mathbf{g}\left(\frac{\partial}{\partial x^i}, \frac{\partial}{\partial x^j}\right). \tag{3.30}$$

The metric product of two vectors $\mathbf{u}$ and $\mathbf{v}$ can then be written

$$\mathbf{g}(\mathbf{u}, \mathbf{v}) = g_{ij}\, u^i v^j. \tag{3.31}$$

It is common to call the metric product the inner product. We will call it the *scalar product*, because its value is a scalar quantity. The scalar product is related to the inner product because the metric gives us a way to associate components of vectors in the tangent space with components of forms in the cotangent space. If we define the operation of lowering an index on a component as

$$u_i = g_{ij}\, u^j, \tag{3.32}$$

then the component u_i with the lowered index could be thought of as a component of a form in the cotangent space. The scalar product of two vectors can then be written in terms of components as

$$\mathbf{g}(\mathbf{u}, \mathbf{v}) = g_{ij}\, u^i v^j = u_i\, v^i. \tag{3.33}$$

Note that this is identical to the result we get from the inner product

$$\langle \mathbf{v}, \bar{\omega}_u \rangle \tag{3.34}$$

using the one form

$$\bar{\omega}_u = u_i\, dx^i. \tag{3.35}$$

This means we can make the assignment

$$\mathbf{g}(\mathbf{u},\ ) = \bar{\omega}_u = u_i\, dx^i, \tag{3.36}$$

making $\bar{\omega}_u$ the one form associated with the vector $\mathbf{u}$ through the metric $\mathbf{g}$.

The operation (3.32) is invertible using the inverse metric, which in a coordinate basis has components that satisfy the equation

$$g^{ij} g_{jk} = \delta^i_k. \tag{3.37}$$

The inverse metric gives us a vector $\mathbf{v}_\omega$ associated with a one form $\tilde{\omega}$

$$\mathbf{v}_\omega = \omega^i \frac{\partial}{\partial x^i}, \qquad \omega^i = g^{ij} \omega_j. \tag{3.38}$$

How is the scalar product of two vectors in the tangent space related to the infinitesimal line element introduced in Chapter 1? A vector $\delta \mathbf{s}$,

$$\delta \mathbf{s} = dx^i \frac{\partial}{\partial x^i}, \tag{3.39}$$

representing an infinitesimal change in coordinates $\{x^i\}$ (where in this case dx^i is not a basis one form, but an infinitesimal change in the x^i coordinate) has the scalar product

$$ds^2 = \mathbf{g}(\delta \mathbf{s}, \delta \mathbf{s}) = g_{ij} \, dx^i dx^j. \tag{3.40}$$

So the metric operating on vectors in the tangent space also gives the line element on the manifold.

Vectors and coordinate transformations

It was common in the past for physics books to define a vector by how it behaves under a change of coordinates

$$\tilde{x}^i = \tilde{x}^i (x^1, x^2, \dots, x^D). \tag{3.41}$$

In that view, a vector $\mathbf{v}$ is represented by v^i, denoting the set of components of $\mathbf{v}$ in a coordinate basis. If an object v^i transforms under (3.41) as

$$\tilde{v}^i = \frac{\partial \tilde{x}^i}{\partial x^j} v^j, \tag{3.42}$$

then v^i is a vector.

In the modern view, the vector $\mathbf{v}$, expanded in a coordinate basis as

$$\mathbf{v} = v^i \frac{\partial}{\partial x^i}, \tag{3.43}$$

is a fundamental geometrical object that remains unchanged by a change in coordinates. The components $\{v^i\}$ of the vector change according to (3.42), and the basis vectors transform via the inverse

$$\frac{\partial}{\partial \tilde{x}^i} = \frac{\partial x^j}{\partial \tilde{x}^i} \frac{\partial}{\partial x^j},$$ (3.44)

so that

$$
\begin{aligned}
\tilde{\mathbf{v}} = \tilde{v}^i \frac{\partial}{\partial \tilde{x}^i} &= \left(\frac{\partial \tilde{x}^i}{\partial x^k} v^k \right) \left(\frac{\partial x^j}{\partial \tilde{x}^i} \frac{\partial}{\partial x^j} \right) \\
&= \left(\frac{\partial \tilde{x}^i}{\partial x^k} \frac{\partial x^j}{\partial \tilde{x}^i} \right) \left(v^k \frac{\partial}{\partial x^j} \right) = \delta^j_k v^k \frac{\partial}{\partial x^j} \\
&= v^j \frac{\partial}{\partial x^j} = \mathbf{v}.
\end{aligned}
$$ (3.45)

Performing a similar operation on a one form $\bar{\omega}$ shows that the basis one forms transform as

$$d\tilde{x}^i = \frac{\partial \tilde{x}^i}{\partial x^j} dx^j,$$ (3.46)

while the components $\{\omega_i\}$ transform as

$$\tilde{\omega}_i = \frac{\partial x^j}{\partial \tilde{x}^i} \omega_j,$$ (3.47)

and the object $\bar{\omega}$ stays the same.

The action of a coordinate transformation on the components of the metric tensor can be deduced from

$$
\begin{aligned}
\tilde{g}_{ij} d\tilde{x}^i \otimes d\tilde{x}^j &= g_{ij} dx^i \otimes dx^j \\
&= \left(g_{ij} \frac{\partial x^i}{\partial \tilde{x}^k} \frac{\partial x^j}{\partial \tilde{x}^l} \right) d\tilde{x}^k \otimes d\tilde{x}^l.
\end{aligned}
$$ (3.48)

The scalar product of two vectors is a *coordinate invariant* object, because the coordinate transformation of the components of the metric cancel the transformation of the components of the two vectors, as

$$
\begin{aligned}
\tilde{\mathbf{v}}^2 = \tilde{g}_{ij} \tilde{v}^i \tilde{v}^j &= \left(g_{kl} \frac{\partial x^k}{\partial \tilde{x}^i} \frac{\partial x^l}{\partial \tilde{x}^j} \right) \left(\frac{\partial \tilde{x}^i}{\partial x^m} \frac{\partial \tilde{x}^j}{\partial x^n} v^m v^n \right) \\
&= \left(\frac{\partial x^k}{\partial \tilde{x}^i} \frac{\partial \tilde{x}^i}{\partial x^m} \right) \left(\frac{\partial x^l}{\partial \tilde{x}^j} \frac{\partial \tilde{x}^j}{\partial x^n} \right) g_{kl} v^m v^n \\
&= \delta^k_m \delta^l_n g_{kl} v^m v^n = g_{kl} v^k v^l = \mathbf{v}^2.
\end{aligned}
$$ (3.49)

This is what it means for the metric tensor $\mathbf{g}$ to be a map $T_p\mathcal{M} \otimes T_p\mathcal{M} \mapsto \mathbf{R}$. Coordinate invariant combinations of vectors, forms and tensors are also called *scalars*, because what they yield is just a real number, which is the same when evaluated in any coordinate system.

Some classes of coordinate transformations leave the metric invariant because they represent symmetries of the spacetime. For example, Euclidean space in any dimension is the same in all directions, at every point. It is isotropic (same in all directions around some point) and homogeneous (the same at every point in some given direction). Isotropy means the Euclidean metric is invariant under rotations of the coordinate system. In $D = 2$ this takes the form

$$\tilde{x} = \cos\theta x - \sin\theta y$$
$$\tilde{y} = -\sin\theta x + \cos\theta y, \tag{3.50}$$

where θ is a constant. Homogeneity means the Euclidean metric is invariant under translations of the coordinates by constants

$$\tilde{x}^i = x^i + c^i. \tag{3.51}$$

In both cases, the line element is the same in the new coordinates as it is in the old coordinates

$$d\tilde{x}^2 + d\tilde{y}^2 = dx^2 + dy^2, \tag{3.52}$$

so the metric components in this basis are the same, $\tilde{g}_{ij} = g_{ij}$.

A coordinate transformation that leaves the metric unchanged is called an *isometry*. Isometries in spacetime give rise to conserved quantities in physics, as we shall see later in this book.

3.3 Vectors in spacetime

In Newtonian physics, the motion of an object is described by a vector in $T_p(\mathbf{E}^3)$, with time appearing as the parameter of evolution in the equations of motion. In relativity, both special and general, time and space are unified into spacetime, so that time has to be accounted for in the geometrical structures in which objects are represented, not just as a parameter in the equations of motion for the objects. In the geometry of spacetime, time is a direction in which a vector can point.

The standard way to represent this in terms of components and bases is to label the timelike direction by the index 0 with basis vector $\partial_0 = \partial_\tau$, and retain the index $\{i \mid 1 \le i \le D\}$ for the standard coordinate basis for $T_p(\mathbf{E}^D)$ with basis vectors $\{\partial_i = \partial_{x^i}\}$. The dimension of spacetime is then $d = D + 1$. Using these conventions, a vector is expanded in $T_p(\mathbf{M}^d)$ as

$$\mathbf{v} = v^\mu \frac{\partial}{\partial x^\mu} = v^\tau \frac{\partial}{\partial \tau} + v^i \frac{\partial}{\partial x^i}, \tag{3.53}$$

where the index i is assumed to be summed over all D space dimensions.

A Lorentz boost $L(\beta)$ with velocity component β^i in the i direction of frame S with coordinates $(x^0, x^1, \ldots, x^D)$ to frame $\tilde{S}$ with coordinates $(\tilde{x}^0, \tilde{x}^1, \ldots, \tilde{x}^D)$ has components

$$
\begin{aligned}
L_0^{\tilde{0}} &= \gamma \\
L_i^{\tilde{0}} &= -\gamma \beta^i \\
L_0^{\tilde{i}} &= -\gamma \beta^i \\
L_j^{\tilde{i}} &= (\gamma - 1) \frac{\beta^i \beta^j}{\beta^2} + \delta^{ij},
\end{aligned}
\tag{3.54}
$$

with the inverse transformation obtained by sending $\beta^i \to -\beta^i$. The components of **v** in the new basis are then

$$
\tilde{v}^\mu = L_\nu^{\tilde{\mu}} \, v^\nu.
\tag{3.55}
$$

The metric tensor **g** of flat spacetime has coordinate basis components $g_{\mu\nu} = \eta_{\mu\nu}$, where

$$
\begin{aligned}
\eta_{00} &= -1 \\
\eta_{0i} = \eta_{i0} &= 0 \\
\eta_{ij} &= \delta_{ij}.
\end{aligned}
\tag{3.56}
$$

The metric serves as a map between the tangent space $T_p(\mathbf{M}^d)$ and the cotangent space $T_p^*(\mathbf{M}^d)$. In a coordinate basis this means index raising and lowering via

$$
\begin{aligned}
v_\mu &= \eta_{\mu\nu} \, v^\nu \\
\omega^\mu &= \eta^{\mu\nu} \, \omega_\nu.
\end{aligned}
\tag{3.57}
$$

The scalar product $\mathbf{u} \cdot \mathbf{v} = \mathbf{g}(\mathbf{u}, \mathbf{v})$ of two vectors **u** and **v** is

$$
\mathbf{u} \cdot \mathbf{v} = \eta_{\mu\nu} v^\mu u^\nu = -v^0 u^0 + \sum_{i=1}^{D} v^i u^i.
\tag{3.58}
$$

Because a Lorentz transformation represents an isometry of the spacetime, a Lorentz transformation of this product yields

$$
\eta_{\mu\nu} v^\mu u^\nu = -v^0 u^0 + \sum_{i=1}^{D} v^i u^i = -\tilde{v}^0 \tilde{u}^0 + \sum_{i=1}^{D} \tilde{v}^i \tilde{u}^i.
\tag{3.59}
$$

The minus sign in the metric presents us with three options for the metric product of a vector with itself:

$$\mathbf{v}^2 = \eta_{\mu\nu} v^\mu v^\nu = -(v^0)^2 + \sum_{i=1}^{D} (v^i)^2 \quad \begin{cases} < 0 & \text{timelike} \\ = 0 & \text{null} \\ > 0 & \text{spacelike.} \end{cases} \qquad (3.60)$$

Since $\mathbf{v}^2$ is a scalar and hence the same for all observers, a timelike vector is timelike, a null vector is null, and a spacelike vector is spacelike, in all coordinate systems and in all inertial frames.

Timelike vectors

A rotation in space of the Euclidean coordinate axes can change the direction in space in which a vector points, so that a vector pointing in the $+x$ direction becomes a vector pointing in the $-x$ direction. But a Lorentz transformation in spacetime cannot change the direction in time in which a timelike vector points, so that a vector pointing into the future becomes a vector pointing into the past.

Let's assume that the direction of increasing coordinate τ is the future. With that convention, a timelike vector

$$\mathbf{v} = v^0 \frac{\partial}{\partial \tau} + v^i \frac{\partial}{\partial x^i}, \qquad \mathbf{v}^2 < 0 \qquad (3.61)$$

we will call *future-pointing* if $v^0 > 0$, and *past-pointing* if $v^0 < 0$. A future-pointing timelike vector cannot be transformed by a continuous Lorentz transformation into a past-pointing timelike vector. Consider the action of a Lorentz boost with spatial velocity $\vec{\beta}$ of the time component of $\mathbf{v}$

$$\begin{aligned} \tilde{v}^0 &= L^{\tilde{0}}_0 v^0 + L^{\tilde{0}}_i v^i \\ &= \gamma v^0 - \gamma \delta_{ij} \beta^i v^j \\ &= \gamma v^0 - \gamma \vec{\beta} \cdot \vec{v}. \end{aligned} \qquad (3.62)$$

The condition for $\tilde{v}^0 \leq 0$ is

$$v^0 \leq \vec{\beta} \cdot \vec{v}, \qquad (3.63)$$

which when squared gives us the inequality

$$(v^0)^2 \leq (\vec{\beta} \cdot \vec{v})^2 \leq |\vec{\beta}|^2 |\vec{v}|^2, \qquad (3.64)$$

where the last term comes from the Schwarz inequality in flat space

$$(\vec{v} \cdot \vec{u})^2 \leq (\vec{v} \cdot \vec{v})(\vec{u} \cdot \vec{u}). \qquad (3.65)$$

The timelike condition $\mathbf{v}^2 < 0$ tells us that

$$(v^0)^2 > |\vec{v}|^2. \tag{3.66}$$

Since $|\vec{\beta}| < 1$,

$$|\vec{\beta}|^2 |\vec{v}|^2 < |\vec{v}|^2, \tag{3.67}$$

and so the inequality (3.63) cannot be satisfied if $\mathbf{v}$ is a timelike vector.

So there is no inertial frame in which a future-pointing timelike vector points to the past, or a past-pointing timelike vector points to the future. The integrated mathematical structure of Minkowski spacetime conspires to keep the past and future separated in a way that preserves the notion of a direction in time for a timelike vector.

The space components of a timelike vector are another matter. A timelike vector can be transformed to point only in the time direction. The space components $\{v^i\}$ of a timelike vector $\mathbf{v}$ can always be transformed away by a Lorentz boost at some velocity $\vec{\beta}$. Proof of this will be left for the reader.

Given a timelike vector $\mathbf{v}$, let's study the set of vectors $\mathbf{u} \in T_p(\mathbf{M}^d)$ that are orthogonal to $\mathbf{v}$, so that $\mathbf{u} \cdot \mathbf{v} = 0$. The coordinate basis components of this vector satisfies

$$u^0 v^0 = \vec{u} \cdot \vec{v}, \tag{3.68}$$

which squares to

$$(u^0)^2 (v^0)^2 = (\vec{u} \cdot \vec{v})^2, \tag{3.69}$$

which by the Schwarz inequality satisfies

$$(u^0)^2 (v^0)^2 = (\vec{u} \cdot \vec{v})^2 \leq |\vec{u}|^2 |\vec{v}|^2, \tag{3.70}$$

or

$$\frac{(u^0)^2 (v^0)^2}{|\vec{u}|^2 |\vec{v}|^2} \leq 1. \tag{3.71}$$

The vector $\mathbf{v}$ is timelike, so $(v^0)^2 > |\vec{v}|^2$. The inequality (3.71) can only be satisfied if $(u^0)^2 < |\vec{u}|^2$, in other words, only if $\mathbf{u}$ is spacelike. Only a spacelike vector can be orthogonal to a timelike vector. A timelike or null vector cannot be orthogonal to a timelike vector.

A corollary to the above result is that the scalar product $\mathbf{u} \cdot \mathbf{v}$ of two timelike vectors satisfies $\mathbf{u} \cdot \mathbf{v} < 0$ if the vectors are pointing the same direction in time (either both to the future or both to the past) and satisfies $\mathbf{u} \cdot \mathbf{v} > 0$ if the two vectors are pointing the opposite direction in time. Proof of this will be left for the reader.

Velocity and momentum in spacetime

The tangent space $T_p(\mathcal{M})$ is constructed using tangent vectors to curves on the manifold $\mathcal{M}$ passing through the point p. So every vector $\mathbf{v} \in T_p(\mathcal{M})$ must be tangent to some curve (more precisely, some family of curves) on the manifold. If $\mathcal{M}$ is flat spacetime in d dimensions, then there exist vectors $\mathbf{u}$ that are timelike, with $\mathbf{g}(\mathbf{u}, \mathbf{u}) < 0$, and are hence tangent to curves representing the world lines of objects traveling through time.

The world line of an object traveling through time in $\mathbf{M}^d$ can be represented by the curve $\mathcal{C}(\lambda) = (\tau(\lambda), x^1(\lambda), \ldots, x^D(\lambda))$, with $D = d - 1$, where the parameter λ is the proper time along world line. At each point on this curve the Lorentz-invariant line element is

$$-d\lambda^2 = -d\tau^2 + \sum (dx^i)^2. \tag{3.72}$$

Dividing both sides by $d\lambda$ gives

$$-\left(\frac{d\tau}{d\lambda}\right)^2 + \sum \left(\frac{dx^i}{d\lambda}\right)^2 = -1. \tag{3.73}$$

We can also write this as

$$\mathbf{u}^2 = u^\mu u_\mu = \eta_{\mu\nu} u^\mu u^\nu = -1, \tag{3.74}$$

where

$$\mathbf{u} = u^\mu \frac{\partial}{\partial x^\mu}, \qquad u^\mu = \frac{dx^\mu}{d\lambda} \tag{3.75}$$

is the spacetime vector tangent to the curve $\mathcal{C}(\lambda)$.

For $d = 4$, which matches the world we live in as far as we can detect experimentally, $\mathbf{u}$ is called the *four-velocity*. For general d, we will call it the spacetime velocity. The space components $\{u^i\}$ of the spacetime velocity $\mathbf{u}$ are related to the usual Newtonian space velocity dx^i/dt by

$$\frac{dx^i}{d\lambda} = \frac{dx^i}{d\tau} \frac{d\tau}{d\lambda}. \tag{3.76}$$

Since λ is the proper time along the curve, then according to relativistic time dilation it must be true that $d\tau = \gamma d\lambda$. The spacetime velocity vector is then revealed to be

$$\mathbf{u} = \gamma \frac{\partial}{\partial \tau} + \gamma \beta^i \frac{\partial}{\partial x^i}, \tag{3.77}$$

where

$$\beta^i = \frac{dx^i}{d\tau} = \frac{1}{c} \frac{dx^i}{dt}. \tag{3.78}$$

If the curve $\mathcal{C}(\lambda)$ is a path of a particle or object in spacetime, then the tangent vector must represent a spacetime generalization of velocity. Normally in Newtonian physics the momentum is $\vec{p} = m\vec{v}$. If we generalize this to flat spacetime in d dimensions, then we should write

$$\mathbf{p} = m\mathbf{u}, \tag{3.79}$$

in which case (3.74) tells us that

$$\mathbf{p}^2 = \eta_{\mu\nu} p^{\mu} p^{\mu} = -(p^0)^2 + |\vec{p}|^2 = -m^2. \tag{3.80}$$

Technically speaking, momentum is a one form, and properly lives in the cotangent space $T_p^*(\mathbf{M}^d)$. However, because the spacetime metric provides an isomorphism between $T_p(\mathbf{M}^d)$ and $T_p^*(\mathbf{M}^d)$, it's usually okay to treat momentum as a vector.

Note that we have absorbed the speed of light into the coordinate system by using $\tau = ct$ as a time coordinate with units of length, so that we can write the time–time component of the Minkowski metric as $\eta_{00} = \eta_{\tau\tau} = -1$ rather than $\eta_{00} = \eta_{tt} = -c^2$, with the inverse being $\eta^{00} = \eta^{\tau\tau} = -1$ rather than $\eta^{00} = \eta^{tt} = -1/c^2$. With this choice of coordinates, spacetime velocity is dimensionless, because the proper time λ comes in units of length as well. However, this convenient choice for the metric components introduces an issue for the units of other physical quantities, as we shall see below.

The time and space components of $\mathbf{p}$ are

$$p^0 = \gamma m$$
$$\vec{p} = \gamma m \vec{\beta}. \tag{3.81}$$

Expanding γ for small β gives

$$p^0 \sim m \left(1 + \frac{1}{2}\beta^2 + \cdots\right) = \frac{1}{c^2}\left(mc^2 + \frac{1}{2}mv^2 + \cdots\right)$$
$$\vec{p} \sim m\frac{\vec{v}}{c} + \cdots \tag{3.82}$$

The time component p^0 of the spacetime momentum looks like the kinetic energy of the object whose world line is $\mathcal{C}(\lambda)$, but there is the extra term mc^2 to account for. This term is present even if the world line is the world line of an object at rest, with $\vec{v} = 0$. This term is called the *rest energy* of the object in question. The time component of the momentum is therefore the relativistic energy of the object, with a contribution from the kinetic energy and a contribution from the mass of the object at rest. This is what lies behind Einstein's famous equation

$$E = mc^2, \tag{3.83}$$

which is what we get in the limit $\beta \to 0$ if we make the assignment $p^0 = E/c^2 = \gamma m$.

The problem is that getting the factors of c out of the metric components ends up putting factors of c into the definitions of energy and space momentum. Rather than deal with all the factors of c that arise, physicists usually adjust the unit system as a whole by setting $c = 1$, in which case the condition $\mathbf{p}^2 = -m^2$ can be written

$$E^2 - |\vec{p}|^2 = m^2. \tag{3.84}$$

This equation is a relationship between the momentum and energy of a massive object called the *mass hyperboloid* or *mass shell*. You will learn more about the mass hyperboloid in Chapter 4 when we discuss particle scattering.

Lorentz boost of velocity

Consider a particle or object traveling in frame S with spacetime velocity $\mathbf{u}$ with Minkowski coordinate components $\{\gamma_u, \gamma_u \vec{\beta}_u\}$, where

$$\gamma_u = \frac{1}{\sqrt{1 - |\vec{\beta}_u|^2}}. \tag{3.85}$$

The components in some frame $\tilde{S}$ moving at velocity $\vec{\beta}$ (as measured in S) relative to S are related to the original components through a Lorentz boost

$$\tilde{u}^\mu = L^{\tilde{\mu}}_\nu u^\nu. \tag{3.86}$$

The time component of $\mathbf{u}$ transforms like

$$\begin{aligned}
\tilde{u}^0 &= L^{\tilde{0}}_0 u^0 + L^{\tilde{0}}_i u^i \\
&= \gamma u^0 - \gamma \vec{\beta} \cdot \vec{u} \\
&= \gamma \gamma_u (1 - \vec{\beta} \cdot \vec{\beta}_u).
\end{aligned} \tag{3.87}$$

Since $\tilde{u}^0 = \tilde{\gamma}_u$, we see that the Lorentz boost rule for γ_u is

$$\begin{aligned}
\tilde{\gamma}_u = \tilde{u}^0 &= \gamma \gamma_u (1 - \vec{\beta} \cdot \vec{\beta}_u) \\
&= \frac{(1 - \vec{\beta} \cdot \vec{\beta}_u)}{\sqrt{1 - |\vec{\beta}_u|^2}\sqrt{1 - |\vec{\beta}|^2}}.
\end{aligned} \tag{3.88}$$

The space components transform in a more complicated manner, with

$$\begin{aligned}
\tilde{u}^i &= L^{\tilde{i}}_0 u^0 + L^{\tilde{i}}_j u^j \\
&= u^i - \gamma \beta^i u^0 + (\gamma - 1)\frac{\vec{\beta} \cdot \vec{u}}{\beta^2} \beta^i.
\end{aligned} \tag{3.89}$$

In this form it's hard to see that this is a Lorentz boost. For simplicity let's work in $d = 3$ with $\vec{\beta} = \beta \hat{e}_x$, so that $\vec{\beta} \cdot \vec{u} = \beta u^x$. We then get

$$
\begin{aligned}
\tilde{u}^x &= u^x - \gamma \beta u^0 + (\gamma - 1) \frac{\vec{\beta} \cdot \vec{u}}{\beta^2} \beta \\
&= \gamma u^x - \gamma \beta u^0 = \gamma \gamma_u ((\beta_u)^x - \beta) \\
\tilde{u}^y &= u^y = \gamma_u (\beta_u)^y,
\end{aligned}
\tag{3.90}
$$

which is the usual formula for a Lorentz boost in one dimension (here in the x direction). The components of the transformed velocity become

$$
\begin{aligned}
(\tilde{\beta}_u)^x &= \frac{\tilde{u}^x}{\tilde{u}^0} = \frac{(\beta_u)^x - \beta}{1 - \vec{\beta} \cdot \vec{\beta}_u} \\
(\tilde{\beta}_u)^y &= \frac{\tilde{u}^y}{\tilde{u}^0} = \frac{(\beta_u)^y}{\gamma (1 - \vec{\beta} \cdot \vec{\beta}_u)}.
\end{aligned}
\tag{3.91}
$$

Notice that although the relative motion between frames S and $\tilde{S}$ is constrained to the x direction, the object's velocity in the y direction is changed by the transformation. In Galilean relativity, the components of velocity orthogonal to the relative motion between the frames are not changed, but in special relativity they are. This is necessary for the speed of light to be preserved by the Lorentz transformation. The equations in (3.91) tell us that in the limit $\beta \to 1$, $(\tilde{\beta}_u)^x \to -1$ and $(\tilde{\beta}_u)^y \to 0$. The component in the y direction ought to vanish if we're boosting the x direction by the speed of light, and the Lorentz transformation guarantees that it does.

Null vectors

The limit $m \to 0$ of (3.84) gives a null momentum vector with time component $p^0 = E = \pm|\vec{p}|$. Null vectors are traditionally labeled by letters from the middle of the alphabet, so let's call this null vector $\mathbf{k}$. A massless object has null momentum, and so travels at the speed of light. A null vector is tangent to the world line of an object traveling at the speed of light. But that world line can only be a straight line, as will be left for the reader to prove as an exercise. So a null vector is a very constrained object, unlike a timelike or a spacelike vector.

A null vector $\mathbf{k}$ is orthogonal to itself, because $\mathbf{k} \cdot \mathbf{k} = 0$. Suppose there is some other null vector $\mathbf{l}$ orthogonal to $\mathbf{k}$. If $\mathbf{k} \cdot \mathbf{l} = 0$, then

$$
|\vec{k}| \, |\vec{l}| = \vec{k} \cdot \vec{l}.
\tag{3.92}
$$

This is true if and only if $\vec{l} = \alpha\vec{k}$, where α is a constant. But if $\vec{l} = \alpha\vec{k}$, then since $l^0 = |\vec{l}|$ and $k^0 = |\vec{k}|$, it is also true that $\mathbf{l} = \alpha\mathbf{k}$. Therefore, two null vectors $\mathbf{l}$ and $\mathbf{k}$ are orthogonal if and only if $\mathbf{l}$ is a constant multiple of $\mathbf{k}$.

We have proven that a timelike vector $\mathbf{v}$ cannot be orthogonal to a null vector, because the inequality (3.71) cannot be satisfied if $(v^0)^2 > |\vec{v}|^2$ and $(u^0)^2 = |\vec{u}|^2$. But this inequality is satisfied automatically if $\mathbf{v}$ is a spacelike vector, so that $(v^0)^2 < |\vec{v}|^2$. Therefore a null vector can be orthogonal to a spacelike vector.

Spacelike vectors

Spacelike vectors are tangent to curves that are *not* the world lines of objects traveling in time. A timelike vector points in a definite direction in time, past or future, but a spacelike vector can point to the past or future depending on the Lorentz frame of the observer. The time component of a spacelike vector $\mathbf{v}$ can be gotten rid of entirely by a Lorentz boost at some velocity $\vec{\beta}$. To get rid of the time component, we need to satisfy the equation

$$v^0 = \vec{\beta} \cdot \vec{v}. \tag{3.93}$$

Squaring this equation and applying the Schwarz inequality tells us that

$$\frac{(v^0)^2}{|\vec{v}|^2} \leq |\vec{\beta}|^2 < 1, \tag{3.94}$$

which can be satisfied if and only if $\mathbf{v}$ is a spacelike vector.

3.4 Tensors and forms

From one forms to p forms

In (3.2) we learned that in addition to the tangent space $T_p(\mathcal{M})$ in which vectors are defined on $\mathcal{M}$ there is also a dual space called the cotangent space $T_p^*(\mathcal{M})$ in which objects called forms are defined. The inner product $\langle \hat{e}_\nu, \bar{e}^\mu \rangle = \delta_\nu^\mu$ shows how to construct the dual basis for $T_p^*(\mathcal{M})$ given the basis for $T_p(\mathcal{M})$. This dual relationship means that a vector $\mathbf{v}$ can be seen as an object that operates on a one form $\bar{\omega} \in T_p^*(\mathcal{M})$ and produces a real number $\langle \mathbf{v}, \bar{\omega} \rangle = v^\mu \omega_\mu$, and a one form $\bar{\omega}$ can be seen as an object that operates on a vector $\mathbf{v} \in T_p(\mathcal{M})$ and produces a real number $\langle \mathbf{v}, \bar{\omega} \rangle = v^\mu \omega_\mu$.

Because of this dual relationship, and the fact that a vector has a direction like an arrow, a one form can be viewed as a kind of surface. The action of a one form on a vector through the inner product $\langle \mathbf{v}, \bar{\omega} \rangle$ gives a real number that could be thought of as the number of surfaces of $\bar{\omega}$ pierced by the arrow $\mathbf{v}$. This provides a heuristic explanation of why momentum is properly treated as a one form, rather than a vector. The phase of a wave is given by the inner product $\langle \mathbf{x}, \bar{k} \rangle$, which yields a number that we can think of as the number of surfaces of equal de Broglie wave momentum $\bar{k}$ pierced by the vector $\mathbf{x}$.

There is an operation defined on one forms that is not defined on vectors, and that is the exterior product (or wedge product). The exterior product is an antisymmetric direct product. The exterior product of two one forms $\bar{\alpha}$ and $\bar{\beta}$ yields a two form $\bar{\omega}$ defined as

$$\bar{\omega} = \bar{\alpha} \wedge \bar{\beta} \equiv \bar{\alpha} \otimes \bar{\beta} - \bar{\beta} \otimes \bar{\alpha}. \tag{3.95}$$

By this definition we can see that

$$\bar{\alpha} \wedge \bar{\beta} = -\bar{\beta} \wedge \bar{\alpha}. \tag{3.96}$$

Therefore if $\bar{\beta} = c\bar{\alpha}$, where c is a constant, then $\bar{\beta} \wedge \bar{\alpha} = c\bar{\alpha} \wedge \bar{\alpha} = 0$.

In a spacetime coordinate basis, if $\bar{\alpha} = \alpha_\mu dx^\mu$ and $\bar{\beta} = \beta_\mu dx^\mu$, then the components of the resulting two form $\bar{\omega}$ are given by

$$\begin{aligned}
\bar{\omega} &= \alpha_\mu \beta_\nu \, (dx^\mu \otimes dx^\nu - dx^\nu \otimes dx^\mu) \\
&= \alpha_\mu \beta_\nu \, dx^\mu \wedge dx^\nu \\
&= \frac{1}{2} (\alpha_\mu \beta_\nu - \alpha_\nu \beta_\mu) \, dx^\mu \wedge dx^\nu \\
&= \frac{1}{2} \omega_{\mu\nu} \, dx^\mu \wedge dx^\nu.
\end{aligned} \tag{3.97}$$

The components of a two form are antisymmetric, $\omega_{\mu\nu} = -\omega_{\nu\mu}$. Note that the most general two form is not a product of two one forms, but any two form $\bar{\omega}$ can be expanded in a coordinate basis as

$$\bar{\omega} = \frac{1}{2} \, \omega_{\mu\nu} \, dx^\mu \wedge dx^\nu, \tag{3.98}$$

with $\omega_{\mu\nu} = -\omega_{\nu\mu}$. A two form that you will become deeply acquainted with in Chapter 5 is the electromagnetic field strength $\bar{F}$, known in spacetime component notation by $F_{\mu\nu}$. Electric and magnetic field vectors in $T_p(\mathbf{E}^3)$ do not give rise to electric and magnetic field vectors in $T_p(\mathbf{M}^4)$. The electric and magnetic fields are instead components of the two form field strength $\bar{F}$. This will be discussed in much greater detail in Chapter 5.

We can keep using the antisymmetric direct product on the coordinate basis one forms $\{dx^\mu\}$ until we run out of coordinates. If we take an antisymmetric product of p basis forms, then we get the basis for a p form, also called a form of degree

p, which can be expanded in this basis as

$$\bar{\omega} = \frac{1}{p!} \omega_{\mu_1 \dots \mu_p} \, dx^{\mu_1} \wedge \dots \wedge dx^{\mu_p}, \tag{3.99}$$

where the components $\omega_{\mu_1 \dots \mu_p}$ are antisymmetric under the exchange of any two indices. The exterior product of a p form $\bar{\alpha}_p$ and a q form $\bar{\beta}_q$ obeys the rule

$$\bar{\alpha}_p \wedge \bar{\beta}_q = (-1)^{pq} \, \bar{\beta}_q \wedge \bar{\alpha}_p. \tag{3.100}$$

Let's count how many linearly independent p forms exist in d dimensions. Each of the indices can take d values, but they must all be different in order that the differential form not vanish. Moreover, changing the order of the indices can (at most) give an overall sign change. Putting these facts together, it is clear that the number of independent p forms in d dimensions is given by the binomial coefficient

$$\binom{d}{p} = \frac{d!}{p!(d-p)!}. \tag{3.101}$$

As this formula indicates, we run out of coordinates in d dimensions when $p > d$, because

$$dx^{\mu} \wedge dx^0 \wedge dx^1 \wedge \dots \wedge dx^D = 0 \tag{3.102}$$

for any value of μ. Forms of degree p do not exist on a manifold of dimension d for $p > d$.

At the limit $p = d$, there is just one possible combination of basis forms,

$$\bar{\epsilon} = dx^0 \wedge dx^1 \wedge \dots \wedge dx^D, \tag{3.103}$$

and this is called the spacetime volume form. The components are

$$\epsilon_{\mu_0 \mu_1 \dots \mu_D} = \begin{cases} 1 & (\mu_0 \mu_1 \dots \mu_D) = \text{ even perm. of } (0\,1 \dots D) \\ -1 & (\mu_0 \mu_1 \dots \mu_D) = \text{ odd perm. of } (0\,1 \dots D) \\ 0 & \text{otherwise.} \end{cases} \tag{3.104}$$

This d form is also called the Levi-Civita permutation symbol, or the Levi-Civita tensor. A tensor is a generalization of vectors and forms that we will get to shortly. The mathematical theory of permutations is discussed in Chapter 8.

Lorentz transformation of forms

A general coordinate transformation $\tilde{x}^{\mu} = \tilde{x}^{\mu}(x^{\nu})$ acts on coordinate basis one forms dx^{μ} as

$$d\tilde{x}^{\mu} = \frac{\partial \tilde{x}^{\mu}}{\partial x^{\nu}} \, dx^{\nu}. \tag{3.105}$$

As a geometric object, a one form $\bar{\omega} = \omega_\mu dx^\mu$ exists independently of any basis used for the cotangent space in which it lives, so the transformation of the components has to cancel the transformation of the basis. Therefore, as discussed earlier, the one form components ω_μ transform as

$$\tilde{\omega}_\mu = \frac{\partial x^\nu}{\partial \tilde{x}^\mu} \, \omega_\nu. \tag{3.106}$$

We have seen that one forms and vectors transform differently under coordinate transformations. This is true, in particular, for a coordinate transformation that is a Lorentz transformation from frame S to frame $\tilde{S}$ moving at velocity $\vec{\beta}$ with respect to the S frame. Instead of the Lorentz components $L^{\tilde{\mu}}_\nu$, which appear in the transformation law of a vector, such as the velocity vector in Eq. (3.86), we need to use the inverse components $L^\nu_{\tilde{\mu}}$,

$$\tilde{\omega}_\mu = L^\nu_{\tilde{\mu}} \, \omega_\nu. \tag{3.107}$$

The inverse components of a Lorentz boost $L^\nu_{\tilde{\mu}}$ are

$$L^0_{\tilde{0}} = \gamma$$
$$L^i_{\tilde{0}} = \gamma \beta^i$$
$$L^0_{\tilde{i}} = \gamma \beta^i$$
$$L^i_{\tilde{j}} = (\gamma - 1) \frac{\beta^i \beta^j}{\beta^2} + \delta^{ij}, \tag{3.108}$$

as can be verified by matrix multiplication. The components $\omega_{\mu_1 \mu_2 \cdots \mu_p}$ of a p form $\bar{\omega}$ transform as

$$\tilde{\omega}_{\mu_1 \mu_2 \ldots \mu_p} = L^{\nu_1}_{\tilde{\mu}_1} L^{\nu_2}_{\tilde{\mu}_2} \ldots L^{\nu_p}_{\tilde{\mu}_p} \, \omega_{\nu_1 \nu_2 \ldots \nu_p}, \tag{3.109}$$

where, as usual, all pairs of repeated upper and lower indices are to be summed over all spacetime dimensions.

What is a tensor?

We can take as many copies as we want of the tangent space $T_p(\mathcal{M})$ and the cotangent space $T_p^*(\mathcal{M})$, take the direct product $T_p(\mathcal{M}) \otimes \ldots \otimes T_p(\mathcal{M}) \otimes T_p^*(\mathcal{M}) \ldots \otimes T_p^*(\mathcal{M})$, and use this as a space for defining the bases of geometrical objects of the manifold $\mathcal{M}$. Such objects are called *tensors*. If we take a direct product of m copies of the tangent space $T_p(\mathcal{M})$ and n copies of the cotangent space $T_p^*(\mathcal{M})$, then we have a space for defining what is called a rank (m, n)

tensor. A vector is a rank $(1, 0)$ tensor and a one form is a rank $(0, 1)$ tensor. A scalar, that is, a number, can be considered to be a rank $(0, 0)$ tensor.

A rank (m, n) tensor $\mathbf{T}$ can be expanded in a coordinate basis as

$$\mathbf{T} = T^{\mu_1 \cdots \mu_m}_{\nu_1 \cdots \nu_n} \frac{\partial}{\partial x^{\mu_1}} \otimes \cdots \otimes \frac{\partial}{\partial x^{\mu_m}} \otimes dx^{\nu_1} \otimes \cdots \otimes dx^{\nu_n}. \tag{3.110}$$

We learned previously that a one form $\bar{\omega} = \omega_\mu \, dx^\mu \in T^*_p(\mathcal{M})$ can be thought of as a map that operates on a vector $\mathbf{v} = v^\mu \partial_\mu \in T_p(\mathcal{M})$ to produce a coordinate invariant scalar $v^\mu \omega_\mu \in \mathbf{R}$. A rank (m, n) tensor $\mathbf{T}$ can be thought of as a map $(T^*_p(\mathcal{M}))^m \otimes (T_p(\mathcal{M}))^n \mapsto \mathbf{R}$, taking as arguments m one forms and n vectors to produce a coordinate invariant scalar

$$\mathbf{T}(\bar{\omega}_1, \ldots, \bar{\omega}_m, \mathbf{v}_1, \ldots, \mathbf{v}_n) = T^{\mu_1 \cdots \mu_m}_{\nu_1 \cdots \nu_n} \omega_{\mu_1} \ldots \omega_{\mu_m} v^{\nu_1} \ldots v^{\nu_n}. \tag{3.111}$$

When there is a metric tensor defined on the manifold, there is an isomorphism between the tangent space and the cotangent space at each point, which is expressed in a coordinate basis through the operation of index raising and lowering. This operation extends naturally from vectors and forms to tensors of any rank. The metric tensor used in this way can turn a rank (m, n) tensor into a rank $(m - 1, n + 1)$ tensor by lowering one of the lower indices. For example,

$$T^{\mu\nu}{}_\kappa = g_{\kappa\lambda} T^{\mu\nu\lambda}. \tag{3.112}$$

The inverse metric operates on a rank (m, n) tensor to produce a rank $(m + 1, n - 1)$ tensor by raising an index, for example

$$T^{\mu\nu\kappa} = g^{\kappa\lambda} T^{\mu\nu}{}_\lambda. \tag{3.113}$$

The metric tensor operates on two vectors to give a coordinate invariant scalar, so it is a map $T_p(\mathcal{M}) \otimes T_p(\mathcal{M}) \mapsto \mathbf{R}$. Therefore the metric tensor can operate on a rank (m, n) tensor to produce a rank $(m - 2, n)$ tensor, for example

$$T^\kappa = g_{\mu\nu} T^{\mu\nu\kappa}. \tag{3.114}$$

The inverse metric yields a rank $(m, n - 2)$ tensor, for example

$$T^\kappa = g^{\mu\nu} T^\kappa{}_{\mu\nu}. \tag{3.115}$$

The study of tensors in full generality is a big subject. In physics we are usually only concerned with certain types of tensors. In spacetime physics, we're concerned with the behavior under Lorentz transformations. A general tensor can be reduced into parts, each of which transforms into itself under a Lorentz transformation. When we can no longer reduce the tensor any further, we say that the individual parts are irreducible tensors. Part of this story involves symmetry and antisymmetry, as will be explained below.

When we discussed p forms, we defined the exterior product of two one forms as their antisymmetric direct product. This is meaningful as a definition because the direct product $T_p^*(\mathcal{M}) \otimes T_p^*(\mathcal{M})$ can be divided into antisymmetric and symmetric subspaces, and this division is coordinate-independent. The antisymmetric subspace of $T_p^*(\mathcal{M}) \otimes T_p^*(\mathcal{M})$ is spanned in a coordinate basis by

$$dx^\mu \wedge dx^\nu = dx^\mu \otimes dx^\nu - dx^\nu \otimes dx^\mu. \tag{3.116}$$

The components of a two form, or antisymmetric tensor of rank $(0, 2)$ are antisymmetric under exchange of indices, so that $T_{\mu\nu} = -T_{\nu\mu}$. Under a coordinate transformation, an antisymmetric tensor remains antisymmetric, so this subspace of the direct product space transforms into itself.

There is also a symmetric subspace of $T_p^*(\mathcal{M}) \otimes T_p^*(\mathcal{M})$, spanned in a coordinate basis by

$$dx^\mu \otimes dx^\nu + dx^\nu \otimes dx^\mu. \tag{3.117}$$

The components of a tensor defined in this subspace are then symmetric under exchange of indices so that $T_{\mu\nu} = T_{\nu\mu}$. Under a Lorentz transformation, a symmetric tensor remains symmetric, so this subspace of the direct product space also transforms into itself.

If we look at tensors as represented by their components in a coordinate basis, then any general rank $(0, 2)$ tensor $T_{\mu\nu}$ is the sum of its symmetric and antisymmetric parts.

$$T_{\mu\nu} = T_{(\mu\nu)} + T_{[\mu\nu]}$$

$$T_{(\mu\nu)} \equiv \frac{1}{2}(T_{\mu\nu} + T_{\nu\mu}) \quad \text{symmetric}$$

$$T_{[\mu\nu]} \equiv \frac{1}{2}(T_{\mu\nu} - T_{\nu\mu}) \quad \text{antisymmetric.} \tag{3.118}$$

But this is not yet the full reduction of the tensor into its irreducible parts. The trace $T = T_{\mu\nu}g^{\mu\nu}$ of a rank $(0, 2)$ tensor is a real number, a coordinate invariant, and hence trivially transforms into itself under a coordinate transformation. The symmetric part of a rank $(0, 2)$ tensor is the sum

$$T_{(\mu\nu)} = T_{\{\mu\nu\}} + \frac{1}{d}\, g_{\mu\nu}T$$

$$T_{\{\mu\nu\}} \equiv \frac{1}{2}(T_{\mu\nu} + T_{\nu\mu}) - \frac{1}{d}\, g_{\mu\nu}\, T \quad \text{traceless symmetric}$$

$$T \equiv g^{\kappa\lambda}T_{\kappa\lambda} \qquad \text{trace} \tag{3.119}$$

of the traceless symmetric part and the trace.

A symmetric tensor has zero antisymmetric part, and an antisymmetric tensor has zero symmetric part. The electromagnetic field strength tensor $\mathbf{F}$ is the primary example of an antisymmetric tensor of rank $(0, 2)$ used in physics, while the metric tensor **g** is the symmetric $(0, 2)$ tensor with which physicists and mathematicians tend to be the most familiar.

The symmetrization and anti-symmetrization process can be extended to tensors of any rank (m, n). A tensor can be symmetric in some pairs of indices and antisymmetric in others, or symmetric or antisymmetric in all pairs of indices.[3] Examples will be left to the reader as an exercise.

Lorentz transformation of tensors

Once you learn how to make a Lorentz transformation of the components of a vector

$$\tilde{v}^\mu = L^{\tilde{\mu}}_\nu v^\nu, \tag{3.120}$$

and a one form,

$$\tilde{\omega}_\mu = L^\nu_{\tilde{\mu}} \omega_\nu, \tag{3.121}$$

transforming the components of a general tensor of rank (m, n) is simple. You just apply vector transformations on the m upper indices, and apply one form transformations on the n lower indices to get

$$\tilde{T}^{\tilde{\mu}_1 \dots \tilde{\mu}_m}_{\tilde{\nu}_1 \dots \tilde{\nu}_n} = L^{\tilde{\mu}_1}_{\kappa_1} \dots L^{\tilde{\mu}_m}_{\kappa_m} L^{\lambda_1}_{\tilde{\nu}_1} \dots L^{\lambda_n}_{\tilde{\nu}_n} T^{\kappa_1 \dots \kappa_m}_{\lambda_1 \dots \lambda_n}. \tag{3.122}$$

3.5 The Principle of Relativity as a geometric principle

Einstein's two postulates don't say anything about a unified spacetime, manifolds, tangent spaces, metrics or tensors. Einstein made two simple, but powerful, proposals:

(i) All physical laws valid in one frame of reference are equally valid in any other frame of reference moving uniformly relative to the first.
(ii) The speed of light (in a vacuum) is the same in all inertial frames of reference, regardless of the motion of the light source.

The rich geometric structure of flat spacetime grows out of those two simple postulates if we follow them to their logical conclusions. In order for all physical laws to be equally valid in frames of reference moving relative to one another, the physical laws have to be expressed in a form that allows such a transformation to

---

[3] There are also more subtle permutation symmetries, which are neither symmetric nor antisymmetric. They are best analyzed using the mathematical theory of the symmetric group.

be defined, and that structure is a manifold with a tangent space and a cotangent space.

In order for the speed of light to be the same in all inertial frames of reference, space and time cannot be absolute and independent. Relativity of simultaneity, time dilation and length contraction are what we get when we follow the second postulate to its logical conclusions. The invariance of the speed of light leads us to a unified picture of spacetime. The spacetime coordinate transformations that leave invariant the speed of light are Lorentz transformations and spacetime translations.

In this chapter we surveyed the fundamentals of geometry in flat spacetime. In Chapter 4 we will put this geometry to work when we examine relativistic mechanics in flat spacetime.

Exercises

3.1   The following pairs of numbers represent the (τ, x) coordinates of events in a spacetime of two dimensions. Using grid paper or your favorite plotting software, plot these events on a spacetime diagram similar to Figure 3.2. Using the invariant interval between each set of events, determine which events have timelike, null or spacelike separations from the other events.

$$E_0 = (0, 0)$$
$$E_1 = (1, 3)$$
$$E_2 = (-2, 5)$$
$$E_3 = (3, 0)$$
$$E_4 = (1, -3).$$

On the same grid, draw a pair of straight lines with slope ± 1 that intersect at event E_0. In what way does this pair of lines relate to the sign of the invariant interval between some other event and E_0? What does this pair of lines represent?

3.2   Suppose that the $\tilde{S}$ frame with coordinates $(\tilde{\tau}, \tilde{x})$ is moving at velocity β relative to the S frame with coordinates (τ, x), with the events $(0, 0)$ coinciding in both frames. On the same grid as in the previous exercise, draw the $(\tilde{\tau}, \tilde{x})$ axes for $\beta = 1/5, 1/2, 4/5$.

3.3   On a (τ, x) coordinate grid, plot the curves

$$-\tau^2 + x^2 = n^2 \qquad \text{(E3.1)}$$

for values $n = 0, 1, 2, 3$, over the region between $\tau = \pm 10$ and $x = \pm 10$. Are these null, timelike or spacelike curves? How would these curves look in coordinates $(\tilde{\tau}, \tilde{x})$, where $(\tilde{\tau}, \tilde{x})$ and (τ, x) are related by a Lorentz transformation at velocity β ?

3.4   On a (τ, x) coordinate grid, plot the curves

$$-\tau^2 + x^2 = -n^2 \tag{E3.2}$$

for values $n = 0, 1, 2, 3$, over the region between $\tau = \pm 10$ and $x = \pm 10$. Are these null, timelike or spacelike curves? How would these curves look in coordinates $(\tilde{\tau}, \tilde{x})$, where $(\tilde{\tau}, \tilde{x})$ and (τ, x) are related by a Lorentz transformation at velocity β ?

3.5   In three spacetime dimensions $(d = 3)$ with coordinates (t, x, y), the Minkowski metric can be written

$$ds^2 = -d\tau^2 + dx^2 + dy^2. \tag{E3.3}$$

(a) Rewrite this metric using coordinates (u, v, y), where $u = \tau + x$ and $v = \tau - x$.

(b) Forgetting about the y direction, on a piece of grid paper or using your favorite computer software, plot the (u, v) axes in (τ, x) coordinates.

(c) What type of world line is represented by a line of constant u or v?

Coordinates such as (u, v) are known as null coordinates, or light-cone coordinates. If one wants to learn string theory, it is a good idea to become familiar with light cone coordinates.

3.6   Consider the pole in the barn scenario discussed in this chapter. Prove that in the case where the pole and barn have the same proper length L_0, the events E_2 and E_3 always have a spacelike separation. Suppose the barn and pole have different proper lengths. Under what conditions, if any, can the separation between events E_2 and E_3 be timelike? In such a case, is there any contradiction in the time ordering between the two events in the pole frame and the barn frame?

3.7   Consider a thin pole of proper length L_0 at rest in the S frame, with one end at $x = 0$, $y = 0$ and the other end at $x = L_0$, $y = 0$. Suppose we are looking at this pole in a universe with three spacetime dimensions with coordinates (τ, x, y).

(a) Find the equation for the world line of each end of the pole in frame S.

(b) Consider an observer in frame $\tilde{S}$ with coordinates $(\tilde{\tau}, \tilde{x}, \tilde{y})$ moving at velocity $\beta^x = 0$, $\beta^y = \beta$ relative to frame S. Using (3.54), compute the equations for the world lines of the ends of the pole in terms of $(\tilde{\tau}, \tilde{x}, \tilde{y})$. What is the length of the pole according to the observer in frame $\tilde{S}$?

(c) Suppose frame $\tilde{S}$ moves instead with velocity $\beta^x = \beta^y = \beta/\sqrt{2}$. Using (3.54), compute the equations for the world lines of the ends of the pole in terms of $(\tilde{\tau}, \tilde{x}, \tilde{y})$ in this case, and compare with your answer above. What is the length of the pole according to the observer in frame $\tilde{S}$?

3.8   Using (3.54), write the Lorentz transformation for four spacetime dimensions $(d = 4)$ as a 4×4 matrix, and calculate the determinant.

3.9   Consider the pole in the barn scenario discussed in this chapter. Imagine a pole with proper length $L_0 = 10$ m heading towards a barn with proper length $L_0 = 10$ m at speed $\beta = 4/5$. Let's call the frame in which the barn is at rest S with coordinates (τ, x), and label the frame in which the pole is at rest by $\tilde{S}$ with coordinates $(\tilde{\tau}, \tilde{x})$. Suppose that the leading edge of the pole passes the front door of the barn at time $\tau = \tilde{\tau} = 0$. Find the total amount of time $\Delta \tau_i$ that the pole is completely inside the barn. What do you learn when you try to calculate the corresponding interval $\Delta \tilde{\tau}_i$?

3.10   Picture a rigid pole at rest according to an observer in frame S with spacetime coordinates (τ, x). One end is at $x = 0$ and the other is at $x = L_0$.
(a) Draw the world sheet of the pole between $\tau = 0$ and $\tau = L$.
(b) On the same plot, draw a line representing the path of a flash of light at $\tau = x = 0$ aimed along the pole in the $+x$ direction.
(c) Suppose the end of the pole at $x = 0$ is sharply tapped at $\tau = 0$. On your plot, identify the set of events on the world sheet of the pole that could possibly be influenced by the tap at the end of the pole, according to special relativity. Identify the set of events on the world sheet of the pole that could not possibly be influenced by the tap at the end of the pole, according to special relativity.
(d) Suppose the tap at the end $x = 0$ is forceful enough to make the pole move at speed $\beta = 1/5$. On a new plot, again using the (τ, x) coordinate system, draw a possible world sheet for the accelerating pole from time $\tau = 0$ to $\tau = 2L_0$. Make sure that this world sheet is consistent with special relativity.
(e) What conclusion would you draw from this exercise about the nature of a rigid body in spacetime?

3.11   The metric for flat space in three dimensions can be written using Euclidean coordinates (x, y, z) as

$$ds^2 = dx^2 + dy^2 + dz^2. \tag{E3.4}$$

Rewrite this metric in (u, v, w) coordinates and describe the curves of constant u, v and w in each case given below
(a) Cylindrical coordinates

$$x = u \cos w$$
$$y = u \sin w$$
$$z = v, \tag{E3.5}$$

(b) Spherical coordinates

$$x = u \cos w \, \sin v$$
$$y = u \sin w \, \sin v$$
$$z = u \cos v. \tag{E3.6}$$

3.12 Consider two flat space dimensions, with Euclidean coordinates (x, y). Let's write the metric in tensor form as

$$\mathbf{g} = dx \otimes dx + dy \otimes dy. \tag{E3.7}$$

(a) Rewrite this metric in polar coordinates $x = r \cos \phi$ and $y = r \sin \phi$.

(b) Write the basis vectors $(\hat{e}_r, \hat{e}_\phi) = (\partial/\partial r, \partial/\partial \phi)$ in terms of the basis vectors $(\hat{e}_x, \hat{e}_y) = (\partial/\partial x, \partial/\partial y)$. Is the basis $(\partial/\partial r, \partial/\partial \phi)$ a coordinate basis?

(c) If some set of basis vectors $\hat{e}_i$ is an orthonormal basis, then $\mathbf{g}(\hat{e}_i, \hat{e}_j) = \delta_{ij}$. Is the basis $(\partial/\partial r, \partial/\partial \phi)$ an orthonormal basis?

(d) Is the basis $(\partial/\partial r, \frac{1}{r}\frac{\partial}{\partial \phi})$ an orthonormal basis? Is it a coordinate basis?

(e) Is the basis $(\partial/\partial x, \partial/\partial y)$ an orthonormal basis? Is it a coordinate basis?

3.13 Consider four flat spacetime dimensions, with Minkowski coordinates (τ, x, y, z). Let's write the metric in tensor form as

$$\mathbf{g} = -d\tau \otimes d\tau + dx \otimes dx + dy \otimes dy + dz \otimes dz. \tag{E3.8}$$

In Minkowski spacetime we define an orthonormal set of basis vectors $\hat{e}_\mu$ by the requirement $\mathbf{g}(\hat{e}_\mu, \hat{e}_\nu) = \eta_{\mu\nu}$.

(a) Rewrite this metric using null coordinates (u, v, y, z), where $u = \tau + x$, $v = \tau - x$.

(b) Write the basis vectors $(\partial/\partial u, \partial/\partial v)$ in terms of the basis vectors $(\partial/\partial\tau, \partial/\partial x)$. Is the basis $(\partial/\partial u, \partial/\partial v, \partial/\partial y, \partial/\partial z)$ a coordinate basis? Is $(\partial/\partial u, \partial/\partial v, \partial/\partial y, \partial/\partial z)$ an orthonormal basis? (In other words, do all of these basis vectors have unit norm and are they all orthogonal to each other?)

(c) Is the basis $(\partial/\partial\tau, \partial/\partial x, \partial/\partial y, \partial/\partial z)$ a coordinate basis? Is it an orthonormal basis?

3.14 Show that in flat spacetime, for any number of dimensions, the only curve whose tangent vector is everywhere lightlike is a straight line.

3.15 Given the following one forms

$$\alpha = \alpha_\mu \, dx^\mu \qquad \beta = \beta_\mu \, dx^\mu$$

$$\gamma = \gamma_\mu dx^\mu \qquad \delta = \delta_\mu \, dx^\mu \tag{E3.9}$$

where, as usual, the repeated Greek index implies a sum over all d spacetime dimensions, compute $\alpha \wedge \beta$, $\alpha \wedge \beta \wedge \gamma$ and $\alpha \wedge \beta \wedge \gamma \wedge \delta$ for $d = 2, 3, 4$.

3.16   Find the coordinate basis components of a rank $\binom{3}{0}$ tensor symmetrized over all pairs of indices.

3.17   Find the coordinate basis components of a rank $\binom{3}{0}$ tensor antisymmetrized over all pairs of indices.

3.18   Consider an antisymmetric rank $\binom{0}{2}$ tensor $F_{\mu\nu}$ in flat spacetime in $d = 4$ with coordinate basis components

$$F_{0i} = -E_i \quad i = x, y, z$$
$$F_{ij} = \epsilon_{ijk} B_k, \tag{E3.10}$$

where ϵ_{ijk} is antisymmetric on all three indices, with $\epsilon_{xyz} = 1$. Calculate the components of this tensor under a Lorentz transformation at speed β in the x direction. This tensor is the electromagnetic field strength tensor, also known as the Faraday tensor. The Lorentz transform of this tensor gives the correct Lorentz transform of the electric and magnetic fields $\vec{E}$ and $\vec{B}$ for the Maxwell equations to be the same in all inertial frames. You will be seeing more of this tensor in Chapter 5.

4

Mechanics in spacetime

To complete this exercise you will need the following:

- A large flat level smooth surface, like the top of a table or desk, and some smooth round glass marbles *or* a pool table and some billiard balls.

The purpose of this exercise is to compare observations in the two coordinate frames most relevant to particle physics – the rest frame of a particle in a collision, and the center of momentum frame of two colliding particles. In the first case a moving marble or billiard ball moving with speed v strikes another one that is initially at rest. Repeat this a few times. Devise a method to determine the final speeds and directions of the two marbles or balls as accurately as you can from the point of view of an observer at rest. For the second case, the center of momentum (CM) frame, there are two possible approaches. You can try to put the two balls or marbles into simultaneous motion on a collision course with equal speeds. The alternative is only to have one ball in motion, as in the first part of the exercise, but to observe the collision in a frame that is moving with speed $v/2$ so that the two marbles or balls appear to have equal and opposite velocities. Try to make repeated observations for each of these two alternatives. Do you find the same results by both methods? Explain why (or why not) they should be equivalent. In what sense does the desk or table provide an absolute frame of reference rather than one that is purely relative?

4.1 Equations of motion in spacetime

Newton's equation and relativity

What are equations of motion and where do they come from? This is a long story, but Newton ultimately wins the prize for inventing the differential equation as a

95

means of computing the motion in space of some object or body as a function of time. This is a book on relativity in which we have just learned that objects move not just in space but in spacetime, and the difference between space and time in spacetime depends on the observer. So we should expect that Newton's equation will have an uneasy confrontation with relativity theory.

Newton's equation can be written

$$\vec{F} = m\vec{a} = m\frac{d^2\vec{x}}{dt^2}. \tag{4.1}$$

The solution of this equation, given a suitable set of initial data, gives the trajectory $\vec{x}(t)$ of some object in response to some force described by the space vector $\vec{F}$. In Newtonian physics, all geometrical objects such as vectors and tensors are defined in D flat space dimensions and evolve dynamically as a function of the absolute universal time t, which is the same for all observers and exists independently of any geometry, being defined in a sense by the differential equations that use it.

Newton believed that the absoluteness of time was a reflection of the absolute power of God. We've learned in the last three chapters that Newton's notion of time doesn't fly if we want physical laws to be the same for all inertial observers, and consistent with the observed constancy of the speed of light. In order to satisfy these conditions, physical laws have to be expressible in terms of vector or tensor equations defined in flat spacetime with $d = D + 1$ dimensions. Therefore both the force and the acceleration have to be expressible as vectors in d spacetime dimensions.

Relativistic acceleration

How do we make a spacetime acceleration vector **a** to replace the space acceleration vector $\vec{a}$ in Newton's equation? Recall from Chapter 3 that the world line $\mathcal{C}(\lambda)$ with coordinates $x^\mu(\lambda)$ has as a tangent vector, the spacetime velocity vector **u**, with coordinate basis components

$$u^\mu = \frac{dx^\mu}{d\lambda}, \tag{4.2}$$

where λ is proper time along the path $\mathcal{C}$. The coordinate basis components a^μ of **a** should therefore be

$$a^\mu = \frac{du^\mu}{d\lambda}, \tag{4.3}$$

which can be written using the chain rule as

$$a^\mu = \frac{du^\mu}{d\lambda} = \frac{\partial u^\mu}{\partial x^\alpha}\frac{dx^\alpha}{d\lambda} = u^\alpha \partial_\alpha u^\mu. \tag{4.4}$$

Let ∇ be the vector associated with the one form with coordinate basis components ∂_α, so that $\nabla^\alpha = \eta^{\alpha\beta} \partial_\beta$. We can then write the acceleration **a** in a basis-independent manner as

$$\mathbf{a} = \mathbf{u} \cdot \nabla \mathbf{u}. \tag{4.5}$$

Since the combination $u^\alpha \partial_\alpha$ is a Lorentz-invariant quantity, the components a^μ of the acceleration **a** transform as a spacetime vector, in other words

$$\tilde{a}^\mu = L^{\tilde{\mu}}_\nu \, a^\nu. \tag{4.6}$$

Spacetime acceleration **a** is always orthogonal to spacetime velocity **u**. Spacetime velocity **u** satisfies $\mathbf{u} \cdot \mathbf{u} = -1$ by definition, and so the metric product $\mathbf{u} \cdot \mathbf{a}$ gives us

$$\mathbf{u} \cdot (\mathbf{u} \cdot \nabla)\mathbf{u} = \eta_{\mu\nu} \, u^\mu \, (u^\alpha \partial_\alpha u^\nu) = \frac{1}{2} u^\alpha \partial_\alpha (\eta_{\mu\nu} u^\mu u^\nu)$$

$$= \frac{1}{2} u^\alpha \partial_\alpha (-1) = 0. \tag{4.7}$$

Since **u** is the tangent vector to the world line $\mathcal{C}(\lambda)$, and $\mathbf{u} \cdot \mathbf{a} = 0$, the spacetime acceleration vector **a** must be the normal vector to the world line $\mathcal{C}(\lambda)$, that is, normal according to the metric of flat spacetime in d dimensions.

It will be left as an exercise for the reader to show that this implies that the spacetime acceleration must be a spacelike, not a timelike or null, vector.

Relativistic force

A spacetime acceleration vector can be defined as a kinematical quantity, but forces have to come from a dynamical principle of nature. In order to be part of an equation of motion that satisfies the two postulates of Einstein, a force has to be expressible as a spacetime vector **F**, the Minkowski basis components $F^\mu(x)$ of which transform under a Lorentz transformation in the same way as the acceleration components in (4.6).

The Lorentz force law in electromagnetism satisfies this constraint, as will be shown in detail in Chapter 5. The strong and weak nuclear forces act over such short ranges that their study requires relativistic quantum mechanics, which will be discussed in Chapter 7.

The most commonly used and experienced force in mechanics is the force of gravity. Unfortunately, the gravitational force cannot be made compatible with special relativity. The incompatibility of gravity with special relativity was the reason Einstein was driven to discover general relativity. In general relativity, the force of gravity comes from the curvature of spacetime. In special relativity, spacetime is flat by definition, so there is no force of gravity.

Because we are considering flat spacetime, where there is no gravity, in our treatment of spacetime mechanics objects have mass but they don't have weight. Since electromagnetism is dealt with in Chapter 5, we will assume for most of the rest of this chapter that there are no forces, and hence no acceleration. The one exception will be the case of uniform acceleration discussed below.

World line of a free particle

A particle traveling at less than the speed of light has a spacetime velocity vector $\mathbf{u}$ tangent to its world line that satisfies the timelike condition $\mathbf{u} \cdot \mathbf{u} = -1$ for all values of proper time λ. If there are no forces, then the equation of motion for the particle is

$$\frac{d\mathbf{u}}{d\lambda} = \mathbf{u} \cdot \nabla \mathbf{u} = \mathbf{0}. \tag{4.8}$$

One obvious solution is $\mathbf{u} = \{1, 0, \ldots, 0\}$, which represents a particle at rest in the frame in which this coordinate basis is defined. To get the equation for the world line of this particle in terms of the Minkowski coordinates $x^\mu(\lambda)$, we need to solve the equation

$$\mathbf{u} = \frac{d\mathbf{x}}{d\lambda}. \tag{4.9}$$

This one vector equation is really a set of d first-order differential equations so we need the values of the d spacetime coordinates at some initial value of λ to completely specify the solution. Let's assume that our particle starts out at proper time $\lambda = 0$ at the origin of the space coordinate frame we are using, so that at coordinate time $x^0 = \tau = 0$, we also have $x^i = 0$ for all $1 \le i \le D$. Then the world line of our particle is

$$x^\mu(\lambda) = \{\lambda, 0, \ldots, 0\}. \tag{4.10}$$

This particle is at rest relative to the coordinate frame we have chosen. It moves forward in time, but it stays at the same location in space.

Now it's time to use a wonderful application of the isometries of flat spacetime in d dimensions. We have one easy solution to (4.8), a particle at rest at the origin of the space coordinates, sitting there as it moves forward in time. This solution can be used to generate all possible timelike world lines, using the translation and Lorentz isometries of flat spacetime. This is because the equations of motion are the same in all coordinate frames related by a coordinate translation or a Lorentz transformation. So a translation or a Lorentz transformation of any one solution will yield another solution.

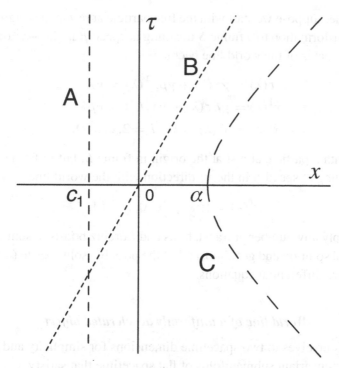

Fig. 4.1. The τ axis can be seen as the world line of an object at rest at $x = 0$. World line A is a translation of an object at rest at $x = 0$ to $x = c_1$. World line B is a Lorentz boost with speed β of an object at rest at $x = 0$. World line C is a uniformly accelerated object that decelerates in the $-x$ direction until it reaches $x = \alpha$, where it stops and turns around and accelerates in the $+x$ direction until it reaches the speed of light again.

We can define a translation vector $\Delta x^\mu = \{c^0, c^1, \ldots, c^D\}$, where all of the c^μ are constants. For any $x^\mu(\lambda)$ that satisfies (4.8), the sum $x^\mu(\lambda) + \Delta x^\mu$ is also a solution. This transformation represents moving the location in spacetime where the particle starts out at $\lambda = 0$. If we apply this translation to our easy solution then we get

$$x^\mu(\lambda) = \{\lambda + c^0, c^1, \ldots, c^D\}. \tag{4.11}$$

This is still a particle at rest, but it rests somewhere else now, because we've moved it.

What about solutions for a particle not at rest? A Lorentz transformation $L^{\tilde{\alpha}}_\mu x^\mu(\lambda)$ of any solution to (4.8) is also a solution. So we can pick any velocity $\vec{\beta}$ that we desire and make a Lorentz transformation of our solution for the particle at rest and get a particle traveling at velocity $\vec{\beta}$. We do this by transforming to a frame $\tilde{S}$ traveling at velocity $-\vec{\beta}$ relative to the frame S in which our particle appears to be at rest.

For example, suppose we start with the free particle at rest at the origin and make a Lorentz transformation to a frame $\tilde{S}$ traveling at speed β in the $-x^1$ direction, so that the coordinates of the world line become

$$\tilde{\tau}(\lambda) = \gamma\tau(\lambda) + \gamma\beta x^1(\lambda) = \gamma\lambda$$
$$\tilde{x}^1(\lambda) = \gamma\beta\tau(\lambda) + \gamma x^1(\lambda) = \gamma\beta\lambda$$
$$\tilde{x}^i(\lambda) = x^i(\lambda) = 0 \quad i = 2, \dots, D. \tag{4.12}$$

We started with a particle at rest at the origin in frame S, but in frame $\tilde{S}$ we see a particle traveling at speed β in the x^1 direction with the world line

$$\tilde{x}^\mu(\lambda) = \{\gamma\lambda, \gamma\beta\lambda, 0, \dots, 0\}. \tag{4.13}$$

We can apply any number of translations and Lorentz boosts at some velocity $\vec{\beta}$ to our original solution and get almost all of the possible solutions to (4.8), without having to solve differential equations.

World line of a uniformly accelerated object

Let's restrict ourselves to two spacetime dimensions for simplicity and look at the set of Lorentz-invariant submanifolds of flat spacetime that satisfy

$$-\tau^2 + x^2 = \alpha^2, \tag{4.14}$$

where $\alpha \in \mathbf{R}$ is a constant. For a given α, the solution to this equation can be written in terms of a parameter λ as

$$\tau(\lambda) = \alpha\sinh(\lambda/\alpha), \quad x(\lambda) = \alpha\cosh(\lambda/\alpha). \tag{4.15}$$

This solution is a timelike curve that passes through the event $(0, \alpha)$ at $\lambda = 0$. The tangent vector to this world line is given by

$$\mathbf{u}(\lambda) = \{\cosh(\lambda/\alpha), \sinh(\lambda/\alpha)\}. \tag{4.16}$$

We can confirm that this world line is everywhere timelike by the fact that $\mathbf{u} \cdot \mathbf{u} = \eta_{\mu\nu}u^\mu u^\nu = -1$ for all values of λ.

According to an observer who measures time and space according to (τ, x), the instantaneous velocity of an object traveling along this world line is

$$\frac{dx}{d\tau} = \frac{dx/d\lambda}{d\tau/d\lambda} = \frac{u^x}{u^0} = \tanh(\lambda/\alpha). \tag{4.17}$$

Therefore this world line represents a particle or object that travels at the speed of light in the $-x$ direction in the infinite past, slows down to a stop at $\lambda = 0$, and speeds up again to end up traveling at the speed of light in the $+x$ direction as $\lambda \to \infty$. The acceleration for this world line is given by

$$\mathbf{a}(\lambda) = \{\frac{1}{\alpha} \sinh(\lambda/\alpha), \frac{1}{\alpha} \cosh(\lambda/\alpha)\}. \tag{4.18}$$

The acceleration is a spacelike vector, as can be seen by the metric product

$$\mathbf{a} \cdot \mathbf{a} = \eta_{\mu\nu} a^\mu a^\nu = \frac{1}{\alpha^2}. \tag{4.19}$$

The magnitude of this acceleration vector is constant, so this world line represents a particle or object undergoing constant acceleration in spacetime.

This world line lives on a timelike Lorentz invariant submanifold of two-dimensional flat spacetime specified by the value of the acceleration α^{-1}. A Lorentz boost by β maps this world line into itself, shifting the world line parameter λ/α by the amount $\theta = \tanh^{-1}\beta$. Proof of this will be left for the reader as an exercise.

In d spacetime dimensions the situation is more complicated. The accelerating world line lives on a larger Lorentz-invariant submanifold of dimension $d - 1$ with the equation

$$-\tau^2 + |\vec{x}|^2 = \alpha^2. \tag{4.20}$$

A Lorentz transformation of a world line on this submanifold gives a world line on this submanifold, but not necessarily in the same direction as the world line we started with.

4.2 Momentum and energy in spacetime

In Chapter 3 we introduced the spacetime momentum vector $\mathbf{p}$, related to the spacetime velocity vector $\mathbf{u}$ by

$$\mathbf{p} = m\mathbf{u}, \qquad \mathbf{p}^2 = \eta_{\mu\nu} p^\mu \; p^\mu = -m^2. \tag{4.21}$$

In a given coordinate frame S, the time and space components of $\mathbf{p}$ can be written

$$p^0 = \gamma m$$
$$\vec{p} = \gamma m \vec{\beta}. \tag{4.22}$$

If we put back the factors of c and expand in powers of $\beta = v/c$, the particle energy is

$$E = c^2 p^0 = mc^2 + \frac{1}{2}mv^2 + \cdots \sim E_0 + T. \tag{4.23}$$

At velocities much smaller than the speed of light, the time component p^0 of the momentum **p** separates into a contribution $T = mv^2/2$ that looks like the nonrelativistic kinetic energy of the object, and a new contribution $E_0 = mc^2$ that is nonzero even when the object is at rest.

This contribution to the energy coming from the object at rest, called the *rest energy*, is very different from what we are used to in Newtonian mechanics. As we shall see later in this chapter, it has profound implications for physics, which is why the equation is well known even to people who have not studied any physics.

Going back to our usual choice of units where $c = 1$, we can write for any velocity

$$E = p^0 = \gamma m = m + (\gamma - 1)m = E_0 + T, \qquad (4.24)$$

where $T = (\gamma - 1)m$ is the relativistic kinetic energy, that is, the part of the relativistic energy that is zero when the object is at rest.

In Newtonian physics, energy and momentum are different quantities. One is a scalar and one is a vector. Under a Galilean transformation, energy remains energy and momentum remains momentum. In relativity, a Lorentz boost mixes energy and momentum, just as it mixes time and space, so that energy and momentum are as relative as time and space. We have learned already that in relativity, the concept of spacetime replaces the separate concepts of space and time from Newtonian physics. The same turns out to be true for energy and momentum. The relationship

$$\mathbf{p}^2 = -E^2 + |\vec{p}|^2 = -m^2 \qquad (4.25)$$

describes a Lorentz-invariant submanifold of a manifold whose coordinates are $\{E, p^1, \ldots, p^D\}$. Each such submanifold is a hyperboloid specified by the value of the mass m. This is why spacetime geometry is considered to be *hyperbolic geometry*. In ordinary space geometry, the surface mapped into itself by a spatial rotation in D dimensions is a $(D - 1)$ dimensional sphere. In spacetime in d dimensions, the surface mapped into itself by a Lorentz boost is a $(d - 1)$ dimensional hyperboloid. When our axes are the Minkowski coordinate basis components of the spacetime momentum **p**, this Lorentz-invariant submanifold is called the *mass hyperboloid*, or the *mass shell*.

Note that since $E = \pm\sqrt{|\vec{p}|^2 + m^2}$, this hyperboloid has two branches, one with $E > 0$ and one with $E < 0$. Only the $E > 0$ branch is shown in Figure 4.2. In classical physics, the negative energy branch can simply be ignored as being

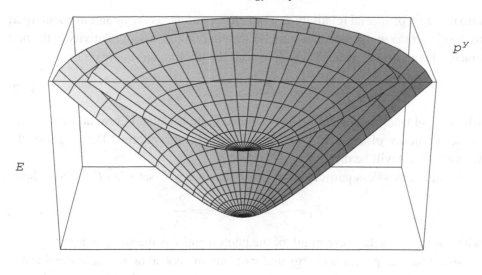

p^y

E

p^x

Fig. 4.2. The mass hyperboloid $E^2 - |\vec{p}|^2 = m^2$ is a Lorentz-invariant submanifold of momentum space. The figure above shows sections of mass hyperboloids for two different values of mass, in three spacetime dimensions with momentum space coordinates $\{E, p^x, p^y\}$. A point on each surface represents a possible momentum and energy for a free particle or object of mass m. A Lorentz transformation maps each hyperboloid into itself. In the massless limit $m \to 0$ the hyperboloid degenerates to a cone with its apex at the origin.

unphysical, but in quantum physics, as shall be shown in Chapter 7, we have to give it a physical interpretation.

Energy and momentum of a massless object

The mass hyperboloid we've just been describing degenerates in the limit $m \to 0$ to

$$\mathbf{p}^2 = -E^2 + |\vec{p}|^2 = 0, \qquad (4.26)$$

which is a cone with its apex at the origin in $\{E, \vec{p}\}$ space. This cone in momentum space is analogous to the light cone of an event in spacetime. Every point on this cone (or hypercone in $d > 3$) represents the possible momentum and energy of some object or particle with zero mass.

After the discovery of the quantum nature of light by Planck and Einstein, the old debate over whether light consists of particles or waves came back to stay. Light started out as a particle, evolved in theory to a wave, but now has come back

into the story of special relativity as a massless particle, as the quantum counterpart
to the electromagnetic field that started Einstein thinking about relativity in the first
place. This massless particle, which we call the photon, has energy

$$E = h\nu = \hbar\omega, \qquad \hbar = \frac{h}{2\pi}, \tag{4.27}$$

where h is Planck's constant, to be discussed in more detail below, and $\nu = \omega/2\pi$
is the frequency of the light wave associated with the photon. The origin of the
formula (4.27) will be discussed in Chapter 7.

Because a massless particle has $\mathbf{p}^2 = 0$, the energy E satisfies $E = |\vec{p}|$, where

$$|\vec{p}| = \frac{h}{\lambda} = \hbar k, \qquad k = \frac{2\pi}{\lambda}, \tag{4.28}$$

where $\lambda = 1/\nu$ is the wavelength of the photon and k is the wave number.

Notice that the photon's energy and momentum depend on frequency and wavelength, which are normally properties of a wave, not a particle. In modern physics,
we no longer debate whether light consists of particles or waves. In quantum
physics, light, along with everything else, has both wave and particle properties.
The dual wave and particle nature of light gives us a quick and easy way to derive
the Doppler shift of light waves using the spacetime momentum of a photon. This
will be left for the reader as an exercise.

Units of mass and energy

Normally we would write $\lambda = c/\nu$, where c is the speed if light, but in the above
formulas for the momentum and energy of a photon, we have come up with $\lambda =
1/\nu$ instead. Why is this? Now is the time to address the question of unit systems
in relativistic physics.

Energy normally has units of ML^2/T^2, where M, L and T represent units of
mass, length and time, respectively. Momentum has units of ML/T, but in this
chapter both energy and momentum end up with units of M alone. In units where
$c = 1$, every 2.998×10^8 m of length are equivalent to 1 s of time. The world line
of an object traveling at the speed of light is easy to draw in these units. However,
it's still important to be able to put back the factors of c when needed, because
that tells us about the scale of measurements in the real world, where we use both
clocks and meter sticks.

Planck's constant h, introduced above, has the value

$$h = 6.63 \times 10^{-34} \text{ J s} = 6.63 \times 10^{-34} \text{ kg m}^2/\text{s}, \tag{4.29}$$

where J stands for joules, the unit of energy in SI units, where the speed of light
is 2.998×10^8 m/s. The more commonly used constant, however, is $\hbar = h/2\pi$. In

SI units, h and $\hbar$ have units of ML^2/T. In units where $c = 1$, they have units of ML. In high-energy physics it is common to make a further rescaling of units by setting $\hbar = 1$ in addition to $c = 1$, so that units of mass become units of inverse length. This unit system is much preferred by high-energy physicists because of the convenience of working only with units of length.

Objects that travel at relativistic speeds tend in practice to be subatomic particles, with masses so small that the kilogram is inconvenient as a unit for mass. The mass unit most commonly used in particle physics is the *electron-volt*, or eV, so named because it represents the energy gained by one electron moving across an electrostatic potential of 1 V. The volt itself is defined in SI units as the electrostatic potential change that imparts 1 J of energy per coulomb of charge. An electron carries a charge

$$q_e = 1.60 \times 10^{-19}\,\text{C}, \tag{4.30}$$

therefore

$$1\,\text{eV} = 1.60 \times 10^{-19}\,\text{J}. \tag{4.31}$$

In its rest frame, an electron has energy

$$E_0 = m_e c^2 = 9.11 \times 10^{-31}\,\text{kg} \times c^2 = 81.9 \times 10^{-14}\,\text{J}. \tag{4.32}$$

Using the above result, the electron rest energy in electron-volts is

$$m_e c^2 = 0.511 \times 10^6\,\text{eV} = 0.511\,\text{MeV}. \tag{4.33}$$

The mass of the electron is derived from the rest energy by $m_e = 0.511\,\text{MeV}/c^2$. In units where $c = 1$, we can say that the electron has a mass of 0.511 MeV, because in units where $c = 1$, energy has units of mass. In this book we started using units with $c = 1$ when we scaled our time coordinate by $\tau = ct$ so that all coordinates have units of length. So unless specifically noted, mass and energy have the same units in this chapter and in the rest of the book.

4.3 Energy and momentum conservation in spacetime

The nonrelativistic case

In both Newtonian and relativistic physics in flat spacetime, a change in the total momentum of a system can only be due to the application of some kind of external force. If no external force is applied, then the total momentum should stay the same. Suppose a system that consists initially of N particles or objects undergoes interactions such that ultimately there are M particles or objects emerging. It doesn't matter what happened during the collision itself. As long as no external forces are applied to the system, the total momentum before the collision must be the same as the total momentum after the collision. In Newtonian physics, momentum conservation means conservation of the Euclidean vector $\vec{p}$, so that the equation

$$\left(\sum_i^N \vec{p}_i\right)_{\text{initial}} = \left(\sum_i^M \vec{p}_i\right)_{\text{final}} \tag{4.34}$$

must always be true.

   The Principle of Relativity says that physical laws should look the same in any inertial frame. If momentum is conserved in some frame S, then it should also be conserved in some other frame $\tilde{S}$ traveling at some constant velocity $\vec{V}$ relative to S. In Newtonian physics, the relevant relativity principle is Galilean relativity. Consider the case of two objects of masses m_1 and m_2 colliding, and then exiting the collision with masses m_3 and m_4. Conservation of momentum tells us that

$$m_1\vec{v}_1 + m_2\vec{v}_2 = m_3\vec{v}_3 + m_4\vec{v}_4. \tag{4.35}$$

In frame $\tilde{S}$ traveling at velocity $\vec{V}$, according to Galilean relativity, the velocity of each object $\vec{v}_i$ appears instead to be $\tilde{\vec{v}}_i = \vec{v}_i - \vec{V}$, so that the equation for momentum conservation is changed in frame $\tilde{S}$ to

$$m_1(\vec{v}_1 - \vec{V}) + m_2(\vec{v}_2 - \vec{V}) = m_3(\vec{v}_3 - \vec{V}) + m_4(\vec{v}_4 - \vec{V}). \tag{4.36}$$

Both equations are true if and only if

$$m_1 + m_2 = m_3 + m_4, \tag{4.37}$$

so that Galilean covariance of momentum conservation requires mass conservation. Although we restricted our analysis to two particles in and out, we could have used any number, and still have obtained this same result.

   In Newtonian mechanics, the kinetic energy of objects with mass is the only energy accounted for in the equations of motion. If the collision involves the conversion of kinetic energy to or from other kinds of energy, such as radiation, heat or plastic deformation, then even though the total energy is conserved in the process, kinetic energy is not conserved. The Q value of a collision or scattering interaction is defined to be the difference between the initial and final kinetic energies of the system, so that

$$Q \equiv T_{\text{final}} - T_{\text{initial}}. \tag{4.38}$$

If $Q > 0$, the collision is called *exothermic*, or *exoergic*, because more kinetic energy leaves the interaction than went into it. If $Q < 0$, the collision is called *endothermic*, or *endoergic*, because there is more kinetic energy going into the system than there is coming out. A collision with $Q = 0$ is called an *elastic* collision

or elastic scattering. In an elastic collision, kinetic energy is conserved by definition, so that

$$\left(\sum_i^N \frac{1}{2} m_i \, |\vec{v}_i|^2 \right)_{\text{initial}} = \left(\sum_i^N \frac{1}{2} m_i \, |\vec{v}_i|^2 \right)_{\text{final}} . \tag{4.39}$$

For the $2 \rightarrow 2$ particle collision described above, assuming that $Q = 0$ and the
scattering is elastic, we have

$$\frac{1}{2} m_1 \, |\vec{v}_1|^2 + \frac{1}{2} m_2 \, |\vec{v}_2|^2 = \frac{1}{2} m_3 \, |\vec{v}_3|^2 + \frac{1}{2} m_4 \, |\vec{v}_4|^2 . \tag{4.40}$$

In frame $\tilde{S}$ this equation becomes

$$\frac{1}{2} m_1 \, |\vec{v}_1 - \vec{V}|^2 + \frac{1}{2} m_2 \, |\vec{v}_1 - \vec{V}|^2 = \frac{1}{2} m_3 \, |\vec{v}_1 - \vec{V}|^2 + \frac{1}{2} m_4 \, |\vec{v}_1 - \vec{V}|^2 . \tag{4.41}$$

Assuming (4.40) is true in frame S, (4.41) can only be true in frame $\tilde{S}$ if momentum
is conserved

$$(m_1 \vec{v}_1 + m_2 \vec{v}_2) \cdot \vec{V} = (m_3 \vec{v}_3 + m_4 \vec{v}_4) \cdot \vec{V} \tag{4.42}$$

and mass is conserved

$$\frac{1}{2} (m_1 + m_2) |\vec{V}|^2 = \frac{1}{2} (m_3 + m_4) |\vec{V}|^2 \tag{4.43}$$

in frame S.

The relativistic case

We've just shown that in Galilean relativity, conservation of mass is necessary for
the covariance of both energy and momentum conservation. This is not the situation in special relativity. Momentum conservation in spacetime does not require
conservation of mass, and as Einstein figured out and as we shall demonstrate in the
next section, that has observable physical consequences of enormous importance.

A Galilean transformation is fundamentally different from a Lorentz transformation. A Galilean transformation takes us from one version of $\mathbf{E}^D$ to another
version moving at velocity V relative to the first. But time is not part of the geometry of $\mathbf{E}^D$, so a Galilean transformation is not a change of coordinates in the space
in which the momentum is defined. A vector $\vec{v}$ is changed to a new vector $\vec{v} - \vec{V}$
by a Galilean transformation. Compare this situation with a rotation of the axes of
$\mathbf{E}^D$. A vector $\vec{v}$ would have different components in the rotated coordinates, but
the vector itself would be the same.

A Lorentz transformation is the spacetime analog of a rotation of the coordinate
axes. A Lorentz transformation operates in spacetime, where time is a coordinate,

and so a transformation to a moving coordinate frame is a coordinate transformation in the space in which the spacetime momentum is defined. Therefore the covariance of spacetime momentum conservation is guaranteed by the fact that we explicitly constructed the spacetime momentum $\mathbf{p}$ as a vector in the tangent space of $\mathbf{M}^d$.

Spacetime momentum conservation can be written in a single equation as

$$\left(\sum_i^N \mathbf{p}_i \right)_{\text{initial}} = \left(\sum_i^M \mathbf{p}_i \right)_{\text{final}}, \tag{4.44}$$

encompassing both energy and momentum conservation in its components. If the components of $\mathbf{p}$ are expanded in a coordinate basis then covariance of spacetime momentum conservation is enforced on the component basis by the fact that the Lorentz transformation is a linear transformation on a vector space, so that if

$$\tilde{p}_i^\mu = L_\nu^{\tilde{\mu}} \, p_i^\nu \tag{4.45}$$

is true for the components of each momentum $\mathbf{p}_i$ in the sum (4.44), then it is also true for the sum. Proof of this will be left for the reader as an exercise.

The time and space components of this equation in some coordinate frame S are

$$\left(\sum_i^N \gamma_i \, m_i \right)_{\text{initial}} = \left(\sum_i^M \gamma_i \, m_i \right)_{\text{final}} \tag{4.46}$$

$$\left(\sum_i^N \gamma_i \, m_i \vec{\beta}_i \right)_{\text{initial}} = \left(\sum_i^M \gamma_i \, m_i \vec{\beta}_i \right)_{\text{final}}, \tag{4.47}$$

with $\gamma_i \equiv 1/\sqrt{1 - |\vec{\beta}_i|^2}$.

Notice that there is no mass conservation being enforced anywhere in the above equations. The conserved quantity appears to be γm instead of m. In nonrelativistic physics, mass conservation seems to be required, but it is not required anywhere in special relativity, and this has important physical consequences. In (4.46), all of the interacting objects were presumed to have mass. But massless particles can be added to the sum as well. Massive particles can collide and produce massless particles, so that the initial and final total masses of a system are not the same. So mass can disappear from a system even though energy and momentum are conserved. Matter can be converted into radiation, and matter can be created out of radiation. In that case, photons will be involved in one or both sides of (4.46), and those terms will contribute energy $E_i = |\vec{p}_i| = h\nu_i$, where ν_i is the frequency of the photon.

What happened to conservation of kinetic energy? The Q value of an interaction is still defined by (4.38), with the relativistic kinetic energy

$$T = (\gamma - 1)m \tag{4.48}$$

in place of the nonrelativistic kinetic energy $T = mv^2/2$. Since we know that γm is conserved by virtue of spacetime momentum conservation, then kinetic energy can only be conserved if mass is conserved in the reaction. This is confirmed by combining (4.38) with (4.46), to get

$$Q = T_{\text{final}} - T_{\text{initial}} = \left(\sum_i^M m_i \right)_{\text{initial}} - \left(\sum_i^N m_i \right)_{\text{final}}. \tag{4.49}$$

We will look at specific applications of conservation of spacetime momentum in the next section.

4.4 Relativistic kinematics

Macroscopic objects ordinarily move at speeds much less than the speed of light and can be reasonably described by nonrelativistic physics. Astronomical objects exist that move near the speed of light, but they are often under the influence of strong gravitational fields and other complicating factors, and so special relativity is not enough to describe their motions. Special relativity is most useful as a tool in understanding collisions and other interactions between subatomic particles. Subatomic particles don't just bounce off of one another; they can be created or destroyed in collisions, fuse together or split apart, or decay into an assortment of other particles. The detailed causes of these interactions can only be studied using specific quantum theories, but the basic motions that can result are revealed through relativistic kinematics, which does not require knowledge of the underlying dynamics.

Photons and charged particles

The interaction of an electromagnetic field with a charged particle of mass m and charge q can be viewed at subatomic distance scales as an infinite series of interactions between the charged particle and the photons that represent collectively the field at the quantum level. Computing and approximating this series to get scattering amplitudes is the business of relativistic quantum field theory, which you will learn a little bit about in Chapter 7. But even without quantum field theory, we can learn an amazing amount about photon interactions with charged particles using spacetime momentum conservation alone.

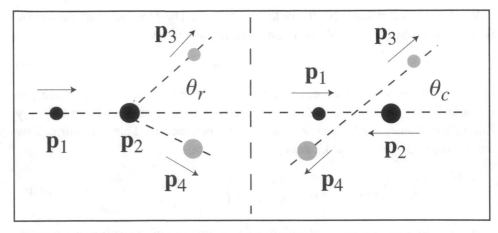

Fig. 4.3. For two particles with mass m and M in an elastic collision, space-time momentum conservation $\mathbf{p}_1 + \mathbf{p}_2 = \mathbf{p}_3 + \mathbf{p}_4$ leads to the Lorentz-invariant equation $\mathbf{p}_2 \cdot (\mathbf{p}_1 - \mathbf{p}_3) = \mathbf{p}_1 \cdot \mathbf{p}_3 - m^2$. The scattering angle θ is defined by $\vec{p}_1 \cdot \vec{p}_3 = |\vec{p}_1||\vec{p}_3|\cos\theta$. The figure on the left shows the collision in the rest frame of particle 2, defined by the condition $\mathbf{p}_2 = (M, \vec{0})$. The figure on the right shows the collision in the center of momentum frame, defined by the condition $\vec{p}_1 + \vec{p}_2 = 0$.

The simplest collision problem has two particles in the initial state and one in the final, which we signify by

$$1 + 2 \rightarrow 3. \tag{4.50}$$

If we label the initial particle momenta $\mathbf{p}_1$ and $\mathbf{p}_2$ and let the final state have momentum $\mathbf{p}_3$, then spacetime momentum conservation tells us that

$$\mathbf{p}_1 + \mathbf{p}_2 = \mathbf{p}_3. \tag{4.51}$$

The mass constraints are

$$\mathbf{p}_1^2 = -m_1^2, \quad \mathbf{p}_2^2 = -m_2^2, \quad \mathbf{p}_3^2 = -m_3^3. \tag{4.52}$$

These conditions combined imply that

$$-2\mathbf{p}_1 \cdot \mathbf{p}_2 = m_3^2 - (m_1^2 + m_2^2). \tag{4.53}$$

It was shown in Chapter 3 that if $\mathbf{p}_1$ and $\mathbf{p}_2$ are null or timelike vectors that are either both future-directed or both past-directed, then the metric product $\mathbf{p}_1 \cdot \mathbf{p}_2 < 0$. Conservation of spacetime momentum therefore requires

$$m_3^2 > m_1^2 + m_2^2. \tag{4.54}$$

If $\mathbf{p}_1$ and $\mathbf{p}_2$ are both past-directed, the reaction is not a collision of two particles giving one but the decay of one particle into two, in which case the stronger mass inequality $m_3 > m_1 + m_2$ also applies.

Since $m_3 = 0$ is not allowed by (4.54) for any real value of m_1 or m_2, then an electron and a positron cannot annihilate into a single photon in the reaction $e^- + e^+ \to \gamma$. The time-reversed process $\gamma \to e^- + e^+$ is also kinematically forbidden. The absorption of a photon by an electron $e^- + \gamma \to e^-$, and the emission of a photon by an electron $e^- \to e^- + \gamma$, are both forbidden by (4.54). In general, a massive particle can neither absorb nor emit a single photon in a manner consistent with spacetime momentum conservation.

Because mass is not conserved in special relativity, it is kinematically allowed for a massive particle to decay into pure light, as in

$$\pi^0 \to \gamma + \gamma, \tag{4.55}$$

where a neutral pi meson (π^0) decays into two photons. The mass constraint allows $m_1 = m_2 = 0$ for any value of $m_3 > 0$. But we don't need to worry that every massive particle is eventually going to vanish into two flashes of light. In addition to conservation of spacetime momentum, such a decay is also constrained by other conservation laws such as those of charge, angular momentum and baryon number. The spontaneous decay of massive particles into light is only allowed for certain electrically neutral particles. Protons and electrons cannot decay in this manner.

Compton scattering

The most basic photon interaction with a charged particle that is allowed by momentum conservation is called Compton scattering

$$q + \gamma \to q + \gamma \tag{4.56}$$

in which a photon collides with a charged particle q and leaves it intact, but changes its energy and momentum. Because the initial and final masses are the same, Compton scattering is an elastic process with $Q = 0$. Let's call our incoming particle and photon momenta $\mathbf{p}_1$ and $\mathbf{k}_1$, respectively, and label the outgoing momenta $\mathbf{p}_2$ and $\mathbf{k}_2$, as shown in Figure 4.4(a). The equations we have to work with are

$$\mathbf{p}_1 + \mathbf{k}_1 = \mathbf{p}_2 + \mathbf{k}_2 \tag{4.57}$$

and

$$\mathbf{p}_1^2 = \mathbf{p}_2^2 = -m^2, \quad \mathbf{k}_1^2 = \mathbf{k}_2^2 = 0. \tag{4.58}$$

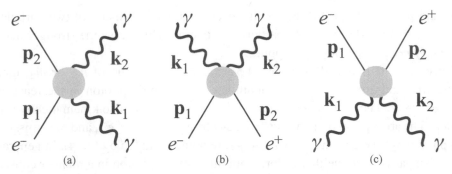

Fig. 4.4. Compton scattering (a), pair annihilation (b) and pair creation (c) are all interactions involving two particles with the same mass, and two massless particles. In this diagram and the ones that follow, the vertical axis is time, the horizontal axis is space, and the round blob in the middle of each interaction represents the details of the interaction that can be ignored when we apply spacetime momentum conservation between the incoming and outgoing particles.

The information we can get from the above equations depends on the question we ask. If we want to know what happens to the outgoing photon, then we can eliminate $\mathbf{p}_2$ from (4.57) and use (4.58) to get

$$\mathbf{p}_1 \cdot (\mathbf{k}_1 - \mathbf{k}_2) = \mathbf{k}_1 \cdot \mathbf{k}_2. \qquad (4.59)$$

Since each term in this equation is the metric product of two spacetime vectors, this equation is valid in any coordinate system that we choose, and therefore in any inertial frame that we desire, so we are free to pick the frame that is the most convenient for the problem we have at hand.

One convenient choice is the rest frame of the initial charged particle, so that the momentum is $\mathbf{p}_1 = \{m, \vec{0}\}$. This is a good choice for studying the scattering of photons from slowly moving charged particles. In this case, the information we put in is the charged particle mass m and the energy of the incoming photon. If we define the scattering angle θ by $\vec{k}_1 \cdot \vec{k}_2 = E_1 E_2 \cos \theta$, then the information we get out is the change in wavelength of the photon as a function of θ

$$\lambda_2 - \lambda_1 = \frac{h}{m}(1 - \cos \theta) = \lambda_c(1 - \cos \theta). \qquad (4.60)$$

The quantity $\lambda_c = h/m$ is known as the Compton wavelength of a particle of mass m. (If one doesn't set $c = 1$, then $\lambda_c = h/mc$.)

Compton's scattering experiment confirmed that light behaves like a massless particle when it scatters from electrons. Compton aimed X-rays with a wavelength $\lambda_1 = 7.11 \times 10^{-11}$ m $= 0.711$ Å at a graphite target and showed that the light in the beam that was scattered from the electrons in the graphite atoms lost momentum from the recoil of the electron just as a massless particle should according to conservation of spacetime momentum. Einstein developed special relativity to make classical mechanics consistent with the classical wave theory of light, but through Compton scattering and other experiments, special relativity turned out to be crucial in understanding the quantum theory of light as well.

Another possible choice of frame for solving (4.59) is a frame in which an electron is heading directly towards a photon at close to the speed of light. In that case $\mathbf{p}_1 = \{E, \vec{p}\}$, with $|\vec{p}| \sim E - m^2/2E$, and $\vec{p} \cdot \vec{k}_1 = -|\vec{p}||\vec{k}_1|$. This is called inverse Compton scattering, and is useful in studying the scattering of high-energy charged particles with low-energy photons, for example the 3×10^{-4} eV photons in the cosmological microwave background. This is called the Sunyaev–Zeldovich effect.

Virtual photons

If the process $q + \gamma \to q$ is strictly forbidden by spacetime momentum conservation, then how can two charged particles interact by exchanging a photon? As we just proved, the absorption and emission are both forbidden as separate interactions. When the two interactions are glued together, somehow it is not only allowed, but forms the foundation of the quantum understanding of how charged particles interact electromagnetically. How can this be true?

Such a photon is what physicists call an *off shell* photon, in reference to the *mass shell*, another name for the mass hyperboloid shown in Figure 4.2. A particle that is off shell does not satisfy the mass hyperboloid constraint $\mathbf{p}^2 = -m^2$. If we demand in (4.53) that the charged particle momentum $\mathbf{p}_1$ be timelike, and $m_3 = m_1$, then the photon momentum $\mathbf{p}_2^2 \neq 0$. The photon serving as the intermediary can only be an off shell photon.

Another name for an off shell particle is a *virtual particle*. Virtual particles are not allowed classically because they are inconsistent with momentum conservation, but quantum mechanics provides a loophole through the Heisenberg Uncertainty Principle. The grey areas in the middle of the reactions shown in Figures 4.4 and 4.5 represent all of the relevant intermediate processes, most of which involve exchange of virtual particles. You will learn more about this topic in Chapter 7.

Pair creation and annihilation

Particle physicists have observed that every particle seems to have an antiparticle, with the same mass and spin but opposite charge. We can apply momentum conservation and mass shell conditions to see under what conditions particle and antiparticle pair creation and annihilation can happen as on shell processes.

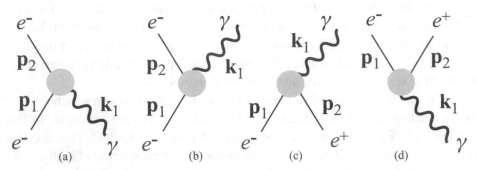

Fig. 4.5. These interactions are kinematically forbidden, that is, forbidden by spacetime momentum conservation. In (a), an electron absorbs a single photon. In (b), an electron emits a single photon. In (c), an electron and positron annihilate to make a single photon. In (d), a single photon decays into an electron and positron. These processes are only allowed to take place if the photon is a virtual photon, with $\mathbf{k}_1^2 \neq 0$, that exists only as a quantum intermediate state in collisions between charged particles, for example $e^- + e^- \to e^- + e^-$ or $e^- + e^+ \to e^- + e^+$.

Pair annihilation is the reaction

$$q + \bar{q} \to \gamma + \gamma, \tag{4.61}$$

where the charged particle is q and the antiparticle $\bar{q}$, as shown in Figure 4.4(b). Momentum conservation tells us

$$\mathbf{p}_1 + \mathbf{p}_2 = \mathbf{k}_1 + \mathbf{k}_2, \tag{4.62}$$

while the mass shell conditions are the same as (4.58) for Compton scattering. Applying these equations and eliminating $\mathbf{k}_2$ yields the equation

$$\mathbf{k}_1 \cdot (\mathbf{p}_1 + \mathbf{p}_2) = \mathbf{p}_1 \cdot \mathbf{p}_2 - m^2. \tag{4.63}$$

A convenient choice of frame when there are two incoming particles of the same mass is the CM frame, defined by $\vec{p}_1 + \vec{p}_2 = 0$. In this frame, the incoming particles have the same energy E, and the outgoing photons have the same energy E_γ. We get the simple relationship

$$E = E_\gamma, \tag{4.64}$$

which can be satisfied for any value of E. We should expect this, because this reaction has $Q = 2m$, making it exothermic. If we shoot a proton and an antiproton at one another with equal energies E, then we should expect radiation with wavelength $\lambda = h/E_\gamma = h/E$. This process is allowed by conservation of energy even if the two particles interact at rest, in which case the light comes out with the Compton wavelength of the proton, $\lambda_{\max} = h/m_\mathrm{p} \sim 10^{-15}$ m.

Pair creation is the opposite reaction

$$\gamma + \gamma \rightarrow q + \bar{q}, \tag{4.65}$$

in which case we can eliminate $\mathbf{p}_2$ to get

$$\mathbf{p}_1 \cdot (\mathbf{k}_1 + \mathbf{k}_2) = \mathbf{k}_1 \cdot \mathbf{k}_2. \tag{4.66}$$

Since pair creation is just pair annihilation running in the opposite direction, we end up with the simple solution

$$E_\gamma = E, \tag{4.67}$$

where this time the photon energy E_γ is the incoming energy that we can control in the lab, and the particle energy E is what comes out as a result.

Pair creation cannot happen unless the incoming photons each have energy $E_\gamma \geq m$ in the CM frame. Since $E_\gamma = h/\lambda$, this means that the wavelength of the incoming light must be smaller than the Compton wavelength of the particles being created, which for electrons and positrons would be $\lambda \leq h/m_e \sim 2.4 \times 10^{-12}$ m.

So now we have a picture of what the Q value of a particle interaction means. If $Q < 0$, as in pair creation, we have to smack the incoming particles together at sufficiently high speed for there to be enough energy for the reaction to take place. If $Q > 0$, as in pair annihilation, even if the incoming particles are at rest relative to one another, there is enough energy for the reaction to take place.

Particle decay

The first confirmation of relativistic time dilation came from observations of the decay

$$\mu^- \rightarrow e^- + \nu_\mu + \bar{\nu}_e \tag{4.68}$$

of a muon into an electron, a muon-type neutrino ν_μ and an electron-type antineutrino $\bar{\nu}_e$. This reaction has $Q \sim m_\mu = 105$ MeV, so it occurs spontaneously in the rest frame of the muon. The lifetime T_μ of the muon is approximately 2.2×10^{-6} s according to clocks at rest in the rest frame of the muon. According to special relativity, if the muon is observed in a laboratory, traveling at some velocity $\vec{\beta}$ relative to the coordinate frame of the lab, the observed lifetime $\tilde{T}_\mu$ of the muon will be

$$\tilde{T}_\mu = \gamma T_\mu, \tag{4.69}$$

where $\tilde{S}$ is the lab frame, and S is the muon rest frame. Since $\gamma \geq 1$, faster muons will decay more slowly than the slower muons according to clocks in the lab frame. For example, a muon moving at $\beta = 0.8$ according to an observer in a lab should have a lifetime $\tilde{T}_\mu = 3.7 \times 10^{-6}$ s according to clocks at rest in the lab, and a muon moving at speed $\beta = 0.994$ should have a lifetime $\tilde{T}_\mu = 2.2 \times 10^{-5}$ s according to clocks at rest in the lab.

The lifetime of an unstable particle is measured by observing the rate of decay Γ of a large number N of such particles. The time dependence of the number of unstable states is given by a decaying exponential factor $e^{-\Gamma t} = e^{-t/T}$, where we have temporarily gone back to using units where $c \neq 1$. Suppose we have some number N_0 of a particle X at some initial time t_0. If at a time T later, there are only $N_0 \times e^{-1} \sim N_0/2.7$ of those particles left, the rest having decayed into their decay products, then we call T the *lifetime* of that particle. The lifetime is the inverse of the decay rate.

The *half life* $t_{1/2}$ is defined in a similar manner, so that $t_{1/2}$ is the time it takes for $N_0/2$ of the unstable X particles to decay. Thus the half life is related to the lifetime by

$$t_{1/2} = T \, \log 2 \sim 0.7T. \tag{4.70}$$

The half life of the muon is then $t_{1/2} = 1.5 \times 10^{-6}$ s. For what follows we will use the half life rather than the lifetime, but they are easily interchangeable since they only differ by a factor of $\log 2$.

The purely relativistic effect of time dilation on the half lives of unstable particles was first confirmed in 1940, and refined in 1963, using cosmic ray muons traveling close to the speed of light, with $\gamma \sim 10$. Neglecting interactions with particles in the Earth's atmosphere, a cosmic ray muon with $\gamma = 10$ can travel an average distance of about $c\tilde{t}_{1/2} = c\gamma t_{1/2} \sim 4500$ m before decaying, according to observers on Earth. In Galilean relativity, the muon would only be able to travel 450 m according to the same observers, because time is not affected by a Galilean transformation.

The physical measurement was made by counting muons at two different locations in Colorado, separated by 1600 m in elevation. According to Galilean relativity, the vast majority of muons that were counted at the first detector should have decayed long before they reached the second detector 1600 m away, and so the muon count at the second detector should be close to zero. According to special relativity, the 1600-m trip takes less than 1 muon half life in the Earth frame, and so the muon count should be less, but not much less, at the second detector than at the first. The relativistic result has been confirmed to a very high precision and relativistic time dilation is routinely taken into account in particle decay experiments today.

The muon has mass $m_\mu = 105$ MeV and the electron has mass $m_e = 0.511$ MeV. The neutrino and antineutrino masses, if they exist, are too small to matter, so the Q value for this reaction is $Q \sim m_\mu$. The muon is sentenced to instability because there exists a particle – the electron – with a much smaller mass and suitable quantum numbers so that the reaction is allowed by quantum selection rules. Almost all elementary particles are unstable and decay very quickly into other particles with smaller masses. Two of the three atomic constituents – the electron and the proton – appear to be stable. A free neutron decays via beta decay

$$n \to p + e^- + \bar{\nu}_e \tag{4.71}$$

into a proton, an electron and an antineutrino, with $Q = 0.78$ MeV, and half life $t_{1/2} = 886$ s. Neutrons bound inside certain nuclei can be stabilized by the extra binding energy. The electron should be absolutely stable because of conservation of charge. The proton is theorized to be unstable, but decays have not yet been observed. According to experiments carried out at the super-Kamiokande detector in Japan, the proton lifetime must be greater than 10^{33} years.

Particle collisions in accelerators

Suppose we want to study the properties of a heavy particle X with mass M that is short-lived and not naturally occurring. A good strategy in that case is to produce it in collisions of other particles. Indeed, there are various high-energy particle accelerator facilities in which collisions are carried out for exactly that purpose. The colliding particles typically are stable charged particles such as electrons, protons or nuclei of atoms. As a specific example, let us consider

$$e^+ + e^- \to X. \tag{4.72}$$

Particle physicists accelerate the electrons and positrons, collide them inside a complicated detector, and keep track of the rate of interactions as the energy of the beams is varied. When the energy is just right to produce a new particle X, the interaction rate increases. This increase can be dramatic in certain cases such as the ψ/J particle ($m_\psi = 3.1$ GeV) and the Z boson ($m_Z = 91$ GeV).

Since X is massive, $e^+ + e^- \to X$ is allowed by spacetime momentum conservation. In the CM frame, in which $\vec{p}_1 + \vec{p}_2 = 0$, we find that $\vec{p}_3 = 0$, and hence $E_3 = M$. Let's call the total energy in this frame $E_{CM} = E_1 + E_2$. Then we have

$$E_{CM} = M. \tag{4.73}$$

So it is the total energy in the CM frame that sets the limit on the mass M of any particle that can be created by colliding two particles.

The example of $e^+ + e^- \to X$ is the simplest example, but it is also true for collisions involving any number of particles in the final state that the total energy in the CM frame sets the kinematic limit on the possible creation of new particles from a collision. For example, in proton–proton collisions (or proton–antiproton collisions) a common strategy is to look for a particular particle of interest (let's again call it X) among all of the particles that emerge from the collision. Thus one considers reactions of the form

$$p + p \rightarrow X + \text{anything.} \tag{4.74}$$

In the CM frame, which is also the lab frame if the proton energies are equal, the collision products have zero total momentum, and hence could in principle all be at rest in this frame. Therefore, in this limiting case, all of the initial energy in this frame goes into the production of new particles, and none goes into the kinetic energy of the new particles.

But this is relativity; how could it matter which inertial frame we're in? It matters if we want to collide particles at high energy using a particle accelerator. The two basic types of particle accelerators used today are colliding-beam accelerators and fixed-target accelerators, and the incoming particle-beam energy gives a different CM energy in each case.

In a colliding-beam accelerator, two beams of particles, or a beam of particles and one of antiparticles, are aimed directly at one another, to collide at some point in space. If the particles in each beam have exactly the same mass m and energy E_b, and collide exactly head on, then the frame in which the accelerator is at rest is also the CM frame for the collision. The CM energy is related to the beam energy by $E_{\text{CM}} = 2E_b$.

In most colliding-beam accelerators, the energies of the two beams are equal, so the lab frame is the CM frame for the collision. However, in recent years colliders have been constructed where the two beam energies are unequal, so that the X particle is in motion in the lab frame. This is advantageous especially if the X particle we're looking for has a very short lifetime. The primary of examples are the B mesons produced at "B factories".

If we did our $p + p \rightarrow X + \text{anything}$ experiment in a fixed-target accelerator, we would fire a beam of protons at some target (such as a thin foil) containing protons (as well as neutrons and electrons) at rest according to observers in the laboratory. The beam energy E_b in this case is related to the CM frame energy E_{CM} by

$$E_{\text{CM}}^2 = -(p_1 + p_2)^2 = 2m(m + E_b). \tag{4.75}$$

The derivation of this formula is left as an exercise. It implies that $E_{\text{CM}} \sim \sqrt{2m E_b}$ for a fixed-target experiment with beam energy $E_b \gg m$. Notice that the effect of the Lorentz transformation to the CM frame gives a much larger cost in useful energy than would be the case using a nonrelativistic Galilean transformation.

The beam energy E_b is determined by real world constraints in the laboratory, including available funding and real estate for construction of the project. Every increase in beam energy in high-energy physics comes at an enormous cost of time and money. For equal beam energies, a colliding-beam accelerator provides more CM energy E_{CM} for particle creation than does a fixed-target accelerator. In a fixed-target accelerator, the CM of the final particles is moving in the laboratory frame, so some of the beam energy E_b has to be used for the kinetic energy of those particles, and is hence unavailable for particle production.

This is why colliding-beam accelerators are favored in high-energy physics for searching for new particles with large masses. Fixed-target accelerators are now mostly used for creating beams of secondary particles such as neutrinos, which can then undergo subsequent collisions.

4.5 Fission, fusion, and $E = Mc^2$

A new world order is born

Einstein's theoretical discovery of special relativity led to a revolution in the observation and understanding of elementary particles, but the public notoriety of the simple equation $E = mc^2$ comes not from particle physics but from nuclear physics, from the dramatic public impact of the energy made available from the conversion of mass into energy as predicted by this simple equation. The impact of this energy was felt in Hiroshima and Nagasaki in nuclear explosions, and it is felt also everywhere in the world where nuclear reactors provide the heat, light and power people use for living day to day. Physicists have felt the impact of Einstein's work in the career choices they have faced regarding whether or not to work on nuclear weapons research.

Consider the collision $1 + 2 \rightarrow 3$, with $m_1 = m_2 = m$ and $m_3 = M$. In the CM frame, where the initial particle velocities satisfy $\vec{\beta}_1 = -\vec{\beta}_2 = \vec{\beta}$, spacetime momentum conservation tells us that $M = 2\gamma m > 2m$, so the new particle is more massive than the two particles that collided to create it. Mass has been created out of the kinetic energy supplied by the initial particles. Mass is not conserved in special relativity. This is so important that redundancy is justified.

Suppose two particles start out bound together into a bound state with mass M, and we want to separate them so that there are two free particles of mass m_1 and m_2 in the final state. In order to separate the two particles, we must supply energy in the form of work done against the attractive force that holds them together. We don't require the details of the dynamics of this attractive force. All we need to know is that we have put some energy, let's call it B, into separating the single bound particle at rest into two free particles at rest. According to conservation of spacetime momentum,

$$M + B = m_1 + m_2, \tag{4.76}$$

so the bound state of two particles is less massive than the two particles when separated. The energy B is called the *binding energy* of the bound state. For example, let's consider a proton and a neutron bound together by the strong nuclear force in a deuteron, which is the nucleus of an atom of deuterium (^{2}H). The proton has a mass $m_p = 938.28$ MeV, and the neutron mass is $m_n = 939.57$ MeV. The mass of the deuteron is $m_d = 1875.63$ MeV. The mass of a proton and a neutron bound together by the strong nuclear force is less by 2.2 MeV than the total mass of a proton and a neutron as free particles. So the binding energy of the deuteron is $B = 2.2$ MeV.

The binding energy of a general bound state of N particles, each with mass m_i, is

$$B = \sum_{i}^{N} m_i - M. \tag{4.77}$$

Notice that by this definition and (4.49), $B = -Q$. A bound object with positive binding energy $B > 0$ being separated by force is like a decay with $Q < 0$, in other words an endothermic decay, with a threshold energy $|Q|$. Likewise, a spontaneous decay with $Q > 0$ is like a bound object with $B < 0$.

Solid objects consist of bound states of atoms. An atom with atomic number Z and atomic mass number A is a bound state of Z electrons and one nucleus, which in turn is a bound state of Z protons and $(A - Z)$ neutrons. An electron is believed to be fundamental, but protons and neutrons have been revealed to be bound states themselves of particles called quarks and gluons. Quarks are bound by the strong nuclear force through the exchange of gluons in such a way that they cannot be observed as free particles in any lab. Therefore the binding energy that holds together a proton or a neutron is technically unmeasurable. The protons and neutrons in the nucleus are bound by the strong forces between their constituent quarks, but these bonds can be created and broken spontaneously in nature and through modern technology. Indeed, the making and breaking of bonds between neutrons and protons inside the Sun is the ultimate energy source for all life on Earth.

The binding energy per nucleon B/A for a nucleus with mass number A and atomic number Z can be determined by subtracting the measured mass M_N of the nucleus from the total mass of the nucleons in the nucleus according to

$$B_N = Z m_p + (A - Z) m_n - M_N. \tag{4.78}$$

The measured binding energy per nucleon B/A as a function of A has an absolute minimum at the deuteron with $B/A \sim 1.4\,\text{MeV}$, rises smoothly to a maximum near ^{56}Fe with $B/A \sim 9$ MeV, and then declines slowly for the heavy elements with $B/A \sim 7.8$ MeV for ^{235}U. The curve on which these points lie is called the *nuclear binding energy curve*. Out of all the elements, those with mass numbers A near iron are the most stable, with the highest binding energy per nucleon. A nucleus with a lower binding energy per nucleon than iron has a higher average mass per nucleon than iron. If that nucleus undergoes a reaction where the reaction products are closer to iron on the binding energy curve than is the initial nucleus, there will be a net reduction of mass in the system, and hence a net release of energy. Nuclei on the lighter side of the binding energy curve can release energy by fusing together, and nuclei on the heavier side of the binding energy curve can release energy by breaking apart. These two kinds of reactions are called *nuclear fusion* and *nuclear fission*, respectively.

Nuclear fission

Heavy nuclei with a high mass number can release energy if they break apart into lighter nuclei through nuclear fission. A typical fission reaction involves a very heavy nucleus such as ^{235}U breaking apart after being hit by a neutron in a reaction such as

$$n + {}^{235}\text{U} \rightarrow {}^{141}\text{Ba} + {}^{92}\text{Kr} + n + n + n, \qquad (4.79)$$

producing one barium nucleus, krypton nucleus and three neutrons. This reaction is exothermic with $Q \sim 175$ MeV.

This is an enormous amount of energy if we consider 1 g of ^{235}U. One gram of anything is equal to 6.02×10^{23} atomic mass units (amu). A ^{235}U nucleus has mass 235 amu, which means there are 2.56×10^{21} nuclei per gram. If the reaction (4.79) has $Q = 175$ MeV per nucleus, then 1 g of uranium could produce about 4.5×10^{23} MeV $= 7.2 \times 10^{10}$ J. In terms of power consumption, since 1 W is equal to 1 J/s, 1 g of ^{235}U could potentially provide 2×10^4 kWh of power, enough to run a 100 W lamp for about 22 years.

This can be compared to a typical combustion reaction, which is an atomic reaction, not a nuclear one, where the binding energies and, hence, the energy released is only on the order of 1 eV. Combustion reactions of fossil fuels or wood require huge amounts of fuel in order to produce energy. We see this fact reflected in the extent of coalmining in industrialized countries, and the extent to which ancient and contemporary societies are now known to have deforested the land around them in order to heat their homes and cook their food.

As you know, nuclear fission has two main applications: providing energy to boil water to drive steam turbines for electric power generation, and providing energy for explosions. A gram of uranium contains a lot of energy but the uranium nuclei rarely undergo fission spontaneously all by themselves. Both applications of the nuclear fission reaction (4.79) require slow neutrons in the fission products to collide with other uranium nuclei and cause them to undergo fission, producing more neutrons to produce still more fission, in a sequence of fission reactions called a *nuclear chain reaction*.

In a nuclear power reactor, engineers use different methods of slowing down neutrons or absorbing them to keep the chain reaction rate high enough to keep the fission process going, yet low enough so that the process does not get out of control and cause a reactor explosion, as happened at the Chernobyl nuclear reactor in the USSR in 1986. The residents of Chernobyl who had to evacuate their homes and never return understood in a very direct manner the power of the relativistic conversion of mass into energy and the importance of controlling it.

Nuclear fusion

Light nuclei with a low mass number can release energy if they fuse together through nuclear fusion into a nucleus with higher mass number. A typical fusion reaction is

$$^2\text{H} + ^3\text{H} \rightarrow ^4\text{He} + n, \tag{4.80}$$

where the nuclei of two different isotopes of hydrogen – deuterium and tritium – fuse together into a helium nucleus and release a neutron. The net kinetic energy from this reaction is $Q = 17.6$ MeV. This is an order of magnitude smaller than $Q = 175$ MeV for uranium fission, but hydrogen is the lightest element in the periodic table, so 1 g of it contains a lot more nucleons than 1 g of ^{235}U. The two reactants have a combined atomic mass number of 5, so 1 g of equal numbers of deuterium and tritium atoms allows for 1.20×10^{23} reactant pairs. With 17.6 MeV released per reaction, 1 g of deuterium and tritium undergoing fusion should produce 2.12×10^{24} MeV $= 3.39 \times 10^{11}$ J, or about 2×10^5 kWh, which is an order of magnitude greater than the energy possible from the nuclear fission of 1 g of uranium.

In addition to producing an order of magnitude more energy per gram, the reactants in nuclear fusion don't have to be mined from the Earth like uranium does. The reaction products of nuclear fusion are nontoxic and nonradioactive, and the fusion reaction does not suffer from the instability of the fission chain reaction as evidenced in the Chernobyl disaster, unless one does this intentionally by constructing a fusion bomb.

So why aren't we using nuclear fusion reactors to generate our power needs today? The problem with fusion is that it occurs through the strong nuclear force, which is a force between quarks, the elementary particles that are bound together to make protons and neutrons. The strong force has a very short range, only 10^{-15} m. Each deuterium and tritium nucleus carries a positive electric charge from the single proton. The coulomb repulsion between the two protons separated by $R \sim 10^{-15}$ m creates a potential energy barrier of magnitude $q^2/R \sim$ 14.4 keV, which the fusing nuclei have to overcome with kinetic energy in order to get close enough for the strong interactions to occur. If this were the average kinetic energy in a thermal distribution of reactants, the temperature would be $T = E/k_B = 14.4\,\text{keV}/(8.617 \times 10^{-5}\,\text{eV/K}) \sim 1.7 \times 10^8$ K, which is a temperature not realizable by current technology on Earth except in bomb form, where the necessary heat is generated by a fission blast.

The problem of generating useful energy from fusion has not yet been solved by humans on Earth, but Nature solved the problem 13 billion years ago, using the gravitational force to compress hydrogen gas to temperatures high enough for a cycle of fusion chain reactions to take place, to form a stable, continuous basis for the energy released by stars. All elements in the Universe heavier than helium are believed to have been made inside stars through nuclear fusion, and then ejected when the stars exploded. The processes of life on Earth are biochemical and only involve nonrelativistic energy exchanges of a few electronvolts per reaction, but the elements that participate in these biochemical reactions were forged inside of stars through the relativistic conversion of mass into energy.

4.6 Rigid body mechanics

In Newtonian physics, time is universal, so there is no problem with defining an extended object by the space it takes up at a particular moment in time. A perfectly rigid body is an extended object where the distance between any two points in the object remains the same regardless of the motion of the object.

In Chapter 3, in the example of the pole and the barn, it was demonstrated that the measured length of a rigid object depends on the observer's notion of time. Time and space are wedded together in special relativity. Two observers in relative motion with respect to one another will measure time differently, and they will measure space differently. So a perfectly rigid object already seems difficult to define in spacetime.

But the death knell for the perfectly rigid object in relativity is the principle of causality. Any object with mass $m^2 > 0$ must travel on a timelike trajectory through spacetime, which means that at every event along its world line the

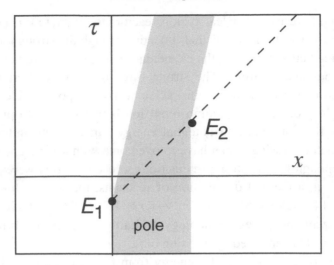

Fig. 4.6. A pole, shown by the shaded area above, starts out at rest with length L in frame S. At event E_1, the left end of the pole is tapped and accelerates almost instantaneously to speed β. The right end of the pole learns that the left end was tapped at some later event E_2. Because the speed of light is the maximum possible speed for any object in spacetime, event E_2 has to occur at or later than the time it takes light, shown by the dashed line, to cross the pole. So it's not possible to have a perfectly rigid object in spacetime.

instantaneous speed must be less than the speed of light. Therefore any signal passing through matter must also travel at less than the speed of light.

The Newtonian notion of a perfectly rigid body requires that information about the displacement of one part of the object travel instantaneously to every other part of the object, so that all points in the object can be displaced at the same time, and the distance between the points can remain the same. This is clearly inconsistent with relativity. In spacetime, there must be a finite speed of displacement propagation for every solid object, and that speed of displacement propagation must be less than the speed of light. An extended object is therefore better described in relativity through the tensor quantities used to describe a continuous deformable system like a fluid, which we shall come to in Chapter 5.

Exercises

4.1   Find the combination of a Lorentz boost and a translation that maps a world line passing through $(\tau, x, y) = (a, 0, 0)$ at $\lambda = 0$ at speed $\vec{\beta}_1 = \beta_1 \partial_x$ to a world line passing through $(\tau, x, y) = (b, 0, 0)$ at $\lambda = 0$ at speed $\vec{\beta}_2 = \beta_2 \partial_y$. Does it matter whether you perform the translation or the boost first?

4.2 Show that acceleration must be a spacelike vector and can never be time-like or null.

4.3 Show that the parameter λ in (4.14) is the proper time measured by an observer traveling along the spacetime path in question.

4.4 A rocket ship is sent out with a crew to take data from the farthest point in the Universe that they can get to and return from in 40 years according to the proper time of the crew. The ship travels at an acceleration equal to one Earth gravity ($9.80\,\mathrm{m/s^2}$) for 10 years, decelerates at the same rate for 10 years, stops briefly to take data, then travels back to Earth by accelerating and decelerating in the same manner. How far can the ship travel according to observers on Earth? How long does the trip take according to observers on Earth? What is the maximum velocity reached by the ship? Suppose the crew is trained to tolerate living under three times the strength of gravity on Earth. In that case, how far can they go, and how long does it take them, in the Earth frame?

4.5 Show that a Lorentz boost with velocity β parallel to the direction of mo-tion of a uniformly accelerating world line in Eq. (4.15) reparametrizes the world line by $\lambda/\alpha \rightarrow \lambda/\alpha - \theta$, where $\theta = \tanh^{-1}\beta$.

4.6 Expand the spacetime momentum conservation equation (4.44) in powers of β_i and show that the first three terms give the nonrelativistic formula for mass, momentum, and energy conservation, respectively.

4.7 Assume that (4.44) is true in some frame S with coordinates $(\tau, \vec{x})$. In some other frame $\tilde{S}$ traveling at velocity $\vec{\beta}$ relative to S, the components of each momentum vector $\mathbf{p}_i$ are related to the components in frame S by a Lorentz transformation

$$\tilde{p}_i^\mu = L^{\tilde{\mu}}_{\nu}\, p_i^\nu. \tag{E4.1}$$

Show that this guarantees that the conservation equation (4.44) valid in frame S is also valid in frame $\tilde{S}$.

4.8 If the space and time axes of a spacetime diagram are measured in meters and seconds, respectively, with time as the vertical axis, then what would be the slope of the world line of a massless particle on that diagram?

4.9 Compute the value of the combination $\hbar c$ in SI units, in units where $c = 1$, and in units where $\hbar = c = 1$.

4.10 Using the dual wave-particle nature of light provides a quick way to de-rive the relativistic Doppler shift for light waves. Suppose a light source at rest in frame S emits light of frequency ν. Using a Lorentz transfor-mation of the components of the photon momentum vector, compute the frequency measured by a receiver in frame $\tilde{S}$ moving at speed β in the x^1

direction relative to S. Compare the labor involved in this derivation with the derivation given in Chapter 3. Which one was simpler?

4.11 Suppose monochromatic blue light is scattered off a beam of charged particles with mass M. What is the value of M in electron-volts required for the light scattered in the backward direction ($\theta = \pi$) to be red? Is there an observed charged particle with this value of mass?

4.12 A high-energy electron ($E_e \gg m_e$) hits a low-energy photon ($E_\gamma \ll m_e$) in the cosmic microwave background. What is the maximum energy that the electron can impart to the photon through the collision?

4.13 Derive the relationship (4.70) between a particle's lifetime T and its half life $t_{1/2}$.

4.14 A beam of muons travels at speed β in the laboratory. In some distance $\Delta \tilde{X}$ in the laboratory, 75 percent of the muons decay. Find the muon speed β as a function of $\Delta \tilde{X}$.

4.15 Derive Eq. (4.75).

4.16 At Fermilab, outside Chicago, protons and antiprotons are collided head on with an energy of 1 TeV in each beam. (Hence the machine is called the Tevatron.) Compute the energy that would be required for an antiproton beam striking protons at rest to produce the same CM energy.

4.17 At the laboratory called DESY, in Hamburg, Germany, a machine called HERA collides 27.5 GeV electrons head-on with 820 GeV protons. Compute the total CM energy as well as the energy that would be required for an electron beam striking protons at rest to produce the same CM energy.

4.18 Compute the value of Q for each of the particle decays listed below. The third decay has not yet been observed, but experiments are looking for it. Relevant particle masses are roughly $m_\pi = 140$ MeV and $m_\rho = 760$ MeV. More precise information can be found in the Particle Data Tables, on the web at http://pdg.lbl.gov/.

$$\rho^0 \to \pi^+ + \pi^-$$
$$\pi^0 \to \gamma + \gamma$$
$$p \to e^+ + \pi^0. \qquad \text{(E4.2)}$$

4.19 Show that the binding energy for the nucleus of an atom with mass number A can be written as $B_N = Z M_1 + (A - Z) m_n - M_A$, where M_1 is the mass of a hydrogen atom, and M_A is the mass of the entire atom. You may neglect the small difference in electron binding energies in the first and third terms.

5

Spacetime physics of fields

Hands-on exercise: the stress tensor of a tub of water

To complete this exercise you will need a lab assistant and the following supplies:

- A 2 or 3 liter tub filled with water.
- A square piece of stiff cardboard not more than a third as wide as the tub on a side.
- A marking pen with waterproof ink.

Fill the tub with water. Let it sit at rest on a table or counter. In your mind's eye, imagine Euclidean coordinate system axes for the tub, with the z axis in the vertical direction, and the x and y axes in the horizontal directions in a right-handed orientation relative to the z axis. These coordinate axes will be referred to as the tub frame.

Using the waterproof pen, draw a large round mark on one side of the cardboard. This will represent the direction normal to the surface of the cardboard. On the same side of the cardboard, draw two arrows, each of which is parallel to one side of the square piece of cardboard. These are the two directions tangential to the surface of the cardboard.

Leaving the tub of water at rest, insert the cardboard into the water so that the normal to the surface points in the positive x direction according to the tub frame. One of the arrows on the surface of the cardboard should be pointing in the positive y direction in the tub frame. Now push very slightly on the cardboard in the positive x direction. The force you feel back from the water is the pressure normal to the surface of the cardboard. Push the cardboard very slightly in the positive y direction, and then in the z direction. Note any resisting force that you feel. Repeat these steps but with the round mark representing the direction orthogonal to the cardboard facing in the positive y direction, with one of the arrows pointing in the positive x direction. Then repeat the process again in the z direction, with the cardboard submerged in the tub with the round mark pointing up and the two arrows both pointing in the $+x$ and $+y$ directions.

The nine forces on this one piece of cardboard make up nine components of a two-index Euclidean tensor that we shall call **T**. The first three components you measured were T^{ix}, where $i = x, y, z$. The cardboard with its normal pointing in the x direction represents a slice of the fluid at some fixed value of x in the tub frame. The second three components you measured were T^{iy}, and the third set of forces gave T^{iz}. The components T^{ii} were all forces that acted orthogonally to the surface of the cardboard and the forces T^{ij}, with $i \neq j$, were forces tangential to the cardboard.

Now imagine the tub of water as an object in Minkowski spacetime rather than Euclidean space. What do the time components of the spacetime version of **T** represent? Consider moving the tub and moving yourself and using Lorentz boosts in the appropriate directions to figure out the significance of the components T^{00}, T^{i0} and T^{0i}.

5.1 What is a field?

The rise of classical field theory

The subject of classical field theory was born in 1687 when Isaac Newton put forward his Law of Universal Gravitation, in which some immaterial agent exists spontaneously in empty space, and leads to a force $\vec{F}_{12}$ acting on an object of mass m_1 located at $\vec{x}_1$ due to the presence of an object with mass m_2 located at $\vec{x}_2$

$$\vec{F}_{12}(x) = -\frac{G\,m_1\,m_2}{|\vec{x}|^2}\,\hat{x}, \qquad \vec{x} \equiv \vec{x}_1 - \vec{x}_2, \qquad \hat{x} = \frac{\vec{x}}{|\vec{x}|}, \qquad (5.1)$$

where Newton's constant G is determined by measurement, and has the value $6.67 \times 10^{-11} (\mathrm{N\,m^2})/\mathrm{kg}^2$ in SI units, and $7.42 \times 10^{-28}\,\mathrm{m/kg}$ in units where $c = 1$.

The gravitational force $\vec{F}_{12}(x)$ is a vector that varies in space, in particular, in flat Euclidean space, the space in which Newtonian physics is defined. A vector that varies in space is called a *vector field*. The immaterial agent put forward by Newton in his law is a vector field carrying the gravitational force. In Newton's time the ordinary derivative was brand new, and physicists didn't understand yet that Newton's immaterial agent was a vector field or that Newton was inventing classical field theory. This was just a law for the manner in which gravity behaved.

It was discovered, by Cavendish in 1771 and then by Coulomb in 1785, that the electrostatic force between two charged objects also obeys the same inverse square law, but with charges q_1 and q_2 replacing the masses m_1 and m_2

$$\vec{F}_{12}(x) = \frac{k\,q_1\,q_2}{|\vec{x}|^2}\,\hat{x},$$ (5.2)

where $\vec{x}$ and $\hat{x}$ are defined as before, and k is a constant determined by measurement, with value $8.99 \times 10^9\,\mathrm{N\,m^2/C^2}$ in SI units, where C stands for coulomb, the SI unit of electric charge. The similarity to Newton's law was inescapable and suggested a unity of mathematical structure in physics that promoted more abstract thinking about forces. This resulted in the birth of the subject that we know as classical field theory.

The above force laws are left unchanged if we make a translation of the origin of the coordinate system by a fixed amount, so let's translate the origin of our coordinate system to the location of object 1. Then we let $\vec{x} = \vec{x}_2$, $m = m_2$ and $M = m_1$. In the case of gravity we can define a new object $\vec{G}(x)$ such that

$$\vec{G}(x) = -\frac{GM}{|\vec{x}|^2}\,\hat{x}, \qquad \vec{F}(x) = m\vec{G}(x).$$ (5.3)

Now we have a new interpretation of the gravitational force. Each massive object has around it a gravitational field $\vec{G}(x)$. When an object with mass m is placed at some location $\vec{x}$ in space it experiences a force $m\vec{G}(x)$ due to the gravitational field of the object of mass M at the origin.

The quantity $\vec{G}(x)$ represents the gravitational field outside a pointlike (or spherically symmetric) object of mass M. If we substitute kQ for GM and q for m, then we get the electrostatic field outside a spherically symmetric object of total charge Q

$$\vec{E}(x) = \frac{kQ}{|\vec{x}|^2}\,\hat{x}.$$ (5.4)

A point charge q placed in this field at point $\vec{x}$ is subject to a force

$$\vec{F}(x) = q\vec{E}(x).$$ (5.5)

The fields $\vec{G}(x)$ and $\vec{E}(x)$ are intrinsic to the objects they surround, which makes them more fundamental as geometrical objects than the forces they induce on other objects.

The Euclidean gradient operator in a Euclidean coordinate basis is $\vec{\nabla} = \partial_x \hat{e}_x + \partial_y \hat{e}_y + \partial_z \hat{e}_z$. The gradient of the function $1/|\vec{x}|$ is

$$\vec{\nabla}\left(\frac{1}{|\vec{x}|}\right) = -\left(\frac{x}{|\vec{x}|^3}\hat{e}_x + \frac{y}{|\vec{x}|^3}\hat{e}_y + \frac{z}{|\vec{x}|^3}\hat{e}_z\right) = -\frac{\hat{x}}{|\vec{x}|^2}.$$ (5.6)

Suppose we define the scalar fields

$$\phi_G(x) = -\frac{GM}{|\vec{x}|}, \qquad \phi_E(x) = \frac{kQ}{|\vec{x}|}. \tag{5.7}$$

Applying (5.6) gives

$$\vec{G}(x) = -\vec{\nabla}\phi_G(x), \qquad \vec{E}(x) = -\vec{\nabla}\phi_E(x). \tag{5.8}$$

The Euclidean gradient operator $\vec{\nabla}$ yields a vector field when operating on the scalar field $\phi(x)$. We can operate with $\vec{\nabla}$ on a vector field $\vec{V}(x)$ to get the divergence $\vec{\nabla} \cdot \vec{V}(x)$. Taking the divergence of the gravitational field $\vec{G}(x)$ yields (for $\vec{x} \neq 0$)

$$\vec{\nabla} \cdot \vec{G} = -GM \left(\hat{e}_x \frac{\partial}{\partial x} + \hat{e}_y \frac{\partial}{\partial y} + \hat{e}_z \frac{\partial}{\partial x} \right) \cdot \left(\frac{\hat{x}}{|\vec{x}|^2} \right)$$

$$= \frac{\partial}{\partial x} \left(\frac{x}{|\vec{x}|^3} \right) + \frac{\partial}{\partial y} \left(\frac{y}{|\vec{x}|^3} \right) + \frac{\partial}{\partial z} \left(\frac{z}{|\vec{x}|^3} \right)$$

$$= 0. \tag{5.9}$$

Since the two fields differ only by a coefficient, it is also true that $\vec{\nabla} \cdot \vec{E} = 0$.

Because the electrostatic and gravitational fields have zero divergence, both the gravitational potential $\phi_G(x)$ and the electrostatic potential $\phi_E(x)$ satisfy the Laplace equation

$$\nabla^2 \phi = \vec{\nabla} \cdot \vec{\nabla}\phi = \frac{\partial^2 \phi}{\partial x^2} + \frac{\partial^2 \phi}{\partial y^2} + \frac{\partial^2 \phi}{\partial z^2} = 0. \tag{5.10}$$

This equation is in effect the equation of motion for these two different classical nonrelativistic fields representing two different forces of nature – gravity and electrostatics. Given some known distribution of charge or mass, one can solve the Laplace equation for the corresponding potential function $\phi(x)$ and compute the electric or gravitational field as the negative gradient $-\vec{\nabla}\phi$ of the potential.

The Laplace equation is only a differential equation in space and not time. Early field theory attracted many strong critics, because the Laplace equation contains no provision for the transmission of information through space about changes in the sources of the field. It requires that one accept the idea of action at a distance, in which a mass M is able to act instantaneously across space to change the gravitational field $\vec{G}(x)$, or a charge Q can act instantaneously to change the electric field $\vec{E}(x)$.

The action-at-a-distance problem for electrostatics was solved by the gradual understanding of the unified dynamics of the electric and magnetic fields. It was realized over the next century that an electric field changing in time induces a magnetic field and a magnetic field changing in time induces an electric field. Maxwell compiled the newly unified theory of electromagnetism into his famous series of equations, and deduced that light propagates as paired fluctuations in electric and magnetic fields. The speed at which these fluctuations in the electromagnetic field propagate in empty space is the speed of light. So the problem of action at a distance was solved in electromagnetic field theory. All fluctuations in the electromagnetic field are limited in their causal influence by the speed of light in empty space.

This is where Einstein arrived on the scene and developed special relativity to make Newtonian mechanics consistent with electromagnetic theory. As shall be shown in this chapter, classical electromagnetism was already a relativistic field theory before special relativity was discovered. The Maxwell equations transform covariantly under a Lorentz transformation, once we correctly identify the geometrical objects that are transforming. The electric and magnetic fields are not elevated from space vectors into spacetime vectors, as happens with velocity and momentum. Instead, they are combined together into an antisymmetric tensor field called the electromagnetic field strength tensor **F**, with coordinate basis components $F_{\mu\nu}(x)$, where x now refers to an event in flat spacetime rather than a location in flat space. A Lorentz transformation in electromagnetism is carried out as a tensor transformation, not a vector transformation. This is why it took time to prove that electromagnetic field theory is a special relativistic field theory.

Despite their similarity in the nonrelativistic limit, the astounding progress in solving the electromagnetic problem was not accompanied by parallel progress in solving the gravitational problem. We will see in the last section of this chapter why the action at a distance problem in Newtonian gravity was not solvable by the application of special relativity, but required Einstein to go a step further and develop general relativity, where spacetime is not flat but curved.

Fields in spacetime

As was shown in Chapter 3, a Euclidean vector $\vec{v}$ is defined in the tangent space to Euclidean space $\mathbf{E}^D$. The Euclidean vector field $\vec{v}(x)$ is a rule for associating a different vector $\vec{v}(x)$ in the tangent space to each point with coordinates $x = \{x^1, \ldots, x^D\}$ in Euclidean space. The gravitational and electric fields in the preceding discussion are two examples of Euclidean vector fields in three dimensions.

A vector field $\mathbf{v}(x)$ in Minkowski spacetime can be defined in an exactly parallel manner as a rule for associating a different spacetime vector $\mathbf{v}(x)$ in the tangent space to $\mathbf{M}^d$ to each point with coordinates $x = \{x^0, x^1, \ldots, x^D\}$ in Minkowski spacetime. Using this definition we can define tensor fields of any rank in spacetime of any dimension.

As was shown in Chapter 3, once we have properly defined a spacetime vector **v** in the tangent space to Minkowski spacetime, the vector itself is unchanged under a change in spacetime coordinates, including a change in coordinates that represents a Lorentz transformation from one inertial frame to another. If we expand the vector in components in a coordinate basis as $\mathbf{v} = v^\mu \hat{x}_\mu$, where $\hat{x}_\mu = \partial_\mu$, then the vector components transform under a Lorentz transformation from an inertial frame S to another inertial frame $\tilde{S}$ by

$$\tilde{v}^\mu = L^{\tilde{\mu}}_\nu v^\nu. \tag{5.11}$$

A spacetime vector field $\mathbf{v}(x)$ can be expanded in a coordinate basis as $\mathbf{v}(x) = v^\mu(x)\, \hat{x}_\mu$. The value of a vector field is a function of the spacetime coordinates, so both the vector components and the coordinates on which they depend undergo a Lorentz transformation when switching from one inertial frame to another. Therefore the coordinate basis components of **v** transform as

$$\tilde{v}^\mu(\tilde{x}) = L^{\tilde{\mu}}_\nu v^\nu(x), \tag{5.12}$$

where

$$\tilde{x}^\mu = L^{\tilde{\mu}}_\nu x^\nu. \tag{5.13}$$

This rule is easily generalized to the coordinate components of any tensor field of any rank in Minkowski spacetime. In particular, a Lorentz-invariant function $f(x)$ of spacetime coordinates can be viewed as a scalar field in spacetime, with $f(\tilde{x}) = f(x)$ as its transformation rule.

5.2 Differential calculus in spacetime

The ordinary derivative came into the world through Newton in the seventeenth century. The Laplacian is a partial differential operator introduced at the turn of the nineteenth century, defined in flat Euclidean spacetime of three dimensions. The beginning of the twentieth century saw the further development of calculus in differential geometry, with the development of the exterior derivative, the Lie derivative, and an understanding of integral calculus in terms of differential forms. All of these topics are important for the subject of relativistic physics.

The Laplace equation in spacetime

The Laplace equation (5.10) gives as a solution the static field potential outside some given configuration of sources at some particular time in the Newtonian universe of fixed absolute space and fixed absolute time. The Laplace equation can be written

$$\nabla^2 \phi(x) = \partial_i \delta^{ij} \partial_j \phi(x) = \partial_i \partial^i \phi(x) = 0, \tag{5.14}$$

where we have used the index raising operation of the Euclidean metric δ_{ij} to write $\partial^i = \delta^{ij} \partial_j$. It is important here to note that the Euclidean metric tensor components $g_{ij} = \delta_{ij}$ in Euclidean coordinates only. In some other coordinate system for Euclidean space, the index of ∂_i is raised by $\partial^i = g^{ij} \partial_j$. This issue will be addressed further in the exercises.

The Laplacian operator ∇^2 is invariant under a rotation $\tilde{x}^i = R^i_j x^j$ of the D coordinates in $\mathbf{E}^D$. A rotation acts on the partial derivative ∂_i through the inverse transformation $\tilde{\partial}_i = R^k_{\tilde{i}} \partial_k$. The result is

$$\tilde{\partial}_i \tilde{\partial}^i \phi(\tilde{x}) = (R^k_{\tilde{i}} \partial_k) \delta^{ij} (R^l_{\tilde{j}}) \partial_l \phi(x)$$
$$= \partial_k \delta^{kl} \partial_l \phi(x) = \partial_k \partial^k \phi(x), \tag{5.15}$$

where we used the fact that $\delta^{ij} R^k_{\tilde{i}} R^l_{\tilde{j}} = \delta^{kl}$. This is the component version of the inverse of $R \cdot I \cdot R^T = I$, the defining equation of a rotation matrix R.

We can do the same thing in flat spacetime by adding time in the most naive way, by extending the index sum in (5.14) over spacetime to get

$$\Box \phi = \partial_\mu \eta^{\mu\nu} \partial_\nu \phi = \partial_\mu \partial^\mu \phi. \tag{5.16}$$

In a coordinate frame S with Minkowski coordinates $(\tau, x^1, \ldots, x^D)$ this expands to

$$\Box \phi = -\frac{\partial^2 \phi}{\partial \tau^2} + \sum_i^D \frac{\partial^2 \phi}{\partial x^{i2}}. \tag{5.17}$$

With this promotion from space to spacetime, the operator $\Box$ is called the *d'Alembertian*, after Jean d'Alembert. D'Alembert did his work on partial differential equations in the early eighteenth century so he had no idea that his name would be given to a Lorentz-invariant differential operator.

Readers should at this point in the book be able to verify by inspection that (5.17), being a hyperbolic version of the Laplacian, is left invariant by hyperbolic rotations in spacetime, just as the Laplacian is left invariant by ordinary rotations in space. A Lorentz transformation $\tilde{x}^\mu = L^{\tilde{\mu}}_\nu x^\nu$ of the d coordinates in $\mathbf{M}^d$ acts on the spacetime partial derivative ∂_μ through the inverse transformation $\tilde{\partial}_\mu = L^\nu_{\tilde{\mu}} \partial_\nu$. A Lorentz transformation leaves invariant the Minkowski coordinate components

$\eta_{\mu\nu}$ of Minkowski metric tensor η. The result is

$$\tilde{\partial}_\mu \tilde{\partial}^\mu \phi(\tilde{x}) = \tilde{\partial}_\mu \eta^{\mu\nu} \tilde{\partial}_\nu \phi(\tilde{x})$$

$$= (L^\rho_{\tilde{\mu}} \partial_\rho) \eta^{\mu\nu} (L^\sigma_{\tilde{\nu}} \partial_\sigma) \phi(x)$$

$$= \partial_\rho \eta^{\rho\sigma} \partial_\sigma \phi(x) = \partial_\rho \partial^\rho \phi(x). \tag{5.18}$$

In this case we have used the relation $\eta^{\mu\nu} L^\rho_{\tilde{\mu}} L^\sigma_{\tilde{\nu}} = \eta^{\rho\sigma}$. This is the component version of the matrix relation $L \cdot \eta \cdot L^T = \eta$, which is the defining property of a Lorentz transformation.

The solutions of $\nabla^2 \phi = 0$ for the Laplacian in space give static field potentials of the form (5.7) for the gravitational or electrostatic field outside some static distribution of mass or charge. The equation $\Box \phi = 0$ in spacetime has plane wave solutions of the form

$$\phi(x) = \phi_0 e^{i\mathbf{k}\cdot\mathbf{x}} = \phi_0 e^{i(\vec{k}\cdot\vec{x} - \omega\tau)}, \qquad k^2 = -\omega^2 + |\vec{k}|^2 = 0, \tag{5.19}$$

with constant amplitude ϕ_0, frequency $\nu = \omega/2\pi$ and wavelength $\lambda = 2\pi/|\vec{k}|$. The metric product $\mathbf{k} \cdot \mathbf{x} = \eta_{\mu\nu} k^\mu x^\nu$ is left invariant by a Lorentz transformation, therefore the field $\phi(x)$ does indeed transform under a Lorentz transformation as a scalar field, with $\tilde{\phi}(\tilde{x}) = \phi(x)$. This is consistent with the fact that $\phi(x)$ solves the Lorentz-invariant differential equation $\partial^\mu \partial_\mu \phi = 0$.

When we add to this scalar wave equation the tensor structure of electromagnetic field theory, the solutions represent electromagnetic waves, in other words, light. Explicit examples will be presented later in this chapter.

The exterior derivative

When Maxwell published his equations in 1873 in *A Treatise on Electricity and Magnetism*, he wrote everything in terms of Euclidean components so that the full set of Maxwell equations looked rather messy and foreboding. It was only later that they were gathered into the more elegant Euclidean vector form that undergraduate physics students learn today. The most economical and beautiful way to understand the Maxwell equations mathematically is not through Euclidean vector analysis, however, but through the mathematics of differential forms, developed by Elie Cartan in the early twentieth century. In this section we shall learn about the differential calculus of forms, and later in the chapter see how they are used to construct the spacetime tensor version of the Maxwell equations.

In Chapter 3 we introduced a geometric object called a p form $\bar{\omega}$, which in a coordinate basis is

$$\bar{\omega} = \frac{1}{p!} \omega_{\mu_1 \ldots \mu_p} dx^{\mu_1} \wedge \ldots \wedge dx^{\mu_p}. \tag{5.20}$$

A p form is defined in the antisymmetric direct product of p copies of the cotangent space $T^*(\mathcal{M})$ of some manifold $\mathcal{M}$. A p form field $\bar{\omega}(x)$ is a p form that takes a different value in $T^*(\mathcal{M})^p$ at every point $x \in \mathcal{M}$. The multiplication operator for a p form $\bar{\omega}$ and a q form $\bar{\sigma}$ is called the exterior product (or wedge product). It gives a $p + q$ form with the symmetry

$$\bar{\omega} \wedge \bar{\sigma} = (-1)^{pq} \bar{\sigma} \wedge \bar{\omega}. \tag{5.21}$$

A function $f(x)$ can be thought of as a p form field with $p = 0$. If x represents the set of coordinates on the manifold on which our cotangent spaces are defined, then the total differential $df(x)$ of some function of those coordinates is

$$df(x) = (\partial_\mu f) \, dx^\mu. \tag{5.22}$$

If we think of the coordinate differentials dx^μ as basis one forms instead of infinitesimal numbers, $df(x)$ becomes a one form field $\bar{\omega}(x)$ with coordinate basis components $\omega_\mu(x) = \partial_\mu f(x)$. The operation df maps a zero form field to a one form field. This operation is called the *exterior derivative*.

Suppose we have a one form field $\bar{\omega} = \omega_\mu dx^\mu$ and we want to apply the exterior derivative operator d. This one form field is a sum of basis one forms dx^μ multiplied by zero forms $\omega_\mu(x)$. If d is a derivative operator (sometimes called a derivation), then it should obey the Leibniz rule of calculus, which for an ordinary derivative in one dimension is

$$\frac{d}{dx}(f(x) \, g(x)) = \frac{df(x)}{dx} \, g(x) + f(x) \frac{dg(x)}{dx}. \tag{5.23}$$

We multiply two forms using the exterior product, and so the Leibniz rule applied to $d\bar{\omega}$ gives us

$$d\bar{\omega} = d(\omega_\mu) \wedge dx^\mu + \omega_\mu \, d(dx^\mu)$$
$$= (\partial_\nu \omega_\mu) \, dx^\nu \wedge dx^\mu. \tag{5.24}$$

We've used $d(dx^\mu) = 0$ because in a coordinate basis, the basis forms are constant. The exterior derivative of a one form $\bar{\omega}$ is a two form. More generally, the exterior derivative of a p form is a $(p + 1)$ form.

What happens if we try to make a two form by operating on our original one form field $df = \partial_\mu f \, dx^\mu$ with the exterior derivative d?

$$d(df) = d^2 f = d(\partial_\mu f) \wedge dx^\mu$$
$$= (\partial_\nu \partial_\mu f) dx^\nu \wedge dx^\mu = -(\partial_\nu \partial_\mu f) dx^\mu \wedge dx^\nu$$
$$= -(\partial_\mu \partial_\nu f) dx^\mu \wedge dx^\nu = -d^2 f = 0. \tag{5.25}$$

The partial derivative operator ∂_μ satisfies $\partial_\mu \partial_\nu f = +\partial_\nu \partial_\mu f$. The exterior product obeys the rule $dx^\mu \wedge dx^\nu = -dx^\nu \wedge dx^\mu$. When we sum over all of the coordinate indices in (5.25), as implied by the repeated index summation convention used in this book, we end up with a sum where every term is matched by its negative, so that the total sum is always exactly zero.

One can use (5.24) and the Leibniz rule to show that the exterior derivative of a p form field $\bar{\omega}$ is the $(p+1)$ form field

$$d\bar{\omega} = \frac{1}{p!} \, \partial_\nu (\omega_{\mu_1...\mu_p}) \, dx^\nu \wedge dx^{\mu_1} \wedge \ldots \wedge dx^{\mu_p}. \tag{5.26}$$

The antisymmetry of the exterior product and the symmetry of the partial derivative guarantee that $d^2\omega = d(d\,\omega) = 0$ for any ω.

Suppose $\bar{\omega}$ is a p form and $\bar{\sigma}$ is a q form. The Leibniz rule consistent with the exterior product (5.21) is

$$d(\bar{\omega} \wedge \bar{\sigma}) = d\bar{\omega} \wedge \bar{\sigma} + (-1)^p \, \bar{\omega} \wedge d\bar{\sigma}. \tag{5.27}$$

Proof of this will be left for the reader as an exercise.

Recall from Chapter 3 that the contraction $\bar{\omega}(\mathbf{v}) \equiv \langle \omega, \mathbf{v} \rangle$ of a vector $\mathbf{v} = v^\mu \hat{e}_\mu$ with a p form $\bar{\omega}$ yields a $(p-1)$ form with components

$$\bar{\omega}(\mathbf{v}) = \frac{1}{(p-1)!} \, v^{\mu_1} \, \omega_{\mu_1\mu_2...\mu_p} \, dx^{\mu_2} \wedge \ldots \wedge dx^{\mu_p}. \tag{5.28}$$

Contraction of a p form with a vector is a linear operation, and obeys the Leibniz rule

$$(\bar{\omega} \wedge \bar{\sigma})(\mathbf{v}) = \bar{\omega}(\mathbf{v}) \wedge \bar{\sigma} + (-1)^p \, \bar{\omega} \wedge \bar{\sigma}(\mathbf{v}). \tag{5.29}$$

Notice the similarity between (5.29) and (5.27). Contraction of a form field with a vector can be viewed as a type of derivative of the form field in the direction of the vector, one that lowers the degree of a form by one, turning a p form into a $(p-1)$ form.

The exterior derivative operates on a p form to yield a $(p+1)$ form. The exterior derivative combined with a contraction would give us a derivative operator that operates on a p form and yields another p form. But do we contract $\mathbf{v}$ with the exterior derivative $d\bar{\omega}$ to get $d\bar{\omega}(\mathbf{v})$, or take the exterior derivative of the contraction $\bar{\omega}(\mathbf{v})$ to get $d(\bar{\omega}(\mathbf{v}))$? The symmetric combination of both operations

$$\mathcal{L}_\mathbf{v} \bar{\omega} = d(\bar{\omega}(\mathbf{v})) + d\bar{\omega}(\mathbf{v}) \tag{5.30}$$

is called the *Lie derivative* of a p form field $\bar{\omega}(x)$ with respect to the vector field $\mathbf{v}(x)$. The Lie derivative is defined for tensors of any rank or symmetry property, and is central in understanding how the geometry of spacetime is reflected in the conservation laws we observe in nature.

The Lie derivative

In relativity, a rate of change of some geometric object such as a tensor can be measured in a coordinate-invariant manner by using the proper time or proper distance $\Delta\lambda$ along some curve $\mathcal{C}(\lambda)$ to which there is some tangent vector field $\mathbf{u}(\lambda)$. The Lie derivative is constructed in this manner. The Lie derivative is a concept from differential geometry that is important for understanding the geometrical basis for physical conservation laws such as conservation of spacetime momentum. This section is a bit more technical than the rest of this chapter and can be omitted in a first reading.

Picture some arbitrary manifold $\mathcal{M}$ as being filled by a set of curves $\{\mathcal{C}(\lambda)\}$ arranged in what is called a smooth congruence, such that every point in the manifold lies on one and only one curve $\mathcal{C}_i(\lambda)$. Each curve in the congruence has a tangent vector $\mathbf{u}_i = (d/d\lambda)_i$. The resulting tangent vector field $\mathbf{u}(x)$ is called the flow of the congruence. Since $\mathbf{u} = d/d\lambda$, the curve $\mathcal{C}_i(\lambda)$ is an integral curve of the tangent vector $\mathbf{u}_i$ taken as a differential operator.

For example, consider flat spacetime in four dimensions, otherwise known as $\mathbf{M}^4$, with coordinates (τ, x, y, z). The set of all straight lines parallel to the τ axis make up one of the possible smooth congruences of curves in $\mathbf{M}^4$. Any event $E_i = (\tau_i, x_i, y_i, z_i) \in \mathbf{M}^4$ can be described as occurring at a proper time $\lambda = \tau_i$ along the curve $\mathcal{C}_i(\lambda) = (\lambda, x_i, y_i, z_i)$ in the congruence, which intersects the $\tau = 0$ plane at the event $(0, x_i, y_i, z_i)$. Each curve in this congruence has a tangent vector $\mathbf{u}_i = d/d\lambda = \partial_\tau$. The resulting tangent vector field $\mathbf{u}(x)$ is a constant vector field, because the congruence of curves to which it is tangent are straight lines. This vector field could represent the velocity field of a fluid or a uniform distribution of objects at rest in the coordinate frame (τ, x, y, z). The curves in the congruence would represent the flow lines of particles in the fluid. In the frame in which the fluid is at rest, each fluid particle P_i traces a path through spacetime that is $\mathcal{C}_i(\lambda) = (\lambda, x_i, y_i, z_i)$.

Suppose we have some generic smooth congruence $\{\mathcal{C}(\lambda)\}$ with tangent vector field $\mathbf{v}$. The Lie derivative $\mathcal{L}_\mathbf{v}$ is a derivative operator that can act on any tensor field $\mathbf{T}(x)$ in spacetime to give the rate of change $\mathcal{L}_\mathbf{v}\mathbf{T}$ of that tensor field along the congruence, the tangent field of which is $\mathbf{v}$.

To take a derivative of a tensor along a curve with parameter λ, we need to compare the tensor field $\mathbf{T}(\lambda)$ at some parameter value $\lambda = \lambda_1$ with its value at $\lambda_2 = \lambda_1 + \Delta\lambda$, and then take the limit $\Delta\lambda \to 0$. But how can we compare a tensor field at two different points on a curve? Recall that a tensor field is defined as a rule for associating a *different* tensor with each point on a manifold. $\mathbf{T}_1(\lambda_1)$ and $\mathbf{T}_2(\lambda_2)$ are two different tensors. It makes no sense to take the difference between two different tensors at two different points. The right thing to do is find a way to drag the second tensor $\mathbf{T}_2(\lambda_2)$ back to the location of the first tensor $\mathbf{T}_1(\lambda_1)$ so that we can take the difference between the two tensors at a single point, and see what happens to that in the limit $\Delta\lambda \to 0$.

The Lie derivative is defined by the manner in which the second tensor is dragged back along the curve to the location of the first tensor for comparison. This operation is called *Lie dragging*. We use the congruence of curves $\{\mathcal{C}(\lambda)\}$ it-self to define the mapping for bringing the second tensor back to the first tensor for comparison. The operation $\lambda \to \lambda + \Delta\lambda$ defines a mapping of the manifold onto itself. We can use this mapping to define a new tensor $\tilde{\mathbf{T}}_2$ by Lie dragging the second tensor $\mathbf{T}_2$ along the curve $\mathcal{C}(\lambda)$ to $\lambda_1 = \lambda_2 - \Delta\lambda$. Then we compare it with $\mathbf{T}_1$ and take the limit $\Delta\lambda \to 0$ to get

$$\mathcal{L}_{\mathbf{v}}\mathbf{T} = \lim_{\Delta\lambda \to 0} \frac{\tilde{\mathbf{T}}_2(\lambda_1) - \mathbf{T}(\lambda_1)}{\Delta\lambda}. \tag{5.31}$$

Notice that by this definition, a tensor field has a vanishing Lie derivative if the tensor field was Lie-dragged to begin with, in other words if $\tilde{\mathbf{T}}_2(\lambda_1) = \mathbf{T}(\lambda_1)$. Therefore Lie dragging must be defined by a vanishing Lie derivative. We shall see this below.

The simplest example of the Lie derivative of a tensor field is the Lie derivative of a scalar field $f(x)$. Along some congruence $\{\mathcal{C}(\lambda)\}$ we can look at $f(x)$ as a function of the parameter λ. At $\lambda = \lambda_1$ we have $f_1 = f(\lambda_1)$ and at $\lambda = \lambda_2$, the field $f_2 = f(\lambda_2)$. The new field $\tilde{f}(\lambda_1)$ is obtained from Lie-dragging f_2 back to λ_1 so that $\tilde{f}(\lambda_1) = f(\lambda_2)$. Equation (5.31) then just gives us

$$\mathcal{L}_{\mathbf{v}}f = \lim_{\Delta\lambda \to 0} \frac{\tilde{f}(\lambda_1) - f(\lambda_1)}{\Delta\lambda} \tag{5.32}$$

$$= \lim_{\Delta\lambda \to 0} \frac{f(\lambda_1 + \Delta\lambda) - f(\lambda_1)}{\Delta\lambda}$$

$$= \frac{df}{d\lambda}. \tag{5.33}$$

In a coordinate basis, with $v^\mu = dx^\mu/d\lambda$, we can use the chain rule of calculus to get

$$\mathcal{L}_{\mathbf{v}}f(x) = v^\mu \frac{\partial f}{\partial x^\mu} \equiv \mathbf{v}(f). \tag{5.34}$$

The Lie derivative of a vector field $\mathbf{w}(x)$ is more tricky. A vector field $\mathbf{w}(x)$ gen-erates its own congruence of curves $\{\mathcal{C}(\sigma)\}$ as integral curves of $\mathbf{w}(x) = \mathbf{w}(\sigma) = d/d\sigma$, and this whole congruence $\{\mathcal{C}(\sigma)\}$ is dragged along the congruence $\{\mathcal{C}(\lambda)\}$ by the map $\lambda \to \lambda + \Delta\lambda$. The set of curves $\{\mathcal{C}(\sigma)\}$ is said to be Lie-dragged along the set of curves $\{\mathcal{C}(\lambda)\}$ if infinitesimal displacements by $\Delta\sigma$ and $\Delta\lambda$ in the

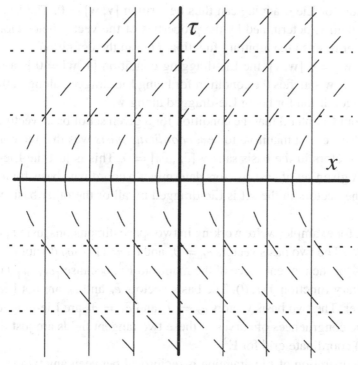

Fig. 5.1. The solid lines parallel to the τ axis belong to the congruence of curves $\{C(\lambda)\}$ with tangent vector field ∂_τ. The dotted lines parallel to the x axis are in the congruence $\{C(\sigma)\}$ with tangent field ∂_x, and they are Lie-dragged along $\{C(\lambda)\}$ by ∂_τ. The curved dashed lines represent the world lines of particles undergoing a constant (or uniform) acceleration in the $+x$ direction. This congruence of curves is not Lie-dragged along $\{C(\lambda)\}$ by ∂_τ, but it is Lie-dragged along $\{C(\sigma)\}$ by ∂_x.

operations $d/d\lambda$ and $d/d\sigma$ make a closed path in the union of the two sets of curves. This condition is illustrated in Figure 5.1.

The mathematical expression of the Lie-dragging condition is

$$\left[\frac{d}{d\lambda}, \frac{d}{d\sigma}\right] f(x) = \frac{d}{d\lambda}\left(\frac{df}{d\sigma}\right) - \frac{d}{d\sigma}\left(\frac{df}{d\lambda}\right) = 0 \tag{5.35}$$

for an arbitrary function $f(x)$ defined on the manifold $\mathcal{M}$. In terms of the vectors $\mathbf{v}$ and $\mathbf{w}$, we say that $\mathbf{w}$ is Lie-dragged along $\mathbf{v}$ if

$$[\mathbf{v}, \mathbf{w}]\, f(x) = v^\mu \frac{\partial}{\partial x^\mu}\left(w^\nu \frac{\partial f}{\partial x^\nu}\right) - w^\mu \frac{\partial}{\partial x^\mu}\left(v^\nu \frac{\partial f}{\partial x^\nu}\right) = 0. \tag{5.36}$$

This equation is satisfied for all $f(x)$ when

$$v^\mu \frac{\partial}{\partial x^\mu} w^\nu - w^\mu \frac{\partial}{\partial x^\mu} v^\nu = 0. \tag{5.37}$$

The condition for Lie dragging can thus be written $[\mathbf{v}, \mathbf{w}] = 0$. The Lie dragging of a vector field is determined by the properties of the vector fields alone, so we will no longer refer to the arbitrary function $f(x)$ in the definition.

Since $[\mathbf{v}, \mathbf{w}] = -[\mathbf{w}, \mathbf{v}]$, the Lie-dragging condition $[\mathbf{v}, \mathbf{w}] = 0$ is a reciprocal relationship. If $\mathbf{w}$ satisfies the condition for being Lie-dragged along $\mathbf{v}$, then $\mathbf{v}$ also satisfies the condition for being Lie-dragged along $\mathbf{w}$.

Recall from Chapter 3 that the condition for a given set of basis vectors $\{\hat{e}_\mu\}$ for the tangent space of a manifold to be a *coordinate basis* was that the commutator of any two vectors in the basis satisfy $[\hat{e}_\mu, \hat{e}_\nu] = 0$. This is just the Lie-dragging condition. A given set of basis vectors determines a coordinate basis for a manifold if each of the vectors in the set is Lie-dragged by all of the other basis vectors in the set.

Suppose, for example, we're working in two space dimensions, using polar coordinates (r, θ). The two basis vectors $\hat{e}_r = \partial_r$ and $\hat{e}_\theta = (1/r)\partial_\theta$ do not form a coordinate basis for the tangent space of the manifold $\mathbf{E}^2$, because $[\hat{e}_r, \hat{e}_\theta] f(r, \theta) \neq 0$ for an arbitrary function $f(r, \theta)$. The basis vectors $\hat{e}_r$ and $\hat{e}_\theta$ are not Lie-dragged by each other. The two basis vectors $\hat{e}_x = \partial_x$ and $\hat{e}_y = \partial_y$ are Lie-dragged by one another. The congruences of curves of these two tangent fields are just the rectangular (x, y) coordinate grid for $\mathbf{E}^2$.

Since the operation of Lie dragging is reciprocal between any two vector fields on a manifold, the Lie derivative of a vector field $\mathbf{u}$ with respect to another vector field $\mathbf{v}$ is just the commutator $[\mathbf{v}, \mathbf{u}]$. In other words,

$$\mathcal{L}_{\mathbf{v}} \mathbf{u} = [\mathbf{v}, \mathbf{u}]. \tag{5.38}$$

If the two vectors are expanded in coordinate basis components so that $\mathbf{v} = v^\mu \partial_\mu$ and $\mathbf{u} = u^\mu \partial_\mu$, the Lie derivative becomes

$$(\mathcal{L}_{\mathbf{v}} \mathbf{u})^\mu = v^\nu \partial_\nu u^\mu - u^\nu \partial_\nu v^\mu. \tag{5.39}$$

The condition for $\mathbf{u}$ to be Lie-dragged by $\mathbf{v}$ (and vice versa) is $\mathcal{L}_{\mathbf{v}}\mathbf{u} = -\mathcal{L}_{\mathbf{u}}\mathbf{v} = 0$. The Lie derivative of a vector field $\mathcal{L}_{\mathbf{v}}\mathbf{u}$ measures the amount by which the two vector fields $\mathbf{u}$ and $\mathbf{v}$ fail to give closed paths in the two congruences of curves that they determine. The relative minus sign between the operations $\mathcal{L}_{\mathbf{v}}\mathbf{u} = -\mathcal{L}_{\mathbf{u}}\mathbf{v}$ arises because the two sets of curves $\{\mathcal{C}(\lambda)\}$ and $\{\mathcal{C}(\sigma)\}$ enter in the opposite order.

The Lie derivative of a one form field $\bar{\omega}(x)$ with respect to a vector field $\mathbf{v}(x)$ can be deduced from the fact that the contraction $\bar{\omega}(\mathbf{u}) = \omega_\mu u^\mu$ is a scalar field on the manifold $\mathcal{M}$. The Leibniz rule applied to the contraction of the two fields gives

$$\mathcal{L}_{\mathbf{v}}(\bar{\omega}(\mathbf{u})) = (\mathcal{L}_{\mathbf{v}}\bar{\omega})(\mathbf{u}) + \bar{\omega}(\mathcal{L}_{\mathbf{v}}\mathbf{u})$$
$$= (\mathcal{L}_{\mathbf{v}}\bar{\omega})_\mu u^\mu + (\mathcal{L}_{\mathbf{v}}\mathbf{u})^\mu \omega_\mu. \tag{5.40}$$

The Lie derivative of the scalar field $\omega_\mu u^\mu$ is

$$
\begin{aligned}
\mathcal{L}_\mathbf{v}(u^\mu \omega_\mu) &= v^\nu \partial_\nu (u^\mu \omega_\mu) \\
&= (v^\nu \partial_\nu u^\mu)\omega_\mu + u^\mu (v^\nu \partial_\nu \omega_\mu) \\
&= (\mathcal{L}_\mathbf{v}\mathbf{u})^\mu \omega_\mu + u^\mu (\omega_\nu \partial_\mu v^\nu + v^\nu \partial_\nu \omega_\mu).
\end{aligned}
\tag{5.41}
$$

The Lie derivative of a one form field is therefore

$$
(\mathcal{L}_\mathbf{v}\bar\omega)_\mu = v^\nu \partial_\nu \omega_\mu + \omega_\nu \partial_\mu v^\nu.
\tag{5.42}
$$

We can write this more elegantly in terms of exterior derivatives and vector contractions as

$$
\mathcal{L}_\mathbf{v}\,\bar\omega = d(\bar\omega(\mathbf{v})) + d\bar\omega(\mathbf{v}).
\tag{5.43}
$$

This is the same as the formula (5.30), applied to the specific case of a one form.

Now let's look at some specific examples of Lie derivatives and what they really do. First consider the vector field $\mathbf{v} = \partial_\tau$ in Minkowski spacetime in four dimensions. The Lie derivative of a scalar field with respect to this vector field is

$$
\mathcal{L}_\mathbf{v} f(x) = \mathbf{v}(f) = \frac{\partial f}{\partial \tau}.
\tag{5.44}
$$

A scalar field is Lie-dragged along the integral curves of this tangent vector field if it is constant in time τ. If we take the Lie derivative of f with respect to the vector field $\mathbf{u} = \gamma \partial_\tau + \gamma\beta \partial_x$, then we get

$$
\mathcal{L}_\mathbf{u} f(x) = \mathbf{u}(f) = \gamma \frac{\partial f}{\partial \tau} + \gamma\beta \frac{\partial f}{\partial x}.
\tag{5.45}
$$

In this case the Lie derivative tells us the rate of change of $f(x)$ along a congruence of curves representing objects or particles traveling at velocity β in the x direction. Since β is constant in this example, $\mathcal{L}_\mathbf{v}\mathbf{u} = [\mathbf{v}, \mathbf{u}] = -\mathcal{L}_\mathbf{u}\mathbf{v} = 0$. The curves to which $\mathbf{u}$ is tangent and the curves to which $\mathbf{v}$ is tangent are Lie-dragged by each other. The two confluences make a grid that is just a skewed version of the rectangular grid implied by Minkowski coordinates (τ, x, y, z).

For an example of two vector fields that are not Lie dragged by one another, consider the tangent field to the world lines of a collection of uniformly accelerated objects traveling in the x direction in $\mathbf{M}^4$, discussed in Chapter 4. Each curve satisfies the equation $-\tau^2 + x^2 = \alpha^2$ for some $\alpha \in \mathbf{R}$. The vector field tangent to this congruence is

$$
\mathbf{w} = \frac{x}{\alpha} \partial_\tau + \frac{\tau}{\alpha} \partial_x.
\tag{5.46}
$$

The two vector fields **v** and **w** have a nonzero Lie derivative

$$\mathcal{L}_\mathbf{v} \mathbf{w} = \frac{1}{\alpha} \partial_x = \frac{\partial \mathbf{w}}{\partial \tau} \neq 0. \tag{5.47}$$

Lie dragging along a curve with tangent vector $\mathbf{v} = \partial_\tau$ is the same as making a translation in coordinate time τ. The vectors **u** and **v** are not Lie-dragged by one another because one of them has a symmetry that the other does not – translation in coordinate time τ. But notice now that we can write these vectors in any coordinates we want, and we have a condition for time translation invariance that applies in all of them.

The Lie derivative of a tensor field of any rank, given in Eq. (5.30), can be deduced from the Lie derivatives of vectors, one forms, and scalars by using vector contraction and the Leibniz rule. For example, the Lie derivative of a rank $\binom{0}{2}$ tensor field $\mathbf{g}(x)$ can be determined from the contraction $\mathbf{g}(\mathbf{u}, \mathbf{v})$ and the Leibniz rule. In coordinate basis components this is

$$(\mathcal{L}_\mathbf{v} \mathbf{g})_{\mu\nu} = v^\rho \partial_\rho \, g_{\mu\nu} + g_{\mu\rho} \, \partial_\nu v^\rho + g_{\rho\nu} \, \partial_\mu v^\rho. \tag{5.48}$$

The proof will be left for the reader as an exercise.

Killing vectors and conservation laws

Suppose the $\binom{0}{2}$ tensor field $\mathbf{g}(x)$ in (5.48) is the metric tensor on a manifold $\mathcal{M}$. The set of vectors $\{\mathbf{k}\}$ for which

$$\mathcal{L}_\mathbf{k} \mathbf{g} = 0 \tag{5.49}$$

are called the *Killing vectors* of **g**. The metric tensor is Lie-dragged along the congruence of curves determined by the Killing vectors $\mathbf{k}_i$. Killing vectors tell us the isometries of a manifold $\mathcal{M}$ with metric tensor **g**. An isometry is a coordinate transformation that maps the metric into itself. Since this book is about special relativity, the metric we care about is the metric for flat spacetime in four dimensions, which in Minkowski coordinates is

$$ds^2 = \eta_{\mu\nu} \, dx^\mu \, dx^\nu = -d\tau^2 + dx^2 + dy^2 + dz^2. \tag{5.50}$$

The Killing vectors of $\mathbf{M}^4$ in these coordinates satisfy the equation

$$\eta_{\mu\rho} \, \partial_\nu k^\rho + \eta_{\rho\nu} \, \partial_\mu k^\rho = 0, \tag{5.51}$$

or, equivalently, using $k_\mu = \eta_{\mu\nu} k^\nu$,

$$\partial_\mu k_\nu = -\partial_\nu k_\mu. \tag{5.52}$$

The simplest class of vector fields that solve this equation are

$$\mathbf{q}_0 = \partial_\tau, \quad \mathbf{q}_1 = \partial_x, \quad \mathbf{q}_2 = \partial_y, \quad \mathbf{q}_3 = \partial_z, \qquad (5.53)$$

which are the basis vectors for Minkowski coordinates. The map generated by the Killing vector ∂_μ is a translation of the coordinate component x^μ by $x^\mu \to x^\mu + a^\mu$, where a^μ are constants.

Flat spacetime is invariant under a translation in any direction, but this isometry is based on a shift in Minkowski coordinates. The metric of flat spacetime is Lie-dragged along Minkowski coordinates, because Minkowski coordinates are adapted to the translation invariance of the spacetime. The Minkowski metric in any other basis is not manifestly invariant under translations, but the Killing vectors will be the same in any coordinate system, and it is the Killing vectors that carry the information about the symmetries of the spacetime.

The set of Killing vectors of a metric constitutes a linear vector space. If $\mathbf{k}_i$ and $\mathbf{k}_j$ are both Killing vectors of a spacetime with some metric $\mathbf{g}$, then

$$\mathcal{L}_{a\mathbf{k}_i + b\mathbf{k}_j}\, \mathbf{g} = a\mathcal{L}_{\mathbf{k}_i}\, \mathbf{g} + b\mathcal{L}_{\mathbf{k}_j}\mathbf{g} = 0, \qquad (5.54)$$

where a and b are constants. So a linear combination of Killing vectors is also a Killing vector. The most important property of Killing vectors is that they form what is called a *Lie algebra*. If $\mathbf{k}_i$ and $\mathbf{k}_j$ are both Killing vectors of $\mathbf{g}$, then the Lie derivative $\mathbf{k}_k = [\mathbf{k}_i, \mathbf{k}_j]$ is also a Killing vector of $\mathbf{g}$. In other words

$$\mathcal{L}_{\mathbf{k}_i}\mathbf{g} = \mathcal{L}_{\mathbf{k}_j}\mathbf{g} = 0 \to \mathcal{L}_{[\mathbf{k}_i, \mathbf{k}_j]}\, \mathbf{g} = 0. \qquad (5.55)$$

The proof will be left for the reader as an exercise.

The algebra of translations is trivial because $[\mathbf{q}_i, \mathbf{q}_j] = 0$ for all of the translation Killing vectors in (5.53). For a nontrivial Lie algebra of Killing vectors, consider the spacelike components of (5.52). Three vectors that satisfy $\mathcal{L}_{\mathbf{k}}\mathbf{g} = 0$ are

$$\mathbf{j}_1 = y\partial_z - z\partial_y$$
$$\mathbf{j}_2 = z\partial_x - x\partial_z$$
$$\mathbf{j}_3 = x\partial_y - y\partial_x. \qquad (5.56)$$

The spacelike vectors $\{\mathbf{j}_1, \mathbf{j}_2, \mathbf{j}_3\}$ are solutions of (5.52) that generate rotations in space around the x, y and z axes, respectively. They also satisfy the algebra

$$[\mathbf{j}_1, \mathbf{j}_2] = -\mathbf{j}_3, \quad [\mathbf{j}_2, \mathbf{j}_3] = -\mathbf{j}_1, \quad [\mathbf{j}_3, \mathbf{j}_1] = -\mathbf{j}_2. \qquad (5.57)$$

This set of Killing vectors forms the Lie algebra for the group of rotations in three space dimensions, which is called $SO(3)$. You will learn more about this in Chapter 8.

We get a third set of Killing vectors when we look at the spacetime components of (5.52). Recall that with the Minkowski metric, in the timelike convention that we have chosen, $k_0 = -k^0$. So the set of Killing vectors that satisfy the $(0i)$ components of (5.52) are

$$k_1 = \tau \partial_x + x \partial_\tau$$
$$k_2 = \tau \partial_y + y \partial_\tau$$
$$k_3 = \tau \partial_z + z \partial_\tau. \tag{5.58}$$

This set of Killing vectors generate the Lorentz boosts that are the crucial isometries of Minkowski spacetime.

Unlike rotations in space, Lorentz boosts in space and time don't make up a closed algebra by themselves. For example, the Lie derivative of a vector field generating a boost in the x direction with respect to a vector field generating a boost in the y direction

$$[\tau \partial_x + x \partial_\tau, \tau \partial_y + y \partial_\tau] = x \partial_y - y \partial_x \tag{5.59}$$

is a rotation around the z axis. In general, the Lie derivative of a Lorentz boost with respect to another Lorentz boost is a rotation around the axis orthogonal to the two Lorentz boosts

$$[k_1, k_2] = j_3, \qquad [k_2, k_3] = j_1, \qquad [k_3, k_1] = j_2. \tag{5.60}$$

The Lie derivatives $[k_i, j_j]$ of boosts with rotations will be left as an exercise for the reader.

The complete set of Lie derivatives, or commutation relations, as they are also called, tells us that the isometry group of flat spacetime is the special orthogonal group in one time and three space dimensions, which is given the label $SO(1, 3)$. Although we have used the case of four spacetime dimensions specifically here, this same basic Lie algebra of Killing vectors occurs in flat spacetime in any dimension d with one time and $D = (d - 1)$ space dimensions, in which case the isometry group is $SO(1, D)$ (or $SO(D, 1)$), also known as the *Lorentz group* in d dimensions. The Lorentz group will be discussed in detail in Chapter 8.

Why are Killing vectors important in physics? Consider the world line of some object with proper time λ and spacetime velocity $u(\lambda)$ that satisfies the Lorentz-covariant version of Newton's equation with no external forces

$$\frac{du}{d\lambda} = u \cdot \nabla u = u^\alpha \partial_\alpha u = 0. \tag{5.61}$$

If $\mathbf{k}$ is a Killing vector, then the quantity $\mathbf{k} \cdot \mathbf{u}$ is a constant along this curve, so that

$$\frac{d}{d\lambda}(\mathbf{k} \cdot \mathbf{u}) = 0. \tag{5.62}$$

To prove this use (5.61) and the Killing condition $\partial_\mu k_\nu + \partial_\nu k_\mu = 0$ to get

$$\frac{d}{d\lambda}(\mathbf{k} \cdot \mathbf{u}) = u^\alpha \partial_\alpha (k_\mu u^\mu)$$

$$= k_\mu u^\alpha \partial_\alpha u^\mu + u^\mu u^\alpha \partial_\alpha k_\mu \tag{5.63}$$

$$= \frac{1}{2} u^\mu u^\alpha (\partial_\alpha k_\mu + \partial_\mu k_\alpha) = 0. \tag{5.64}$$

As discussed earlier, the d basis vectors $\{\partial_\mu\}$ for Minkowski coordinates $(\tau, x^1, \ldots, x^D)$ of flat spacetime are d Killing vectors $\{\mathbf{q}\}$. Let the spacetime momentum of an object with mass m be $\mathbf{p} = m\mathbf{u}$. Then the theorem in the previous paragraph implies that

$$\frac{d}{d\lambda}(\mathbf{q}_i \cdot m\mathbf{u}) = \frac{d}{d\lambda}(\mathbf{q}_i \cdot \mathbf{p}) = 0. \tag{5.65}$$

Spacetime momentum is conserved in every direction in which there is a Killing vector representing translation invariance, which in the case of flat spacetime means all directions, including time. Translation invariance in the time direction leads to conservation of the timelike component of the spacetime momentum, which is the energy E. In a spacetime with a metric $\mathbf{g}$ that is changing in time, conservation of energy would not be enforced along the paths of objects with zero acceleration.

But time is just a coordinate in the frame of reference of some observer. In relativity we have to define things in a coordinate-invariant manner. Minkowski coordinates $(\tau, x^1, \ldots, x^D)$ are the coordinates in which the components of the metric tensor are all equal to ± 1. So in Minkowski coordinates, translation invariance is manifest in every direction. But no matter what coordinate system we use to describe flat spacetime, we will end up finding the same timelike Killing vector, and conservation of energy will still apply. The Lie derivative and Killing vector construction give us a coordinate-invariant manner to define the symmetries of the spacetime in question, and to understand the relationships between them.

The rotational Killing vectors $\{\mathbf{j}\}$ represent the rotational invariance of flat spacetime. Conservation of angular momentum is encoded in the value of $\mathbf{j}_i \cdot \mathbf{p}$ remaining constant along the world line to which $\mathbf{p} = m\mathbf{u}$ is tangent. Notice that if

$\vec{r}$ is the vector $x\,\partial_x + y\,\partial_y + z\,\partial_z$, then the components of $\mathbf{j}_i \cdot \mathbf{p}$ can be written

$$\mathbf{j}_1 \cdot \mathbf{p} = y\,p_z - z\,p_y = (\vec{r} \times \vec{p})^x$$
$$\mathbf{j}_2 \cdot \mathbf{p} = z\,p_x - x\,p_z = (\vec{r} \times \vec{p})^y$$
$$\mathbf{j}_3 \cdot \mathbf{p} = x\,p_y - y\,p_x = (\vec{r} \times \vec{p})^z. \tag{5.66}$$

The conserved values of $\mathbf{j} \cdot \mathbf{p}$ are the spacelike components of the Newtonian angular momentum $\vec{J} = \vec{r} \times \vec{p}$, which in this case represent the angular momentum about the origin of the coordinate system. Angular momentum in Newtonian physics is treated as a vector but properly transforms under Lorentz transformations as a tensor. The angular momentum tensor will be discussed in the next section.

The case for a geometric understanding of the Universe is especially strong when one sees the close relationship between geometry and conservation laws for physical quantities like energy and momentum. The Killing vector is central to this understanding and becomes even more important when one graduates from flat to curved spacetime in the study of general relativity.

5.3 Integral calculus in spacetime

Integration is defined as the inverse of differentiation in the fundamental theorem of calculus, which for an indefinite integral can be written

$$\int df = f + c, \tag{5.67}$$

where c is a constant. It's tempting to look at the ordinary differential df as a differential form. In one dimension, on the manifold $\mathbf{E}^1$, the exterior derivative of a scalar field $f(x)$ can be written

$$df(x) = \frac{\partial f}{\partial x}\,dx. \tag{5.68}$$

If we consider dx to be the basis one form for the cotangent space of $\mathbf{E}^1$, then the differential df is a one form in this cotangent space as well. Integrating the one form df over a connected region $\mathcal{R}$ of $\mathbf{E}^1$ with coordinate x and boundary points (a, b) gives

$$\int_{\mathcal{R}} df = \int_a^b \frac{\partial f}{\partial x}\,dx = f|_a^b = f(b) - f(a). \tag{5.69}$$

The definite integral of the one form df is equal to the zero form f evaluated at the boundaries (a, b) of the region of integration $\mathcal{R} \subset \mathbf{E}^1$. This simple result can be generalized to higher dimensions using differential forms of higher degree. The generalization is known as Stokes's theorem and will be discussed below.

Any definite integral can be written as the integral of a differential form. In special relativity, where the difference between time and space depends on the observer, forms give us a Lorentz-invariant way to define integrations over regions of space. This is very important for understanding electromagnetism as a classical relativistic field theory, because we need to be able to integrate the Maxwell equations in any frame.

Volumes and forms

Consider two vectors $\vec{u} = u^x \partial_x + u^y \partial_y$ and $\vec{v} = v^x \partial_x + v^y \partial_y$ in Euclidean space in two dimensions with coordinates (x, y). When placed tail to tail, the two vectors determine a parallelogram with area $A = |\vec{u}||\vec{v}| \sin \theta$, where θ is the angle between $\vec{u}$ and $\vec{v}$. Using $\vec{u} \cdot \vec{v} = |\vec{u}||\vec{v}| \cos \theta$, we get

$$
\begin{aligned}
A &= |\vec{u}||\vec{v}| \sin \theta \\
&= \pm (|\vec{u}|^2 |\vec{v}|^2 - (\vec{u} \cdot \vec{v})^2)^{1/2} \\
&= u^x v^y - v^x u^y. \\
&= \epsilon_{ij} u^i v^j \\
&= \bar{\epsilon}(\vec{u}, \vec{v}),
\end{aligned}
\tag{5.70}
$$

where $\bar{\epsilon}$ is the two form

$$
\bar{\epsilon} = dx \wedge dy = \frac{1}{2!} \epsilon_{ij} dx^i \wedge dx^j
\tag{5.71}
$$

with nonzero components $\epsilon_{xy} = -\epsilon_{yx} = 1$. The two form $\bar{\epsilon}$, with the two vectors $\vec{u}$ and $\vec{v}$ as arguments, returns the area of the parallelogram made by the two vectors. Area is what we call volume in two space dimensions, and so the two form $\bar{\epsilon} = dx \wedge dy$ is called the *volume form* for $\mathbf{E}^2$.

If the components of the vectors $\vec{u}$ and $\vec{v}$ make up the columns of the matrix

$$
\mathbf{A} = \begin{pmatrix} u^x & v^x \\ u^y & v^y \end{pmatrix},
\tag{5.72}
$$

then the area A of the parallelogram made by $\vec{u}$ and $\vec{v}$ is equal to the matrix determinant of $\mathbf{A}$

$$
A = \epsilon_{ij} u^i v^j = u^x v^y - v^x u^y = \det \mathbf{A}.
\tag{5.73}
$$

According to (5.77) and (5.72), the area A can be negative. The volume form $\bar{\epsilon}$ is antisymmetric under exchange of $\vec{u}$ and $\vec{v}$, with $\bar{\epsilon}(\vec{u}, \vec{v}) = -\bar{\epsilon}(\vec{v}, \vec{u})$. The sign of A tells us the *orientation* of the parallelogram made by $\vec{u}$ and $\vec{v}$. The volume form for $\mathbf{E}^2$ is chosen by convention to be $dx \wedge dy$ and not $dy \wedge dx$. The coordinate axes

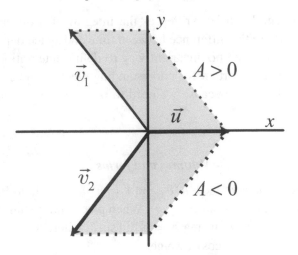

Fig. 5.2. Two vectors $\vec{u}$ and $\vec{v}$ determine a parallelogram when their tails are placed together, with area $A = u^x v^y - u^y v^x = \bar{\epsilon}(\vec{u}, \vec{v})$, where $\bar{\epsilon} = dx \wedge dy$. In the figure above, $\bar{\epsilon}(\vec{u}, \vec{v}_1) > 0$, but $\bar{\epsilon}(\vec{u}, \vec{v}_2) < 0$. The area is positive when the two vectors obey the "right hand rule" and negative when they do not. This difference in sign comes about because we chose an orientation for the volume form by using the convention $\bar{\epsilon} = dx \wedge dy$.

for $\mathbf{E}^2$ are chosen by convention to be as shown in (5.77) and not the opposite. Those choices determine the conditions under which $A > 0$ and $A < 0$.

The ability to choose an orientation for coordinate axes consistently everywhere on a manifold $\mathcal{M}$ means that $\mathcal{M}$ is an *orientable* manifold. In special relativity, the manifold in question is flat spacetime. Flat space and flat spacetime can be twisted topologically to make *unorientable* versions that are still flat, but don't allow a globally consistent choice of orientation for the volume form. The Möbius strip in two dimensions and the Klein bottle in three dimensions are examples of unorientable manifolds. We shall assume unless specified that every space or spacetime under consideration in this text is orientable.

We can easily extend the analysis from two dimensions to arbitrary D to get the volume form in $\mathbf{E}^D$

$$\bar{\epsilon} = dx^1 \wedge dx^2 \wedge \ldots \wedge dx^D$$
$$= \frac{1}{D!} \epsilon_{i_1 i_2 \ldots i_D} \, dx^{i_1} \wedge dx^{i_2} \wedge \ldots \wedge dx^{i_D}, \tag{5.74}$$

with Euclidean components $\epsilon_{i_1 i_2 \ldots i_D}$ that are antisymmetric on any pair of indices and take the values

$$\epsilon_{i_1 i_2 \ldots i_D} = \begin{cases} 1 & (i_1 i_2 \ldots i_D) = \text{even perm. of } (1\,2\ldots D) \\ -1 & (i_1 i_2 \ldots i_D) = \text{odd perm. of } (1\,2\ldots D) \\ 0 & \text{otherwise.} \end{cases} \tag{5.75}$$

The volume of the n-dimensional parallelepiped determined by $n \leq D$ vectors is given by a $(D - n)$ form. If $n = D$ then the volume is equal to the determinant of the $D \times D$ matrix with the D vectors as columns. For example, if we enlarge our space in the previous example from $\mathbf{E}^2$ to $\mathbf{E}^3$ with Euclidean coordinates (x, y, z), the volume form becomes $\bar{\epsilon} = dx \wedge dy \wedge dz$. The area of the parallelogram made by $\vec{u}$ and $\vec{v}$ is a one form

$$\bar{a} = \bar{\epsilon}(\vec{u}, \vec{v},) = \epsilon_{ijk} u^i v^j dx^k = (u^x v^y - v^x u^y) dz. \tag{5.76}$$

The metric of the manifold $\mathbf{E}^3$ provides a one to one map between one forms and vectors through raising and lowering the component indices using $a^i = \delta^{ij} a_i$. The vector $\vec{a}$ derived from the one form $\bar{a}$ in (5.76) is

$$\vec{a} = (u^x v^y - v^x u^y) \partial_z = \vec{u} \times \vec{v}. \tag{5.77}$$

The antisymmetry of the Euclidean vector cross product $\vec{u} \times \vec{v}$ stems from the antisymmetry of the volume form on the manifold. The vector cross product is crucial in describing electromagnetism through the Maxwell equations, so you will meet this relation again later.

Suppose there is a third vector $\vec{w}$ that does not lie in the plane defined by $\vec{u}$ and $\vec{v}$, so that $\vec{w} \cdot \vec{a} = \bar{a}(\vec{w}) \neq 0$. The three vectors $(\mathbf{u}, \mathbf{v}, \mathbf{w})$ then form a parallelepiped with volume $V = \vec{w} \cdot (\vec{u} \times \vec{v})$. As the reader should expect by now, we can write

$$V = (u^x v^y - v^x u^y) w^z = \epsilon_{ijk} u^i v^j w^k = \bar{\epsilon}(\vec{u}, \vec{v}, \vec{w}). \tag{5.78}$$

As in two dimensions, if the components of the vectors make up the columns in the matrix

$$V = \begin{pmatrix} u^x & v^x & w^x \\ u^y & v^y & w^y \\ u^z & v^z & w^z \end{pmatrix}, \tag{5.79}$$

then the volume of the parallelepiped they make is $V = \det \mathbf{V} = \epsilon_{ijk} u^i v^j w^k$.

What happens if we change coordinates? In two dimensions if we define new coordinates $(\tilde{x}(x, y), \tilde{y}(x, y))$ then the new volume form $d\tilde{x} \wedge d\tilde{y}$ is related to the old one $dx \wedge dy$ by

$$dx \wedge dy = \left(\frac{\partial x}{\partial \tilde{x}} \frac{\partial y}{\partial \tilde{y}} - \frac{\partial x}{\partial \tilde{y}} \frac{\partial y}{\partial \tilde{x}} \right) d\tilde{x} \wedge d\tilde{y} = J d\tilde{x} \wedge d\tilde{y}, \tag{5.80}$$

where J is the Jacobian of the coordinate transformation. If we represent the coordinate transformation as a matrix Γ, then

$$\Gamma = \begin{pmatrix} \frac{\partial x}{\partial \tilde{x}} & \frac{\partial y}{\partial \tilde{x}} \\ \frac{\partial x}{\partial \tilde{y}} & \frac{\partial y}{\partial \tilde{y}} \end{pmatrix}, \qquad J = \det \Gamma. \tag{5.81}$$

The components g_{ij} of the Euclidean metric in the new coordinates $(\tilde{x}, \tilde{y})$ are

$$g_{ij} = \frac{\partial x^m}{\partial \tilde{x}^i} \delta_{mn} \frac{\partial x^n}{\partial \tilde{x}^j}. \tag{5.82}$$

It will be left for the reader as an exercise to show that the Jacobian $J = \sqrt{g}$, where $g = \det g_{ij}$ is the determinant of the metric components g_{ij} in the new coordinate basis. If the coordinate transformation is a rotation, then the Jacobian determinant satisfies $J = \sqrt{g} = 1$ and the volume form is unchanged.

This result extends to D dimensions

$$\begin{aligned}\bar{\epsilon} &= dx^1 \wedge dx^2 \wedge \ldots \wedge dx^D \\ &= \sqrt{g} d\tilde{x}^1 \wedge d\tilde{x}^2 \wedge \ldots \wedge d\tilde{x}^D,\end{aligned} \tag{5.83}$$

where $(x^1, \ldots, x^D)$ are the usual Euclidean coordinates and $\sqrt{g}$ is the determinant of the metric tensor components g_{ij} in the coordinate basis $(\tilde{x}^1, \ldots, \tilde{x}^D)$ given in (5.82). As in the case with $D = 2$, if the new coordinates are related to the old coordinates by a rotation, then $J = \sqrt{g} = 1$, and the volume form is left unchanged.

The presence of the metric determinant $\sqrt{g}$ in the transformation law means that the coordinate basis components $\epsilon_{\mu_1 \ldots \mu_D}$ of the volume form do not transform as the components of a tensor under a coordinate transformation. They transform instead as a *tensor density*. When we see the volume form appear in a physical equation, then the quantity being represented is a density to be integrated over space. The presence of the metric determinant in the transformation law means that the quantity belongs inside an integral over the volume of the manifold.

To extend the analysis from space to spacetime, add a time direction to the wedge product in (5.74) to get

$$\begin{aligned}\bar{\epsilon} &= d\tau \wedge dx^1 \wedge dx^2 \wedge \ldots \wedge dx^D \\ &= \frac{1}{d!} \epsilon_{\mu_0 \mu_1 \ldots \mu_D} dx^{\mu_0} \wedge dx^{\mu_2} \wedge \ldots \wedge dx^{\mu_D},\end{aligned} \tag{5.84}$$

where we have used $\tau = x^0$. The Minkowski components $\epsilon_{\mu_0 \mu_1 \ldots \mu_D}$ of the spacetime volume form $\bar{\epsilon}$ are antisymmetric on all indices just like the Euclidean components in (5.75), except in the spacetime case the coordinate ordering goes from x^0 to x^D. If we switch from Minkowski coordinates $(\tau, x^1, \ldots, x^D)$ to some other coordinates $(\tilde{\tau}, \tilde{x}^1, \ldots, \tilde{x}^D)$, the spacetime volume form is

$$\bar{\epsilon} = \sqrt{|g|} \, d\tilde{\tau} \wedge d\tilde{x}^1 \wedge d\tilde{x}^2 \wedge \ldots \wedge d\tilde{x}^D, \tag{5.85}$$

where now we need to use $\sqrt{|g|}$ because $g < 0$ for a spacetime metric. The spacetime coordinate transformation that leaves the spacetime metric unchanged is a

Lorentz transformation, so it follows from (5.85) that the spacetime volume form as defined above is Lorentz-invariant.

What does it mean to have a volume element with a time dimension? Let's compare the volume form $\bar{\epsilon}_s = dx \wedge dy \wedge dz$ in $\mathbf{E}^3$, with three space dimensions, with the volume form $\bar{\epsilon}_{st} = d\tau \wedge dx \wedge dy$ in $\mathbf{M}^3$, with one time and two space dimensions. When contracted with vectors $\vec{u} = u^x \partial_x + u^y \partial_y$ and $\vec{v} = v^x \partial_x + v^y \partial_y$, the volume form in space yields

$$\bar{a} = \bar{\epsilon}_s(\vec{u}, \vec{v}) = A \, dz, \quad \vec{a} = A \partial_z. \tag{5.86}$$

The vector associated with the one form is pointing in the z direction, with magnitude equal to the area of the parallelogram.

If we make $\vec{u}$ and $\vec{v}$ into spacetime vectors $\mathbf{u} = (0, \vec{u})$ and $\mathbf{v} = (0, \vec{v})$, then the contraction with the spacetime volume form yields

$$\bar{a} = \bar{\epsilon}_{st}(\mathbf{u}, \mathbf{v}) = A \, d\tau, \quad \mathbf{a} = -A \partial_\tau, \tag{5.87}$$

where the minus sign comes about because $a^0 = \eta^{00} a_0 = -A$. In this case the spacetime vector $\mathbf{a}$ associated with the one form $\bar{\epsilon}_{st}(\mathbf{u}, \mathbf{v})$ is tangent to the world line of an object at rest in the (τ, x, y) coordinate frame.

In some other inertial frame traveling at velocity $\vec{\beta}$ relative to the (τ, x, y) frame, the vectors $\mathbf{u}$ and $\mathbf{v}$ will have components that are different from those in the (τ, x, y) frame. But the volume form $\bar{\epsilon}_{st}$ will be the same, because the Jacobian determinant $J = \sqrt{|g|} = 1$ when the two spacetime coordinate frames are related by a Lorentz transformation.

Integration of forms

Consider the d form field $\bar{\omega} = f(x) \, d\tau \wedge dx^1 \wedge \ldots \wedge dx^D$, where by $f(x)$ we mean some scalar function of the Minkowski coordinates for $\mathbf{M}^d$. The vectors $(\delta \mathbf{x}_0 = \Delta x^0 \, \partial_0, \, \delta \mathbf{x}_1 = \Delta x^1 \, \partial_1, \ldots, \, \delta \mathbf{x}_D = \Delta x^D \, \partial_D)$ describe an infinitesimal volume ΔV of $\mathbf{M}^d$. The contraction of these vectors with the d form $\bar{\omega}$ gives

$$\bar{\omega}(\delta \mathbf{x}_0, \ldots, \delta \mathbf{x}_D) = f(x) \, \Delta x^0 \ldots \Delta x^D = f(x) \, \Delta V. \tag{5.88}$$

Adding up all contributions from regions with volume ΔV inside a total spacetime volume V and then taking the limit $\Delta V \to 0$ gives

$$\int_V \bar{\omega} = \int f(x) \, dV = \int f(x) \, d^d x. \tag{5.89}$$

If we make a coordinate transformation to coordinates $(\tilde{\tau}, \tilde{x}^1, \ldots, \tilde{x}^D)$, the integral of the d form $\bar{\omega}$ becomes

$$\int_V \bar{\omega} = \int f(\tilde{x}) \, d\tilde{V} = \int f(\tilde{x}) \sqrt{|\tilde{g}|} \, d^d\tilde{x}. \qquad (5.90)$$

It wasn't relevant to this derivation whether we're in spacetime or space, so the relation (5.89) should apply to definite integrals of functions of real numbers in any dimension. In this sense, all definite integrals are integrals of differential forms.

Stokes's theorem

If you learned to do integrals perfectly well without using differential forms, then why should you bother to start using them now? Because very important integrals in physics are multidimensional and involve integrating fields such as electric and magnetic fields over volumes and surfaces of sources. The divergence theorem of Gauss

$$\int_V \vec{\nabla} \cdot \vec{E} \, dV = \oint_S \vec{E} \cdot d\vec{S} \qquad (5.91)$$

equates the integral of the divergence of the electric field $\vec{E}$ over a three-dimensional volume V to the integral of $\vec{E}$ over the two-dimensional closed surface S bounding the volume V. Stokes's theorem

$$\int_S (\vec{\nabla} \times \vec{B}) \cdot d\vec{S} = \oint_\ell \vec{B} \cdot d\vec{\ell} \qquad (5.92)$$

equates the integral of curl of the magnetic flux field $\vec{B}$ over a two-dimensional surface S to the integral of $\vec{B}$ around a one-dimensional closed path ℓ bounding S. These two theorems from electromagnetic field theory are both special cases of a more simple and elegant general theorem on integrating differential forms, known as Stokes's theorem.

Stokes's theorem for differential forms says that the integral of a p form $d\bar{\omega}$ over a p-dimensional compact volume V is equal to the integral of the $p-1$ form $\bar{\omega}$ over the $(p-1)$-dimensional boundary ∂V of the volume V. Thus Stokes's theorem takes the elegant form

$$\int_V d\bar{\omega} = \int_{\partial V} \bar{\omega}. \qquad (5.93)$$

This simple formula contains (5.91) and (5.92) as special cases, as shall be shown later.

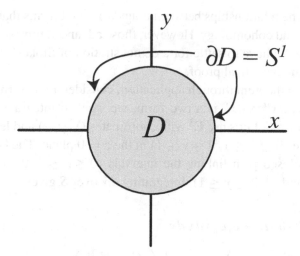

Fig. 5.3. The shaded area labeled D is a circular disk of unit radius, which is defined as the set of all points in $\mathbf{E}^2$ that satisfy $x^2 + y^2 \leq 1$. The boundary of this region, labeled ∂D, is the set of points that satisfy $x^2 + y^2 = 1$, which is just the circle S^1. A circle has no boundary, hence $\partial S^1 = \partial(\partial D) = 0$. This is a simple example of the general fact that the boundary of a boundary is zero.

What we mean by ∂V in the integral above is the set of all points in a manifold $\mathcal{M}$ that lie on the boundary of some region V, where V is a compact open set in $\mathcal{M}$. The boundary operator ∂ is similar to the exterior derivative operator d in that operating with it twice always yields zero. By the antisymmetry of differential forms, if a differential form $\bar{\alpha}$ satisfies the condition $\bar{\alpha} = d\bar{\omega}$, then it is true that $d\bar{\alpha} = d(d\bar{\omega}) = 0$. If a surface $S = \partial V$ is the boundary of a volume V, then

$$\partial S = \partial(\partial V) = 0. \tag{5.94}$$

In other words, *the boundary of a boundary is zero.*

A surface S that satisfies $\partial S = 0$ is called a *closed* surface, and a surface that satisfies $S = \partial V$ is called a *boundary*. All surfaces that are boundaries are closed, but not all closed surfaces are boundaries. The study of closed surfaces that are not boundaries is called *homology*. An exactly parallel situation exists with forms. If a form $\bar{\alpha}$ satisfies $d\bar{\alpha} = 0$ then $\bar{\alpha}$ is called a *closed* form, and if $\alpha = d\omega$ for some form ω, then α is called an *exact* form. All exact forms are closed, but not all closed forms are exact. The study of closed forms that are not exact is called *cohomology*.

As one might expect from their definitions and names, cohomology and homology are closely related subjects in topology. Stokes's theorem is best proved in theory by using the relationships between boundaries and forms that are explained using homology and cohomology. However, those advanced topics will not be presented here, so we'll have to settle for a demonstration of Stokes's theorem rather than a rigorous mathematical proof.

To test Stokes's theorem through application, consider a one form $\bar{\alpha} = \alpha_\mu \, dx^\mu$. Since $d\alpha = \partial_\mu \alpha_\nu \, dx^\mu \wedge dx^\nu$ is a two form, we want to integrate it over a two-dimensional surface S. Let's use $\mathbf{E}^3$ with coordinates (x, y, z) and let S the interior of the unit square $(0 \le x \le 1, 0 \le y \le 1)$ in the $z = 0$ plane. The boundary of the square ∂S is a closed path linking the intervals $(0 \le x \le 1, 0)$, $(1, 0 \le y \le 1)$, $(0 \le x \le 1, 1)$ and $(0, 0 \le y \le 1)$. Integrating $d\bar{\alpha}$ over S gives

$$
\begin{aligned}
\int_S d\bar{\alpha} &= \int_0^1 \int_0^1 (\partial_x \alpha_y - \partial_y \alpha_x) \, dx \, dy \\
&= \int_0^1 \left(\int_0^1 \partial_x \alpha_y \, dx \right) dy - \int_0^1 \left(\int_0^1 \partial_y \alpha_x \, dy \right) dx \\
&= \int_0^1 \alpha_x(x, 0) \, dx + \int_0^1 \alpha_y(1, y) \, dy + \int_1^0 \alpha_x(x, 1) \, dx + \int_1^0 \alpha_y(0, y) \, dy \\
&= \int_{\partial S} \alpha.
\end{aligned}
\tag{5.95}
$$

Stokes's theorem is thus verified in the case of a two-dimensional square S in Euclidean space.

If instead of S we apply (5.95) to an infinitesimal square δS of locally Euclidean coordinates on a larger curved surface S, then we could patch the squares δS together across the larger surface and prove the theorem for a general one form and a general two-dimensional surface S embedded in $\mathbf{E}^3$ by taking the limit $\delta S \to 0$. If we choose a consistent orientation for integrating $\bar{\alpha}$ around all of the $\partial \delta S$, then the contributions from segments of $\partial \delta S$ that overlap have the opposite sign and cancel each other out, leaving only the integral of $\bar{\alpha}$ over the total boundary ∂S. So the fact that Stokes's theorem works for this unit square of Euclidean space is enough to show that it works in general.

If we use the metric of Euclidean space to make a vector $\vec{\alpha} = \alpha^i \, \partial_i$ whose components are $\alpha^i = \delta^{ij} \alpha_j$, then (5.95) is equivalent to

$$
\int_S (\vec{\nabla} \times \vec{\alpha}) \cdot d\vec{S} = \oint_\ell \vec{\alpha} \cdot d\vec{\ell},
\tag{5.96}
$$

where $d\vec{S} = dx \, dy \, \partial_z$ is the area element normal to the unit square S and $d\vec{\ell}$ is tangent to the path ℓ that goes along the boundary of the square S. This is the familiar form of Stokes's theorem taught in Euclidean vector calculus.

To apply Stokes's theorem to a volume V in three space dimensions requires that $d\bar{\alpha}$ be a three form, therefore $\bar{\alpha}$ is a two form. In $\mathbf{E}^3$ with coordinates (x, y, z)

$$\bar{\alpha} = \frac{1}{2}\alpha_{ij}\, dx^i \wedge dx^j$$
$$= \alpha_{xy}\, dx \wedge dy + \alpha_{yz}\, dy \wedge dz + \alpha_{zx}\, dz \wedge dx. \qquad (5.97)$$

The exterior derivative $d\bar{\alpha}$ is

$$d\bar{\alpha} = \left(\partial_x\alpha_{yz} + \partial_y\alpha_{zx} + \partial_z\alpha_{xy}\right) dx \wedge dy \wedge dz. \qquad (5.98)$$

Define the volume V as a unit cube in the same way we defined the unit square above. The integral of $d\bar{\alpha}$ over V is

$$\int_V d\bar{\alpha} = \int_0^1 \int_0^1 \int_0^1 \left(\partial_x\alpha_{yz} + \partial_y\alpha_{zx} + \partial_z\alpha_{xy}\right) dx\, dy\, dz$$

$$= \int_0^1 \int_0^1 \left(\int_0^1 \partial_x\, \alpha_{yz}\, dx\right) dy\, dz + \int_0^1 \int_0^1 \left(\int_0^1 \partial_y\, \alpha_{zx}\, dy\right) dz\, dx$$

$$+ \int_0^1 \int_0^1 \left(\int_0^1 \partial_z\, \alpha_{xy}\, dz\right) dx\, dy$$

$$= \int_0^1 \int_0^1 \left(\alpha_{yz}(1, y, z) - \alpha_{yz}(0, y, z)\right) dy\, dz$$

$$+ \int_0^1 \int_0^1 \left(\alpha_{zx}(x, 1, z) - \alpha_{zx}(x, 0, z)\right) dz\, dx$$

$$+ \int_0^1 \int_0^1 \left(\alpha_{xy}(x, y, 1) - \alpha_{xy}(x, y, 0)\right) dx\, dy$$

$$= \int_{\partial V} \alpha, \qquad (5.99)$$

where the changes of sign in the integral reflect the orientation of the coordinate axes determined by the volume form $dx \wedge dy \wedge dz$. For example, the faces $x = 1$ and $x = 0$ of the cube are on opposite sides of the cube, therefore the integrals over those faces have the opposite orientation. Either they contribute to the surface integral with a relative minus sign, or the limits of integration are the opposite in one of the coordinates on the face.

As with Stokes's theorem applied to a unit square, we can take this result to be for an infinitesimal volume δV and add up all of the volume elements to prove that Stokes's theorem works for three-dimensional volumes V that are not cubes. The contributions from the overlapping boundary pieces $\partial\delta V$ cancel due to relative minus signs, so the sum of the surface integrals for each piece gives the integral over the total surface $S = \partial V$.

To get back the familiar Euclidean vector version of this integral, we need to derive a vector $\vec{\alpha} = \alpha^i \, \partial_i$ from the two form $\bar{\alpha}$. The volume form

$$\bar{\epsilon} = dx \wedge dy \wedge dz = \frac{1}{3!} \epsilon_{ijk} \, dx^i \wedge dx^j \wedge dx^k, \tag{5.100}$$

with components ϵ_{ijk} given by (5.75) for $D = 3$, maps a vector $\vec{\alpha}$ to a two form $\bar{\alpha}$ through the contraction $\bar{\alpha} = \epsilon(\vec{\alpha})$. The components α_{ij} of the two form $\bar{\alpha}$ are related to the components α^i of $\vec{\alpha}$ by $\alpha_{ij} = \epsilon_{ijk} \, \alpha^k$. We can use the inverse Euclidean metric tensor components δ^{ij} to raise the indices on ϵ_{ijk} to get $\epsilon^{ijk} = \epsilon_{ijk}$. It will be left for the reader as an exercise to show that

$$\vec{\alpha} = \frac{1}{2} \epsilon^{ijk} \alpha_{jk} \partial_i, \quad \bar{\alpha} = \frac{1}{2} \epsilon_{ijk} \alpha^i \, dx^j \wedge dx^k, \tag{5.101}$$

and that we can rewrite (5.99) as

$$\int_V \vec{\nabla} \cdot \vec{\alpha} \, dV = \oint_S \vec{\alpha} \cdot d\vec{S}, \tag{5.102}$$

which is the formula taught in calculus classes as the divergence theorem of Gauss.

Equation (5.102) could be the electric integral (5.91) if the electric field $\vec{E}$ were derived from a Euclidean two form $\bar{E}$, and Eq. (5.96) could be the magnetic integral (5.92) if the magnetic field $\vec{B}$ were derived from a Euclidean one form $\bar{B}$. However, satisfying Einstein's two postulates of special relativity requires that we work with geometric objects defined in spacetime rather than just space.

As has been shown previously in this text, two observers in relative motion will not agree on whether two given events occur at the same time. This relativity also holds for electric and magnetic fields. Two observers in relative motion will not agree on the measurement of a magnetic or electric field. Neither an electric nor a magnetic field can be described as an independent geometric object in spacetime. The magnetic and electric fields are bundled together into a spacetime two form $\bar{F} = 1/2 \, F_{\mu\nu} \, dx^\mu \wedge dx^\nu$ that you will meet later in this chapter. The derivations of the specific magnetic and electric field versions of Stokes's theorem (5.92) and Gauss's theorem (5.91) from the generic differential form version in (5.93) will be shown at that point.

5.4 Continuous systems in spacetime

Fields are used in physics to describe continuous systems in nature. A continuous system can be a thin cloud of dust, made of some number N of noninteracting particles, which can be represented in a continuum limit $N \to \infty$ by a field. Or a continuous system can be a fluid made of particles bound by mutual forces that make representing the system as a collection of individual particles uselessly complicated, in which case the fluid is much better represented by a continuous field. Another continuous system is the electromagnetic field, which at the classical level can't be said to be made up of anything; it's just represented by its own geometrical objects, which are completely fundamental objects from the point of view of the classical theory.

How do we work with continuous systems in special relativity? First we need to define the relevant physical quantities as geometric objects in spacetime. Are they scalar, vector, or tensor? We need to know what conservation laws apply, how the conserved quantities are represented as geometric objects, and how the conservation laws are expressed in terms of differential equations.

Energy and momentum

The energy and momentum of a single particle is described by a spacetime vector **p** that is tangent to the worldline of the particle. A continuous distribution of matter or energy described by a field, like the electromagnetic field, doesn't have a unique worldline with a unique tangent vector. The energy and momentum of a continuous system like a fluid or some other field must be described by something other than a vector, but it should give back a vector if we go back to the limit of one particle. In fact, the energy and momentum of a continuous system must be represented by a tensor.

What kind of tensor? Let's work in three space dimensions with Euclidean coordinates (x, y, z) first and then graduate to spacetime. Pressure is defined as force per unit area, with units M/LT^2. Those are the same units as energy density, or energy per unit volume. A fluid can store energy density in the form of pressure. Consider a three-dimensional cube of fluid with volume V. One way to measure the pressure is to divide the fluid into two portions by a two-dimensional plane, for example the surface $z = 0$, and measure the force $\vec{F}$ over a unit of area $\Delta \bar{A} = \tilde{\epsilon}(\Delta \vec{x}, \Delta \vec{y}) = \Delta A dz$. If we call our desired tensor **T**, then we require that $\mathbf{T}(\Delta \bar{A}) = \vec{F}$. The tensor we want takes a one form as an argument and returns a vector. Therefore **T** is a rank $\binom{2}{0}$ tensor, referred to as the *energy–momentum tensor*, but also called the *stress–energy tensor*, or more simply the *stress tensor*.

The same argument applies to a hypercube of matter in D dimensions. The force across a $(D - 1)$-dimensional hypersurface is a vector in the tangent space of $\mathbf{E}^D$ and the area element on that surface is the D-dimensional volume form contracted with $(D - 1)$ vectors, leaving again a one form. So in any number of dimensions in space, the stress tensor **T** is a rank $\binom{2}{0}$ tensor.

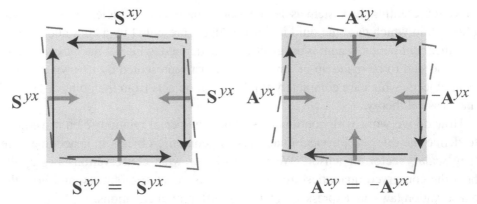

Fig. 5.4. The two figures above show the forces on a block of fluid from the symmetric (**S**) and antisymmetric (**A**) portions of the stress tensor **T**. The short grey arrows pointing inward are the normal vectors to the surfaces that bound the cube, and the long black arrows are the forces on the sides of the cube from the surrounding fluid. The symmetric forces from **S** create a deformation of the fluid called *shear*, shown by the dashed lines around the shaded block on the left. The forces from the antisymmetric portion **A** give torques that would lead to the creation of vortices and the nonconservation of angular momentum. Therefore it must be true that **A** = 0, and the stress tensor is a symmetric tensor.

To promote this object to spacetime, think of the example above in $\mathbf{E}^D$ as being a measurement occurring in the rest frame of a volume of fluid that occurs over some $(D-1)$-dimensional surface ΔA and over some time interval $\Delta \tau$. We still get back a one form for the resulting object, but a one form defined in $\mathbf{M}^d$ rather than $\mathbf{E}^D$, where $d = D + 1$. The force vector across this spacetime hypersurface is likewise a vector in $\mathbf{M}^d$ rather than $\mathbf{E}^D$. Therefore the resulting stress tensor **T** is an object defined in the direct product of two copies of the tangent space of $\mathbf{M}^d$, which makes it a rank $\binom{2}{0}$ tensor in spacetime.

What are the symmetry properties of this tensor? Suppose we decompose the components of **T** into symmetric and antisymmetric portions

$$\mathbf{T} = \mathbf{S} + \mathbf{A}, \tag{5.103}$$

with coordinate basis components

$$S^{\mu\nu} \equiv \frac{1}{2}\left(T^{\mu\nu} + T^{\nu\mu}\right), \qquad A^{\mu\nu} \equiv \frac{1}{2}\left(T^{\mu\nu} - T^{\nu\mu}\right). \tag{5.104}$$

If **T** is a symmetric tensor, then we should be able to prove by physical argument that **A** = 0 and therefore **T** = **S**. In Chapter 3 it was shown that the Lorentz transform of a symmetric tensor is symmetric, and the Lorentz transform of an antisymmetric tensor is antisymmetric. Therefore if we prove $\mathbf{A} = 0$ in the rest frame of the fluid, it will be true in all other frames.

Let's employ a convention where the second index of $T^{\mu\nu}$ corresponds to normal to the surface across which the force is measured, while the first index corresponds to the direction of the measured force. The coordinate component T^{ii} therefore gives the force in the i direction across the surface, the normal of which is in the i direction as well. The component T^{ii} describes hydrostatic pressure in a fluid, which always exerts a force normal to the surface on which it is being measured. If $\mathbf{S}$ were zero, then the stress tensor couldn't describe normal forces like hydrostatic pressure. Therefore $\mathbf{S} \neq 0$.

Now we will argue that $\mathbf{A} = 0$. Let's compare T^{ij} with T^{ji} for $i \neq j$, and examine the forces in the (xy) plane on a small block of fluid. Let the block have sides of equal length L. The component $T^{xy} = \mathbf{T}(\hat{e}^x, \hat{n}^y)$, where $\hat{n}^y$ is the unit one form normal to the surface across which the force is being transmitted, and $\hat{e}^x$ is the basis one form in the x direction. At the surface $y = -L/2$, $\hat{n}^y = \hat{e}^y$, but at the surface $y = L/2$, $\hat{n}^y = -\hat{e}^y$. Therefore $T^{xy}(x, L/2) = -T^{xy}(x, -L/2)$. The same arguments apply for T^{yx}, with $T^{yx}(L/2, y) = -T^{yx}(-L/2, y)$. These conditions apply for all components of $\mathbf{T}$. For $\mathbf{S}$ we use the condition that $S^{xy} = S^{yx}$, while for $\mathbf{A}$ we use the condition $A^{xy} = -A^{yx}$. The resulting force diagrams for $\mathbf{S}$ and $\mathbf{A}$ are shown in Figure 5.4.

As shown in the right of Figure 5.4, the forces from $\mathbf{A}$ add up to create a net torque on the block of fluid in the z direction, which vanishes if and only if $A^{xy} = 0$. If $A^{0y} \neq 0$ and $A^{xy} = 0$ in frame S, then we can make a Lorentz transformation in the x direction to another frame $\tilde{S}$ where $\tilde{A}^{xy} = L_0^{\tilde{x}} A^{0y} \neq 0$. It must then be true that all of the components of $\mathbf{A}$ vanish in every frame. The stress tensor is therefore a purely symmetric tensor, $\mathbf{T} = \mathbf{S}$, with coordinate basis components $T^{\mu\nu} = T^{\nu\mu}$.

The stress tensor for a perfect fluid

The stress tensor is not a fundamental geometric object, but rather one that is defined in terms of more fundamental geometric objects to describe a given distribution of matter or energy. Therefore its tensor structure must derive from that of fundamental geometric objects on the spacetime manifold. So what geometric objects can we use to construct $\mathbf{T}$ for a given physical system?

Electromagnetism has its own set of fields that are vectors and tensors in $\mathbf{M}^4$, from which $\mathbf{T}$ for an electromagnetic field can be constructed. To describe a distribution of matter that is not a separate field in nature but just a very large collection of interacting particles taken to the continuum limit, we can use the fields already defined in spacetime to make the symmetric rank $\binom{2}{0}$ tensor $\mathbf{T}$. In the limit we consider a perfect (or ideal) fluid that is incompressible, has no viscosity, and does not exchange heat with its surroundings. Such a fluid is characterized in its local rest frame by a pressure P and a mass density ρ.

The Minkowski metric tensor η is a tensor of rank $\binom{0}{2}$, and its inverse

$$\eta^{-1} = \eta^{\mu\nu}\partial_\mu \otimes \partial_\nu \tag{5.105}$$

is a symmetric tensor with the desired rank of $\binom{2}{0}$. The spacetime velocity field $\mathbf{u}$ is a vector, which is a $\binom{1}{0}$ tensor, and so the direct product $\mathbf{u} \otimes \mathbf{u}$ makes a symmetric rank $\binom{2}{0}$ tensor as well. So the stress tensor for a continuous distribution of matter, such as a perfect fluid in flat spacetime, should take the form

$$\mathbf{T} = A\mathbf{u} \otimes \mathbf{u} + B\eta^{-1}. \tag{5.106}$$

for a suitable choice of the coefficients A and B.

Now consider a volume of a perfect fluid in its rest frame, with velocity field $\mathbf{u} = \partial_\tau$. The only nonzero components of $\mathbf{T}$ as defined above are T^{00} and T^{ii}. The component T^{00} represents the force component in the time direction, across a surface, the normal of which is also in the time direction. In the rest frame of a particle, the momentum in the time direction is the just mass of the particle. Therefore $T^{00} = \rho c^2$, where ρ is the mass density of the fluid. For what follows we shall use units where $c = 1$ where mass density and energy density have the same units of M/L^D, so we'll just write ρ rather than ρc^2. The component T^{ii} represents the force component in the i direction, across a surface whose normal is also in the i direction. Therefore $T^{ii} = P$, where P is the hydrostatic pressure in the fluid.

Applying this physical reasoning to the formula (5.106) gives $B = P$ and $A = P + \rho$. We finally arrive at the stress tensor $\mathbf{T}$ of a perfect fluid in spacetime

$$\mathbf{T} = (P + \rho)\mathbf{u} \otimes \mathbf{u} + P\eta^{-1}. \tag{5.107}$$

The coordinate basis components are

$$T^{\mu\nu} = (P + \rho)u^\mu u^\nu + P\eta^{\mu\nu}. \tag{5.108}$$

The tensor $\mathbf{h} = \eta^{-1} + \mathbf{u} \otimes \mathbf{u}$, with components $\eta^{\mu\nu} + u^\mu u^\nu$, is a projection operator. Any spacetime vector $\mathbf{w}$ can be written as $\mathbf{w} = a\mathbf{u} + \mathbf{v}$, with $\mathbf{u} \cdot \mathbf{v} = 0$, and $\mathbf{u}^2 = -1$. The metric product $\mathbf{h} \cdot \mathbf{w}$ is

$$\mathbf{h} \cdot \mathbf{w} = (\eta^{-1} + \mathbf{u} \otimes \mathbf{u}) \cdot (a\mathbf{u} + \mathbf{v}) = \mathbf{v}. \tag{5.109}$$

From this example we see that $\mathbf{h}$ projects away the vector component parallel to the spacetime velocity $\mathbf{u}$, leaving the component that is orthogonal to $\mathbf{u}$. As was shown in Chapter 3, if $\mathbf{u}$ is a timelike vector and $\mathbf{v} \cdot \mathbf{u} = 0$, then $\mathbf{v}$ is a spacelike vector. Therefore the tensor $\mathbf{h}$ projects out the spacelike portion of any given vector $\mathbf{w}$. The stress tensor $\mathbf{T}$ can be written as

$$\mathbf{T} = \rho \mathbf{u} \otimes \mathbf{u} + P\mathbf{h}. \tag{5.110}$$

The hydrostatic pressure P is thus defined in a Lorentz-covariant manner as the spacelike portion of the stress tensor.

Stress tensor for pointlike particles

A particle that is pointlike can be treated as a continuous system in spacetime by describing the particle as a highly localized distribution given by a Dirac delta function. The Dirac delta function $\delta(x)$ is defined as a limit of sharply peaked functions, with unit area under the peak. (For example, the large α limit of $\sqrt{\alpha/\pi} \exp(-\alpha x^2)$.) In the limit

$$\delta(x - a) = \begin{cases} 0 & x \neq a \\ \infty & x = a. \end{cases} \tag{5.111}$$

It satisfies the integral property

$$f(a) = \int_{-\infty}^{\infty} \delta(x - a) f(x) dx. \tag{5.112}$$

The Dirac delta function in a d-dimensional spacetime can be defined as the product of d one-dimensional delta functions, with the result

$$f(\mathbf{a}) = \int_{\mathcal{U}} f(\mathbf{x}) \delta^d(\mathbf{x} - \mathbf{a}) d^d x, \tag{5.113}$$

where $\mathbf{a}$ refers to an event in $\mathbf{M}^d$, and the integration region $\mathcal{U}$ is meant to be all of spacetime. Note that by this definition, the delta function in d dimensions carries implied units of L^{-d}, to balance the dimensions L^d of $d\mathcal{U} = d^d x$.

The d-dimensional delta function $\delta^d(\mathbf{x} - \mathbf{a})$ describes a zero-dimensional point in spacetime that only exists for the instant $x^0 = a^0$ and then is gone. The world line of a particle in spacetime is one-dimensional and can be written as $\mathbf{x} = \mathbf{x}(\lambda)$. If the particle is massive, then λ can be chosen to be the proper time along the world line. In frame S with coordinates $(\tau, \vec{x})$, we consider

$$\delta^d(\mathbf{x} - \mathbf{x}(\lambda)) = \delta(\tau - \tau(\lambda)) \delta^D(\vec{x} - \vec{x}(\lambda)). \tag{5.114}$$

We can make a one-dimensional distribution out of this by adding up all of the instants where $\tau = \tau(\lambda)$ along the world line, which means doing an integral. But what should we integrate over? If we want the answer to be the path in space as a function of coordinate time τ, rather than the world line proper time λ, then we have to integrate over the proper time λ

$$\delta^D(\vec{x} - \vec{x}(\tau)) = \int_{-\infty}^{\infty} \delta^d(\mathbf{x} - \mathbf{x}(\lambda)) \frac{d\tau(\lambda)}{d\lambda} d\lambda. \tag{5.115}$$

The derivative $d\tau/d\lambda$ is the time component of the spacetime velocity $\mathbf{u}$ of the particle. If we have N such particles, and each particle has spacetime velocity $\mathbf{u}_i$, then the number density of the N particles in $\mathbf{M}^d$ is given by the vector

$$\mathbf{n}(x) = \sum_{i=1}^{N} \mathbf{u}_i(\tau) \delta^D(\vec{x} - \vec{x}_i(\tau)) = \sum_{i=1}^{N} \int \mathbf{u}_i(\lambda_i) \delta^d(\mathbf{x} - \mathbf{x}_i(\lambda_i)) d\lambda_i, \tag{5.116}$$

where λ_i is the parameter along the world line of the ith particle with velocity $\mathbf{u}_i$. It will be left for the reader as an exercise to show that the components of $\mathbf{n}$ transform as a spacetime vector under a Lorentz transformation.

The stress tensor for N particles with spacetime momentum $\mathbf{p}_i = m_i \mathbf{u}_i$ is

$$\mathbf{T}(x) = \sum_{i=1}^{N} \mathbf{p}_i \otimes \mathbf{u}_i \, \delta^D(\vec{x} - \vec{x}_i(\tau))$$

$$= \sum_{i=1}^{N} \int_{-\infty}^{\infty} \mathbf{p}_i \otimes \mathbf{u}_i \, \delta^d(\mathbf{x} - \mathbf{x}_i(\lambda_i)) d\lambda_i. \tag{5.117}$$

The above form for the stress tensor shows very plainly that stress tensor measures the flux of momentum $\mathbf{p}$ across a surface with normal vector $\mathbf{u}$.

If our collection of N particles has some type of scalar property, such as electric charge, then the current density of that charge is given by

$$\mathbf{J}(x) = \sum_{i=1}^{N} q_i \mathbf{u}_i(x) \delta^D(\vec{x} - \vec{x}_i(\tau)) = \sum_{i=1}^{N} \int q_i \mathbf{u}_i(\lambda_i) \delta^d(\mathbf{x} - \mathbf{x}_i(\lambda_i)) d\lambda_i, \tag{5.118}$$

where q_i is the charge carried by ith particle.

To define the continuum limit, suppose all N particles are headed in the same direction at the same speed for the most part, so that the $\mathbf{u}_i$ are all about equal, and the charges q_i are equal. In that case we can take the velocity, momentum and charge out of the sums for $\mathbf{n}$, $\mathbf{T}$ and $\mathbf{J}$, and the number density, stress tensor and current can be written

$$\mathbf{n} = n(x)\mathbf{u}$$
$$\mathbf{T} = mn(x)\mathbf{u} \otimes \mathbf{u} = \rho(x)\mathbf{u} \otimes \mathbf{u} = \mathbf{p} \otimes \mathbf{n}$$
$$\mathbf{J} = qn(x)\mathbf{u} = \rho_e(x)\mathbf{u} = q\,\mathbf{n}, \tag{5.119}$$

where the functions $n(x)$, $\rho(x)$ and $\rho_e(x)$ are the particle number density, mass density and charge density, as measured in the frame where $\mathbf{u} = \partial_\tau$. (In the definition of $\mathbf{T}$, the mass density $\rho(x) \to \rho(x)c^2$ in units where $c \neq 1$.)

Stokes's theorem and momentum conservation

The stress tensor $\mathbf{T}$ gives a spacetime vector $\xi = \mathbf{T}(\,,\bar{\omega}) = T^{\mu\nu}\omega_\nu\,\partial_\mu$ when we evaluate it on a one form field $\bar{\omega} = \omega_\mu\,dx^\mu$. This vector can be contracted with the spacetime volume form to give a $(d-1)$ form $\bar{\alpha} = \bar{\epsilon}(\xi)$. The exterior derivative $d\bar{\alpha}$ is a d form. If we integrate this d form over a d-dimensional spacetime volume U and then apply Stokes's theorem for differential forms (5.95), we get

$$\int_U d(\bar{\epsilon}(\mathbf{T}(\,,\bar{\omega}))) = \oint_{\partial U} \bar{\epsilon}(\mathbf{T}(\,,\bar{\omega})). \tag{5.120}$$

Let's look at a toy example in $\mathbf{M}^3$ and figure out what this strange expression could mean. Let's use coordinates (τ, x, y) and choose our one form to be $\bar{\omega} = d\tau$. Using the symmetry of $\mathbf{T}$ we can write the vector $\mathbf{T}(\,,\bar{\omega})$ as

$$\mathbf{T}(,\bar{\omega}) = T^{00}\partial_0 + T^{0x}\partial_x + T^{0y}\partial_y. \tag{5.121}$$

The spacetime volume form is $\bar{\epsilon} = d\tau \wedge dx \wedge dy$, therefore

$$\bar{\epsilon}(\mathbf{T}(\,,\bar{\omega})) = T^{00}\,dx \wedge dy + T^{0x}dy \wedge d\tau + T^{0y}d\tau \wedge dx. \tag{5.122}$$

Taking the exterior derivative gives

$$d(\bar{\epsilon}(\mathbf{T}(\,,\bar{\omega}))) = (\partial_0\,T^{00} + \partial_x\,T^{0x} + \partial_y\,T^{0y})\,d\tau \wedge dx \wedge dy. \tag{5.123}$$

The terms in the exterior derivative can be integrated separately, as in the previous section, to get the value of $\mathbf{T}$ on the boundary ∂U of the spacetime volume U. After some index manipulation, (5.120) reduces to

$$\int_U \partial_\nu T^{0\nu}d\tau\,dx\,dy = \oint_{\partial U} T^{0\nu}d^2\Sigma_\nu, \tag{5.124}$$

where $\{d^2\Sigma_\nu\}$ are the coordinate components of the two form area element on the two-dimensional surface ∂U that bounds the given three-dimensional spacetime volume U.

To examine this expression in more detail, we need to choose a spacetime volume U over which to integrate. Suppose we want to look at the energy flowing in and out of a two-dimensional box with sides of length L, and we want to integrate from time $\tau = \tau_0$ to time $\tau = \tau_1$. The left-hand side of (5.124) becomes

$$\int_0^L \int_0^L \int_{\tau_0}^{\tau_1} (\partial_\nu T^{0\nu}) \, d\tau \, dx \, dy. \tag{5.125}$$

Taking into account the orientation of the area form $d^2\Sigma_\nu$ on the boundary ∂U, the right hand side can be written

$$\oint_{\partial U} T^{0\nu} d^2\Sigma_\nu = \int_0^L \int_0^L (T^{00}(\tau_2, x, y) - T^{00}(\tau_1, x, y)) \, dx \, dy$$

$$+ \int_{\tau_0}^{\tau_1} \int_0^L (T^{0x}(\tau, L, y) - T^{0x}(\tau, 0, y)) \, dy \, d\tau$$

$$+ \int_{\tau_0}^{\tau_1} \int_0^L (T^{0y}(\tau, x, L) - T^{0y}(\tau, x, 0)) \, dx \, d\tau. \tag{5.126}$$

The first term in the above sum is the net change in the total energy density T^{00} contained in the box between times τ_0 and τ_1. The second term is the net change in the x component of the momentum density T^{0x} flowing through the faces $x = 0$ and $x = L$ between times τ_0 and τ_1. The third term is the net change in the y component of the momentum density T^{0y} flowing through the faces $y = 0$ and $y = L$ between times τ_0 and τ_1.

If there are no sources of energy inside the box, then any net change in time in the energy density inside the box can only come from a net momentum density crossing the boundary of the box, either entering or leaving. Therefore, if the total spacetime momentum of this system is conserved, then the integral in (5.126) must be zero. This physical argument is independent of the precise boundaries of the box and times of initial and final measurement, and is also ultimately independent of the spacetime dimension. Therefore if the total energy and momentum are conserved for a general continuous system, it must be true that

$$\partial_\nu T^{0\nu} = 0. \tag{5.127}$$

The spacelike components of (5.120) can be reduced to the form (5.126) in a parallel manner. This result can be extended to arbitrary spacetime dimension by using the linearity of the tangent space on a manifold. Stokes's theorem for the stress tensor $\mathbf{T}$ in Minkowski spacetime in d dimensions can be written

$$\int_U \partial_\nu T^{\mu\nu} d^d x = \oint_{\partial U} T^{\mu\nu} d^D \Sigma_\nu. \tag{5.128}$$

Hence conservation of spacetime momentum for a continuous system in spacetime with stress tensor $\mathbf{T}$ in any number of dimensions is enforced by the equation

$$\partial_\nu T^{\mu\nu} = 0.$$
(5.129)

This is a Lorentz-covariant equation, so it is true in any inertial frame.

The total momentum $\mathbf{P} = P^\mu \, \partial_\mu$ crossing a D-dimensional surface with area element $d\Sigma = d^D \Sigma_\nu \, dx^\nu$ is

$$P^\mu = \int_S T^{\mu\nu} d^D \Sigma_\nu.$$
(5.130)

If the system in question has the energy contained within some finite volume for which a rest frame can be defined, then the total mass of the system is given by

$$M = P^0 = \int_S T^{00} d^D x,$$
(5.131)

where S is some given spacelike volume contained in $\mathbf{M}^d$ that represents all of space inside the volume at some time τ_0. Note that in units where $c \neq 1$, this formula gives the energy Mc^2, not the mass M.

Energy conservation for a perfect fluid

Let's explore the consequences of the conservation equation (5.129) for a perfect fluid (5.108). The spacetime velocity field $\mathbf{u}$ has components $u^0 = \gamma$ and $u^i = \gamma \beta^i$. If we make the definition $\alpha \equiv \gamma^2 (\rho + P)$, then the timelike component of (5.129) can be written

$$\frac{\partial \alpha}{\partial \iota} + \vec{\nabla} \cdot (\alpha \, \vec{\beta}) - \frac{\partial P}{\partial \tau} = 0.$$
(5.132)

This is the relativistic continuity equation. The spacelike components of (5.129) can be written

$$\alpha \frac{\partial \vec{\beta}}{\partial \tau} + \alpha \, (\vec{\beta} \cdot \vec{\nabla}) \, \vec{\beta} + \vec{\nabla} P + \vec{\beta} \frac{\partial P}{\partial \tau} = 0.$$
(5.133)

This is the relativistic version of the Euler equation for hydrodynamic flow.

It will be left for the reader as an exercise to show that in the nonrelativistic limit in units where $c \neq 1$, these equations reduce to the usual hydrodynamic continuity equation

$$\frac{\partial \rho}{\partial t} + \vec{\nabla} \cdot (\rho \, \vec{v}) = 0$$
(5.134)

and the nonrelativistic form of Euler's equation

$$\rho \frac{\partial \vec{v}}{\partial t} + \rho \, (\vec{v} \cdot \vec{\nabla}) \, \vec{v} + \vec{\nabla} P = 0.$$
(5.135)

Energy conservation as a vanishing Lie derivative

Recall that the Lie derivative of a p form $\bar{\omega}$ field with respect to a vector field ξ is

$$\mathcal{L}_\xi \bar{\omega} = d(\bar{\omega}(\xi)) + d\bar{\omega}(\xi). \tag{5.136}$$

Since the exterior derivative of a d form in a d-dimensional spacetime is identically zero, the Lie derivative of the volume form $\bar{\epsilon}$ satisfies $\mathcal{L}_\xi \bar{\epsilon} = d(\bar{\epsilon}(\xi))$. If the vector $\xi = \mathbf{T}(\ , \bar{\sigma})$, where $\bar{\sigma}$ is a one form, then

$$d(\bar{\epsilon}(\xi)) = \mathcal{L}_\xi \bar{\epsilon}, \tag{5.137}$$

and the differential form of energy conservation for a continuous system in spacetime (5.129) can be rewritten as

$$\mathcal{L}_\xi \bar{\epsilon} = 0. \tag{5.138}$$

The energy and momentum of a continuous system are conserved when the volume form of spacetime is Lie-dragged by the momentum flux density vector $\xi = \mathbf{T}(\ , \bar{\sigma})$.

The momentum flux field ξ is a vector field that determines a congruence of curves that are the flow lines of energy transported through the volume. If the volume form $\bar{\epsilon}$ is Lie-dragged along such a congruence of flow lines, then the same density of flow lines enters an infinitesimal volume as leaves it. This is equivalent to saying that as much energy and momentum enter an infinitesimal volume as leave it. Hence conservation of spacetime momentum has a natural interpretation as the vanishing Lie derivative of the volume form with respect to the momentum flux in the system.

This is a purely geometrical view of what energy conservation means for a fluid or some other field or set of fields on spacetime. All conservation laws are related to symmetries and all symmetries come from geometry. There is a geometrical way of expressing any conservation law that is observed in physics. Modern theoretical physics has become more and more a subject about geometry. The story of special relativity is just the tip of the geometric iceberg.

Angular momentum

Consider a rotational Killing vector in $\mathbf{M}^4$ representing a rotation about the origin in the z direction

$$\mathbf{j}_3 = x\partial_y - y\partial_x, \tag{5.139}$$

which satisfies the conservation equation

$$\frac{d}{d\lambda}(\mathbf{j}_3 \cdot \mathbf{p}) = \frac{d}{d\lambda}(xp_y - yp_x) = \frac{d}{d\lambda}J^z = 0, \tag{5.140}$$

where λ is the proper time along the world line of a free particle or object with spacetime momentum $\mathbf{p}$, and

$$\vec{J} = \vec{r} \times \vec{p} \tag{5.141}$$

is the Newtonian angular momentum with respect to the origin of the coordinate system.

Angular momentum conservation for a particle or object with momentum $\mathbf{p}$ has a direct analog in a continuous system with stress tensor $\mathbf{T}$. As one might expect by now, this conservation law will come from applying Stokes's theorem to a vector that we make from a physically relevant tensor. In this case, the relevant tensor is the angular momentum density tensor with coordinate basis components

$$\mathcal{J}^{\mu\nu\rho} = x^\mu T^{\nu\rho} - x^\nu T^{\mu\rho}. \tag{5.142}$$

One problem inherent in this definition is that $\mathbf{x}$ is not a true vector on the manifold. The quantity $\mathbf{x}$ represents the coordinate displacement between a specified event in Minkowski spacetime and some event we've chosen to call $\mathbf{0}$. Because we're in flat spacetime, we can overlook this, because the tangent space on a flat manifold is the same at every point.

Since the coordinate components $\mathcal{J}^{\mu\nu\rho} = -\mathcal{J}^{\nu\mu\rho}$ are antisymmetric on the first two indices, then it seems logical to make a vector from this tensor by evaluating it on a two form $\bar{\sigma}$, which represents the plane of rotation. If we define the spacetime vector $\zeta = \mathcal{J}(\bar{\sigma},)$, then we can make a three form by contracting this vector with the spacetime volume form $\bar{\epsilon}$ and taking the exterior derivative. Then we can apply Stokes's theorem as before.

For example, let's evaluate $\mathcal{J}$ on the two form $dx \wedge dy$, so that we're looking at a rotation around the z axis

$$\zeta = \mathcal{J}(dx \wedge dy,) = \mathcal{J}^{xy0}\partial_\tau + \mathcal{J}^{xyx}\partial_x + \mathcal{J}^{xyy}\partial_y + \mathcal{J}^{xyz}\partial_z, \tag{5.143}$$

with $\mathcal{J}^{xy\rho} = xT^{y\rho} - yT^{x\rho}$. The contraction of this vector with the spacetime volume form $\bar{\epsilon} = d\tau \wedge dx \wedge dy \wedge dz$ is

$$\begin{aligned}\bar{\epsilon}(\zeta) = \mathcal{J}^{xy0}dx \wedge dy \wedge dz &- \mathcal{J}^{xyx}d\tau \wedge dy \wedge dz \\ + \mathcal{J}^{xyy}d\tau \wedge dx \wedge dz &- \mathcal{J}^{xyz}d\tau \wedge dx \wedge dy.\end{aligned} \tag{5.144}$$

Taking the exterior derivative gives

$$d(\bar{\epsilon}(\zeta)) = \partial_\rho (x\, T^{y\rho} - y\, T^{x\rho})\, d\tau \wedge dx \wedge dy \wedge dz.$$
$$= (T^{yx} - T^{xy})\, d\tau \wedge dx \wedge dy \wedge dz, \tag{5.145}$$

where we have assumed that energy and momentum are conserved, so that $\partial_\rho T^{x\rho} = \partial_\rho T^{y\rho} = 0$.

Extending the above example from the (xy) plane of $\mathbf{M}^4$ to the $(\mu\nu)$ plane in $\mathbf{M}^d$, with $d = D + 1$, we can write Stokes's theorem applied to angular momentum as

$$\int_U \partial_\rho \mathcal{J}^{\mu\nu\rho} d^d x = \int_U (T^{\mu\nu} - T^{\nu\mu})\, d^d x = \oint_{\partial U} \mathcal{J}^{\mu\nu\rho}\, d^D\Sigma_\rho. \tag{5.146}$$

The integral of $\partial_\rho \mathcal{J}^{\mu\nu\rho}$ over a spacetime volume U is equal to the integral over the boundary ∂U of $\mathcal{J}^{\mu\nu\rho}\, d^D\Sigma_\rho$, which is the net flux of spacetime angular momentum across the boundary. If there are no sources or sinks for angular momentum inside the volume, that is, no way for angular momentum to be created or destroyed inside the volume, then the net flux of angular momentum crossing the boundary of any spacetime volume U must be zero. So it must be true that $d(\bar{\epsilon}(\zeta)) = 0$, and therefore $T^{\mu\nu} = T^{\nu\mu}$. The condition for conservation of angular momentum in this system is just the symmetry of the stress tensor itself. This example verifies what we learned previously: that antisymmetry in the stress tensor is associated with a singular rotation density in the fluid in question, not just a bulk rotation.

The geometric way to say that angular momentum is conserved is to say that the Lie derivative of the volume form $\bar{\epsilon}$ with respect to the angular momentum flux in a given plane $\bar{\sigma}$ vanishes

$$\mathcal{L}_\zeta\, \bar{\epsilon} = 0, \qquad \zeta = \mathcal{J}(\bar{\sigma},). \tag{5.147}$$

The angular momentum flux in a given plane determines a congruence of curves that are the flow of lines of angular momentum in that plane. If the volume form is Lie-dragged along those flow lines, then angular momentum is conserved. If the stress tensor in the fluid or matter in question is not symmetric but carries a singular angular momentum density of its own, then flow lines can begin or end inside an infinitesimal volume element. In that case, a different density of flow lines can enter a volume element than leave it and angular momentum is not conserved.

The total angular momentum $\mathbf{J}$ of an isolated system has components

$$\mathbf{J}^{\mu\nu} = \int_{\mathcal{V}} \mathcal{J}^{\mu\nu\rho} d^D\Sigma_\rho, \tag{5.148}$$

where $\mathcal{V}$ is the volume of the system in question. In the rest frame of the system this becomes

$$J^{\mu\nu} = \int_{\mathcal{V}} \mathcal{J}^{\mu\nu0}\, d^D x = \int_{\mathcal{V}} (x^\mu T^{\nu\,0} - x^\nu T^{\mu 0})d^D x. \qquad (5.149)$$

This tensor represents the angular momentum of some system about the origin of the coordinate frame and is not invariant under a translation of the coordinate system $\mathbf{x} \to \mathbf{x} + \mathbf{a}$.

In quantum theory (discussed in Chapter 7) a particle can have intrinsic angular momentum, which is called its spin. The spin S of a particle with total angular momentum tensor $\mathbf{J}$ can be defined as the $(d - 3)$ form

$$\bar{S} = \frac{1}{2}\bar{\epsilon}(\mathbf{J}, \mathbf{u}), \qquad (5.150)$$

with coordinate basis components in $d = 4$

$$S_\mu = \frac{1}{2}\epsilon_{\mu\nu\alpha\beta} J^{\nu\alpha} u^\beta. \qquad (5.151)$$

The corresponding vector $\mathbf{S}$, with components $S^\mu = \eta^{\mu\nu} S_\nu$, has $\mathbf{S} \cdot \mathbf{u} = 0$. Therefore the spin vector is a spacelike vector, the time component of which can be made to vanish by a Lorentz transformation.

5.5 Electromagnetism

Einstein came to the theory of special relativity after he grappled with the inconsistency between the successful electromagnetic field theory of the Maxwell equations, and the successful vector field theory of Newtonian mechanics. A light wave, according to Maxwell theory, has no consistent description according to the equations of motion in a frame traveling at the speed of light. Yet Newtonian mechanics is consistent in any frame moving at any constant velocity. Einstein had the courage to resolve this crisis with his two postulates, which rearranged the rigid and absolute Newtonian notions of space and time into a unified spacetime in which the difference between space and time is flexible and depends upon the motion of the observer.

How did it turn out that way? Electromagnetic waves don't have a rest frame because electromagnetic theory was already a relativistic theory, even though the people who developed it had no idea of spacetime at all. Their observations of nature at work in electromagnetic phenomena detected and described Minkowski spacetime, but because all those scientists knew was Euclidean space and Newtonian time, they wrote everything in that language. When written in terms of

Euclidean vectors in $\mathbf{E}^3$ and Newtonian time t, the Maxwell equations are

$$\vec{\nabla} \cdot \vec{E} = 4\pi\rho$$

$$\vec{\nabla} \times \vec{B} - \frac{1}{c}\frac{\partial \vec{E}}{\partial t} = \frac{4\pi}{c}\vec{J}$$

$$\vec{\nabla} \cdot \vec{B} = 0$$

$$\vec{\nabla} \times \vec{E} + \frac{1}{c}\frac{\partial \vec{B}}{\partial t} = 0, \qquad (5.152)$$

where ρ is the static electric charge density and $\vec{J}$ is the electric current. Please note that we are using Gaussian units in this chapter. Gaussian units in electromagnetism will be discussed in more detail below.

In a region of space where there is no charge or current, the four equations imply that

$$\nabla^2\vec{E} - \frac{\partial^2\vec{E}}{\partial \tau^2} = \partial^\mu\partial_\mu\vec{E} = 0$$

$$\nabla^2\vec{B} - \frac{\partial^2\vec{B}}{\partial \tau^2} = \partial^\mu\partial_\mu\vec{B} = 0, \qquad (5.153)$$

where we've used $\tau = ct$. These two equations are manifestly Lorentz-invariant wave equations for the electric and magnetic fields, respectively. So evidence of a unified spacetime is not far from the surface in electromagnetism, once one knows what to look for. It will turn out that the electric and magnetic fields also transform under Lorentz transformations, so we will still have to reexamine the Lorentz covariance of these equations more carefully.

The Maxwell equations yield a manifestly Lorentz-invariant wave equation defined in spacetime, but the vectors $\vec{E}$ and $\vec{B}$ are defined in regular old Euclidean space. How can we know that the equations (5.152) are really Lorentz-covariant? The best solution to this dilemma is to find a way to define the electric and magnetic fields as objects in $\mathbf{M}^4$ rather than $\mathbf{E}^3$, as was done with Newtonian momentum in Chapter 4.

Newtonian mechanics is "relativized" by adding a time component p^0 to the Euclidean momentum vector $\vec{p} = p^i\,\partial_i$ to make vector $\mathbf{p} = p^\mu\partial_\mu = p^0\partial_\tau + p^i\partial_i$ that is defined in Minkowski spacetime rather than Euclidean space. In Newtonian mechanics energy E is a Euclidean scalar, and it just turns out that setting $p^0 = E$ is what solves the problem of finding the relativistic version of Newtonian mechanics.

However, there are no such scalar quantities defined for electricity and magnetism that can convert the Euclidean vectors $\vec{E}$ and $\vec{B}$ into spacetime vectors $\mathbf{E}$ and $\mathbf{B}$. The Maxwell equations are a complete system of equations, with no dynamical quantities missing that need to be provided by a relativistic theory. We know that Maxwell's equations for electromagnetism unified what were believed to be two separate forces of nature – electricity and magnetism – into a single theory, albeit a single theory of two related fields. So perhaps the vectors $\vec{E}$ and $\vec{B}$ should properly be considered as components derived from a single unified field.

When counted together as a single geometric object, the $\vec{E}$ and $\vec{B}$ fields in $\mathbf{E}^3$ have a total of six independent components. A p form in d spacetime dimensions has

$$C_p^d = C_{d-p}^d = \frac{d!}{p!\,(d-p)!} \tag{5.154}$$

independent components. The solution to $\binom{4}{p} = 6$ is $p = 2$. Therefore the Euclidean $\vec{E}$ and $\vec{B}$ fields have enough independent components to be combined into a spacetime two form, which is called the *electromagnetic field strength tensor*, or the *Faraday tensor*

$$\bar{F} = \frac{1}{2} F_{\mu\nu}\,dx^\mu \wedge dx^\nu. \tag{5.155}$$

Faraday was a nineteenth-century experimentalist who arrived at the concept of an electromagnetic field by drawing lines of force measured in his own experimental work. Faraday wasn't good at mathematics himself and never heard of differential forms, which weren't developed until the twentieth century. His visual, hands on, understanding of electric and magnetic fields motivated the abstract mathematical understanding achieved by Maxwell. Faraday sought to understand electricity and magnetism as a unified field theory, so the Faraday tensor represents the realization of his life's quest.

The Faraday tensor

The six components of $\vec{E}$ and $\vec{B}$ are bundled into the Faraday tensor by using the three components of $\vec{E}$ as the $(0i)$ components of the two form $\bar{F}$ and putting the three components of $\vec{B}$ to work as the (ij) components. The Minkowski coordinate basis components of $\bar{F}$ can thus be represented by the matrix

$$(F_{\mu\nu}) = \begin{pmatrix} 0 & -E^x & -E^y & -E^z \\ E^x & 0 & B^z & -B^y \\ E^y & -B^z & 0 & B^x \\ E^z & B^y & -B^x & 0 \end{pmatrix}. \tag{5.156}$$

The Faraday tensor is a tensor and so as a geometric object in spacetime exists independently of the coordinates in which we describe it. It is important for understanding relativity to keep in mind at all times that these are only the components of the tensor in the Minkowski basis (τ, x, y, z). This tensor can be described with equal validity in many other choices of coordinates for $\mathbf{M}^4$. Finding the components of this tensor in other coordinate bases for $\mathbf{M}^4$ will be left as an exercise for the reader.

Taking the exterior derivative $d\bar{F}$ gives the three form field

$$d\bar{F} = \frac{1}{2} \partial_\kappa F_{\mu\nu} \, dx^\kappa \wedge dx^\mu \wedge dx^\nu, \tag{5.157}$$

which expanded in coordinates is

$$\begin{aligned}
d\bar{F} = {} & (\partial_x F_{yz} + \partial_y F_{zx} + \partial_z F_{xy}) \, dx \wedge dy \wedge dz \\
& + (\partial_\tau F_{xy} + \partial_x F_{y0} + \partial_y F_{0x}) \, d\tau \wedge dx \wedge dy \\
& + (\partial_z F_{0x} + \partial_\tau F_{xz} + \partial_x F_{z0}) \, dz \wedge d\tau \wedge dx \\
& + (\partial_y F_{z0} + \partial_z F_{0y} + \partial_\tau F_{yz}) \, dy \wedge dz \wedge d\tau.
\end{aligned} \tag{5.158}$$

If we decide to set $d\bar{F} = 0$ then we get four independent equations

$$\begin{aligned}
\partial_x B^x + \partial_y B^y + \partial_z B^z &= 0 \\
\partial_x E^y - \partial_y E^x + \partial_\tau B^z &= 0 \\
\partial_z E^x - \partial_x E^z + \partial_\tau B^y &= 0 \\
\partial_y E^z - \partial_z E^y + \partial_\tau B^x &= 0.
\end{aligned} \tag{5.159}$$

If we use $\tau = ct$ then these four equations obtained by setting $d\bar{F} = 0$ are recognizable as the two source-free Maxwell equations

$$\begin{aligned}
\vec{\nabla} \cdot \vec{B} &= 0 \\
\vec{\nabla} \times \vec{E} + \frac{1}{c} \frac{\partial \vec{B}}{\partial t} &= 0.
\end{aligned} \tag{5.160}$$

The Faraday tensor $\bar{F}$ is defined on the direct product space of two copies of the cotangent space of Minkowski spacetime, and so it is by definition a geometric object in spacetime and exists independently of the coordinate frame of any observer in spacetime. The components of the tensor are what changes under a transformation from one frame to another. The Minkowski basis components of the Faraday tensor transform from frame S to frame $\tilde{S}$ under a Lorentz transform as

$$\tilde{F}_{\mu\nu} = L_{\tilde{\mu}}^\alpha F_{\alpha\beta} L_{\tilde{\nu}}^\beta. \tag{5.161}$$

Suppose there is a constant electric field in frame S, so that $F_{x0} = -F_{0x} = E^x = \text{constant}$ and all of the other components of $\bar{F}$ are zero. What does this field look like according to an observer in frame $\tilde{S}$ traveling at velocity β in the $+y$ direction? The nonzero Lorentz boost components are $L_0^0 = L_y^y = \gamma$, $L_x^x = L_z^z = 1$, and $L^0_y = L^y_0 = \gamma\beta$, so the nonzero components of (5.161) are

$$\tilde{E}^x = \tilde{F}_{x0} = (L^x_x L^0_0 - L^0_x L^x_0) F_{x0}$$
$$= \gamma E^x$$
$$\tilde{B}^z = \tilde{F}_{xy} = (L^x_x L^0_y - L^0_x L^x_y) F_{x0}$$
$$= \gamma\beta E^x. \tag{5.162}$$

To the observer in frame S, there was only an electric field. To an observer in frame $\tilde{S}$, there is both an electric field and a magnetic field. The electric field points in the same direction as in the other frame, but has a greater magnitude because $\gamma \geq 1$. The magnetic field is pointing in a direction at right angles to both the electric field and the direction of motion between the two frames. Notice that in both S and $\tilde{S}$ it is true that $\vec{E} \cdot \vec{B} = 0$.

One can use (5.161) to show that a general electric and magnetic field transform under a Lorentz boost as

$$\vec{E}_{\tilde{S}} = \gamma\vec{E} + \gamma\vec{\beta} \times \vec{B} - \frac{\gamma^2}{\gamma+1}\vec{\beta}\,(\vec{\beta} \cdot \vec{E})$$
$$\vec{B}_{\tilde{S}} = \gamma\vec{B} - \gamma\vec{\beta} \times \vec{E} - \frac{\gamma^2}{\gamma+1}\vec{\beta}\,(\vec{\beta} \cdot \vec{B}). \tag{5.163}$$

If $\vec{E} = E^x\,\partial_x$ and $\vec{B} = 0$, (5.163) confirms the previous result that

$$\vec{E}_{\tilde{S}} = \gamma E^x\partial_{\tilde{x}}, \qquad \vec{B}_{\tilde{S}} = \gamma\beta E^x\partial_{\tilde{z}}. \tag{5.164}$$

The field that an observer in frame S sees as a pure electric field, appears to an observer in frame $\tilde{S}$ to be a combined magnetic and electric field. The difference between electricity and magnetism is relative, not absolute.

These transformations confirm Faraday's belief that the electric and magnetic fields were not separate dynamical fields at all, but different components of the same dynamical field. The distinction between an electric and magnetic field depends on the observer. Therefore there can only be one physical field that is a fundamental field, and the electric and magnetic field must be derived from that field somehow. We will see what that field is below, but first we need to show how to derive the other two Maxwell equations in (5.152) from a tensor equation.

The Maxwell tensor

Now we want to derive the remaining two Maxwell equations in (5.152) from the Faraday tensor. An electric current consists of moving electric charge, which is a hint that the source charge density $\rho_e(x)$ and current density $\vec{J}$ should be combined into a spacetime current vector. But current carries a unit of time in the denominator. In units where $c \neq 1$, we need to set $J^0 = \rho_e c$ so that

$$\mathbf{J} = J^\mu \partial_\mu = \rho_e c \, \partial_\tau + J^i \partial_i \tag{5.165}$$

to make the spacetime vector $\mathbf{J}$ dimensionally balanced with units of current density in all components. The unit systems used in electromagnetism will be addressed in more detail below. In the discussion below we will not use units with $c = 1$, but we will continue absorbing a factor of c in the time coordinate with the assignment $\tau = ct$, so that the τ axis continues to measure length and the coordinate system and metric tensor components are dimensionally balanced.

There are two spacetime operations that we can perform on the two form $\bar{F}$ in order to reduce the Maxwell equations with source to a version as simple as $d\bar{F} = 0$. The first operation is to create a tensor $\mathbf{F} = F^{\mu\nu}\partial_\mu \otimes \partial_\nu$, the components of which are related to the two form components $F_{\mu\nu}$ by raising the indices with the Minkowski metric

$$F^{\mu\nu} = \eta^{\mu\alpha}\eta^{\nu\beta}F_{\alpha\beta}. \tag{5.166}$$

In matrix form this gives

$$(F^{\mu\nu}) = \begin{pmatrix} 0 & E^x & E^y & E^z \\ -E^x & 0 & B^z & -B^y \\ -E^y & -B^z & 0 & B^x \\ -E^z & B^y & -B^x & 0 \end{pmatrix}. \tag{5.167}$$

At this stage in the game we can write the remaining Maxwell equations as

$$\partial_\nu F^{\mu\nu} = \frac{4\pi}{c} J^\mu, \tag{5.168}$$

as the reader will be asked to show in an exercise.

We can turn the tensor $\mathbf{F}$ back into a two form by contracting it with the volume form of spacetime $\bar{\epsilon} = d\tau \wedge dx \wedge dy \wedge dz$. The resulting tensor is called the dual of $\bar{F}$ and is written as

$$*\bar{F} \equiv \frac{1}{2}\bar{\epsilon}(\mathbf{F}) = \frac{1}{4}\epsilon_{\mu\nu\alpha\beta}F^{\alpha\beta}dx^\mu \wedge dx^\nu. \tag{5.169}$$

This two form is called the *Maxwell tensor*, with matrix form

$$(*F_{\mu\nu}) = \begin{pmatrix} 0 & B^x & B^y & B^z \\ -B^x & 0 & E^z & -E^y \\ -B^y & -E^z & 0 & E^x \\ -B^z & E^y & -E^x & 0 \end{pmatrix}. \tag{5.170}$$

We can perform the same operation to the spacetime current vector, creating a three form

$$* \bar{J} \equiv \bar{\epsilon}(\mathbf{J}) = \frac{1}{3!} J^\alpha \epsilon_{\alpha\mu\nu\kappa} dx^\mu \wedge dx^\nu \wedge dx^\kappa \tag{5.171}$$

that is the dual form to the spacetime current vector $\mathbf{J}$. For example, consider a static charge distribution $\rho_e(x)$ with $\mathbf{J} = \rho_e c \, \partial_\tau$. The three form $* \bar{J}$ dual to this vector is

$$* \bar{J} = \bar{\epsilon}(\rho c \partial_\tau) = \rho_e c \, dx \wedge dy \wedge dz. \tag{5.172}$$

The exterior derivative of the Maxwell tensor is

$$d * \bar{F} = \frac{1}{4} \partial_\lambda F^{\alpha\beta} \epsilon_{\alpha\beta\mu\nu} dx^\lambda \wedge dx^\mu \wedge dx^\nu. \tag{5.173}$$

Using the component expansion as we did with the stress tensor in (5.123), it can be shown that

$$d * \bar{F} = \frac{1}{3!} (\partial_\lambda F^{\alpha\lambda}) \epsilon_{\alpha\beta\mu\nu} dx^\beta \wedge dx^\mu \wedge dx^\nu. \tag{5.174}$$

The complete set of Maxwell equations in (5.152) can thus be written

$$d\bar{F} = 0$$
$$d * \bar{F} = \frac{4\pi}{c} * \bar{J}. \tag{5.175}$$

Since $d(d\bar{\omega}) = 0$ for any p form $\bar{\omega}$, then we know that $d(d * \bar{F}) = 0$. But this also enforces the condition $d * \bar{J} = 0$, which in coordinate components yields the usual equation for current conservation

$$\partial_\mu J^\mu = \frac{\partial \rho_e}{\partial t} + \vec{\nabla} \cdot \vec{J} = 0. \tag{5.176}$$

Gaussian units and relativity

So far we've only addressed the microscopic Maxwell equations, where the only sources of field energy are static or flowing distributions of charged particles. Macroscopic media can be polarized and/or magnetized by external electric and magnetic fields. The resulting polarization and magnetization fields appear in the Maxwell equations through two constants that reflect bulk properties of macroscopic media, called the permittivity constant ϵ and the permeability constant μ. (In anisotropic media these would be tensors.)

In Chapter 2 we gave the vacuum Maxwell equations in what are called rationalized mksA units, or just mksA units for short. By "vacuum" we mean in a region of spacetime where there is no charge or current present. The two Maxwell equations with source terms are properly written in terms of the displacement field $\vec{D}$ and magnetic field $\vec{H}$.[1] In mksA units

$$\vec{D} = \epsilon_0 \vec{E} + \vec{P}, \qquad \vec{H} = \frac{1}{\mu_0} \vec{B} - \vec{M}, \tag{5.177}$$

where $\vec{P}$ is the polarization of the macroscopic medium, and $\vec{M}$ is the magnetization. The microscopic Maxwell equations with $\vec{P} = \vec{M} = 0$ are

$$\epsilon_0 \vec{\nabla} \cdot \vec{E} = \rho \tag{5.178}$$

$$\frac{1}{\mu_0} \vec{\nabla} \times \vec{B} - \epsilon_0 \frac{\partial \vec{E}}{\partial t} = \vec{J}$$

$$\vec{\nabla} \cdot \vec{B} = 0$$

$$\vec{\nabla} \times \vec{E} + \frac{\partial \vec{B}}{\partial t} = 0. \tag{5.179}$$

These are the Maxwell equations in mksA units. The two constants ϵ_0 and μ_0 are called the permittivity and permeability of free space, respectively. These two constants appear because Maxwell, like almost everyone else in his era, thought of empty space as a mechanical medium that could be polarized and/or magnetized like a bulk material in the laboratory. The concept of an electric field existing in empty space without some kind of polarizable matter serving as an intervening mechanical medium didn't yet exist in the collective imagination of physicists.

As was pointed out in Chapter 2, Maxwell came to the realization that the traveling wave solutions to the vacuum equations (5.178) could describe the propagation of light because the squared speed of these traveling waves was $v_w = (\mu_0 \epsilon_0)^{-1/2}$, and the measured value of $(\mu_0 \epsilon_0)^{-1/2}$ was equal to the measured value of the speed of light. And he was exactly right in his speculation. Today we know that these two constants meant to represent the assumed bulk properties of space are just standing in for the speed of light, which wasn't thought to have anything to do with electricity or magnetism before Maxwell made his theoretical observation. In fact, today they are no longer treated as true constants of nature and (as a definition of units) just given the values

$$\epsilon_0 = \frac{10^7}{4\pi c^2}, \qquad \mu_0 = 4\pi \times 10^{-7}. \tag{5.180}$$

Only the product is independent of conventions.

In Gaussian units, one effectively sets $\epsilon_0 = 1/4\pi$, and then the symbols ϵ_0 and μ_0 can be discarded entirely. The displacement and magnetic fields are defined

---

[1] The field $\vec{B}$ is commonly referred to as the magnetic field but is properly called the magnetic induction.

to be

$$\vec{D} = \vec{E} + 4\pi \vec{P}, \qquad \vec{H} = \vec{B} - 4\pi \vec{M}. \tag{5.181}$$

The microscopic Maxwell equations with $\vec{P} = \vec{M} = 0$ are those given in (5.152). The appearance of the factors of c in the Maxwell equations in Gaussian units are in the denominators of the time derivatives, and so can be absorbed in the time coordinate definition $\tau = ct$, as was done above.

Note that in Gaussian units the fields $\vec{E}$ and $\vec{B}$ both have the units Q/L^2, where Q represents units of electric charge. As one can see from (5.178) the same relationship does not hold in mksA units, where the units of $\vec{E}$ and $\vec{B}$ differ by a factor of L/T.

The vector potential

Since the Faraday tensor satisfies the condition $d\bar{F} = 0$, we can define a one form $\bar{A}$ such that $\bar{F} = d\bar{A} = A_\mu dx^\mu.$[2] This one form field historically was first defined as a vector $\vec{A}$ through the Euclidean version of the exterior derivative $\vec{B} = \vec{\nabla} \times \vec{A}$. For this reason the field is called the *vector potential* for the electromagnetic field. This parallels the situation with spacetime momentum **p**, which is more properly defined as a one form $\bar{p}$ but is used in practice as a spacetime vector.

In Minkowski spacetime we have a metric tensor to provide a one-to-one map between one forms and vectors; it doesn't matter whether we work with the one form field $\bar{A}$ or the vector field **A**. However, it is important to note that the traditional definition of $\vec{A}$ as a vector means that the "relativized" spacetime version is

$$\mathbf{A} = A^\mu \partial_\mu = \phi \, \partial_\tau + A^i \partial_i, \tag{5.182}$$

where $\phi(x)$ is the coulomb electrostatic potential discovered in the seventeenth century by Cavendish and in the eighteenth century by Coulomb. Thus the one form field $\bar{A}$ ends up being

$$\bar{A} = A_\mu dx^\mu = -\phi d\tau + A_i dx^i, \tag{5.183}$$

where the minus sign in the time component comes from the map $A_\mu = \eta_{\mu\nu} A^\nu$ that takes the vector components to the components of the associated one form.

Setting $\bar{F} = d\bar{A}$ yields a two form with coordinate basis components

$$F_{\mu\nu} = \partial_\mu A_\nu - \partial_\nu A_\mu. \tag{5.184}$$

---

[2] This fundamental theorem of differential geometry is strictly true only in regions that are topologically trivial. The nontrivial example of a magnetic monopole will be discussed later.

In terms of electric and magnetic fields this gives

$$\vec{E} = -\vec{\nabla}\phi - \frac{\partial \vec{A}}{\partial \tau}$$

$$\vec{B} = \vec{\nabla} \times \vec{A}. \tag{5.185}$$

Gauge invariance

Consider a one form field $\bar{A}$ for which $\bar{F} = d\bar{A}$. Now consider another one form field $\bar{A}' = \bar{A} + d\psi$, where $\psi(x)$ is some scalar function of the spacetime coordinates. Since $d(d\bar{\omega}) = 0$ for any p form $\bar{\omega}$, then

$$\bar{F}' = d\bar{A}' = d\bar{A} + d(d\psi) = d\bar{A} = \bar{F}. \tag{5.186}$$

Since the function $\psi(x)$ was completely arbitrary, this result means that the Maxwell equations don't care whether we use $\bar{A}$ or $\bar{A}'$. We get the same electric and magnetic fields and equations in either case. Therefore the vector potential is not defined uniquely, but is only unique up to the addition of the gradient of a scalar field. This lack of uniqueness in the definition of the field $\bar{A}$ is called *gauge invariance*.

Since the function $\psi(x)$ is completely arbitrary and is invisible to the Maxwell equations, we can use this freedom to bring the vector potential to a special form that is convenient for the problem that we are working on. Making a specific choice of this type is called *choosing a gauge*.

Choosing a gauge in electromagnetism is somewhat analogous to choosing a coordinate system in a scattering problem. The momentum conservation condition is good in any inertial frame, but we have to choose some particular frame in order to solve an actual problem. We can choose the rest frame of one of the initial particles, or we can choose the center of momentum frame of all the initial particles. The components in one frame are related to the components in the other frame by a Lorentz transformation, so therefore both ways of looking at the problem are equally valid.

The two most common choices of gauge in electromagnetism are called *Lorentz gauge* and *Coulomb gauge*. The Lorentz gauge consists of the condition $\partial_\mu A^\mu = -\partial_\tau \phi + \vec{\nabla} \cdot \vec{A} = 0$. This condition is enforced on a gauge field that obeys $\partial_\mu \partial^\mu \phi = 0$ by the Maxwell equations. The Coulomb gauge (also called the radiation gauge or transverse gauge) is defined by $\vec{\nabla} \cdot \vec{A} = 0$. The Lorentz gauge condition $\partial_\mu A^\mu = 0$ is manifestly Lorentz-invariant.[3] The Coulomb gauge condition

---

[3] This statement assumes that A^μ transforms as a vector under Lorentz transformations, as the notation would seem to suggest. However, this is strictly true only up to a gauge transformation.

$\vec{\nabla} \cdot \vec{A} = 0$, on the other hand, is not Lorentz-invariant and leads to apparent (but not actual) violations of causality. The topic of causality will be addressed in detail in Chapter 6.

The electromagnetic stress tensor

The stress tensor **T** for an electromagnetic field is made from the Faraday two form $\tilde{F}$ and its rank $\binom{2}{0}$ version **F**. The coordinate basis components of **T** in Gaussian units are

$$T^{\mu\nu} = \frac{1}{4\pi} \left(F^{\mu\alpha} F^{\nu\beta} \eta_{\alpha\beta} - \frac{1}{4} \eta^{\mu\nu} F_{\alpha\beta} F^{\alpha\beta} \right). \tag{5.187}$$

In Minkowski coordinates (τ, x, y, z) the components of the stress tensor are

$$T^{00} = \frac{1}{8\pi} \left(|\vec{E}|^2 + |\vec{B}|^2 \right)$$

$$T^{0i} = \frac{1}{4\pi} \left(\delta^{ij} \epsilon_{jkm} E^k B^m \right) = \frac{1}{4\pi} (\vec{E} \times \vec{B})^i$$

$$T^{ij} = \frac{1}{8\pi} \left((|\vec{E}|^2 + |\vec{B}|^2) \delta^{ij} - 2(E^i E^j + B^j B^i) \right). \tag{5.188}$$

The component T^{00} is recognizable as the electromagnetic energy density. The component $T^{0i} = T^{i0}$ gives the momentum flux density of the electromagnetic field

$$\vec{P} = \frac{c}{4\pi} (\vec{E} \times \vec{B}), \tag{5.189}$$

a quantity sometimes referred to as the *Poynting vector*. The spacelike diagonal components with $i = j$ are

$$T^{ii} = \frac{1}{8\pi} \left\{ (|\vec{E}|^2 + |\vec{B}|^2) - 2((E^i)^2 + (B^i)^2) \right\}, \tag{5.190}$$

which give the pressure normal to the face of a small cube of electromagnetic field. The off-diagonal spacelike components with $i \neq j$

$$T^{ij} = -\frac{1}{4\pi} (E^i E^j + B^i B^j) \tag{5.191}$$

give the shear stresses that act parallel to the faces of an infinitesimal region of the electromagnetic field.

As a simple example of electromagnetic stress energy, consider a constant electric field in the x direction, with $\vec{E} = E \partial_x$. The nonzero components of **T** are

$$T^{00} = \frac{E^2}{8\pi}, \qquad T^{xx} = -\frac{E^2}{8\pi}, \qquad T^{yy} = T^{zz} = \frac{E^2}{8\pi}. \tag{5.192}$$

The forces on the cubic region of electromagnetic field with volume L^3 centered at the origin consist of tension on the faces at $x = \pm L/2$, and pressure on the faces at $y = \pm L/2$ and $z = \pm L/2$. The electric field pulls on the cube in the direction parallel to the field and pushes on the cube in the two directions orthogonal to the field.

Now consider an observer at rest in some frame $\tilde{S}$ that is moving at velocity $\vec{\beta} = \beta \partial_y$ relative to frame S. It will be left for the reader as an exercise to show that this observer measures the energy and momentum

$$\tilde{T}^{00} = \frac{E^2}{8\pi}\gamma^2(1 + \beta^2), \qquad \tilde{T}^{0y} = \tilde{T}^{y0} = -\frac{E^2}{4\pi}\gamma^2\beta \qquad (5.193)$$

and forces

$$\tilde{T}^{xx} = -\frac{E^2}{8\pi}, \qquad \tilde{T}^{zz} = \frac{E^2}{8\pi}, \qquad \tilde{T}^{yy} = \frac{E^2}{8\pi}\gamma^2(1 + \beta^2). \qquad (5.194)$$

In the relativistic limit $\gamma \gg 1$, the pressure $\tilde{T}^{yy}$ on the faces at $\tilde{y} = \pm L/2$ is the dominant pressure in the field and dwarfs the tiny amount of tension on the faces $\tilde{x} = \pm L/2$.

Electromagnetic waves

Special relativity was developed by Einstein to address the curious lack of a consistent rest frame for light waves as described by electric and magnetic fields through the Maxwell equations. Since then we have learned that the propagation of light can be described in two ways in relativity. In Chapter 4 we described light as a massless particle that travels along null paths through spacetime with spacetime momentum satisfying the massless condition $\mathbf{p}^2 = 0$. Now we are going to describe light as an oscillating solution of the Maxwell equations in empty space, and explore the connection between this description and the description that we used in Chapter 4.

In Lorentz gauge $\partial_\mu A^\mu = 0$, the Maxwell equations with source can be reduced to one equation for the coordinate components of the one form potential $\bar{A} = A_\mu dx^\mu$

$$\partial_\alpha \partial^\alpha A_\mu = \frac{4\pi}{c} J_\mu, \qquad (5.195)$$

where $J_\mu = \eta_{\mu\nu} J^\mu$. In empty spacetime outside any region containing charge or current, this equation becomes

$$\partial_\alpha \partial^\alpha A_\mu = 0. \qquad (5.196)$$

Suppose we look for a wavelike oscillating solution of the form

$$\bar{A} = A_\mu(x)dx^\mu = A \exp(iS(x))\varepsilon_\mu \, dx^\mu. \tag{5.197}$$

The function $S(x)$ is called the *phase* of the electromagnetic wave and the components ε_μ are called the *polarization*. The constant A is the amplitude of the wave. Let's assume for what follows that $S(x)$ is a real function, and the polarization components ε_μ are constant. Define the one form $\bar{k}$ as the exterior derivative of the phase function $\bar{k} \equiv dS = \partial_\mu S dx^\mu$. The spacetime vector $\mathbf{k}$ with components $k^\mu = \eta^{\mu\nu}k_\mu$ is then normal to surfaces of constant phase S. This means that $\mathbf{k}$ points in the direction in which the wave is traveling, so we can consider it to be the propagation vector of the wave. The Lorentz gauge condition $\partial_\mu A^\mu = 0$ yields the condition

$$i(\partial_\mu S)A^\mu = i\mathbf{k} \cdot \mathbf{A} = i(\mathbf{k} \cdot \boldsymbol{\varepsilon})A \exp(iS(x)) = 0, \tag{5.198}$$

where the vector $\boldsymbol{\varepsilon}$ has components $\varepsilon^\mu = \eta^{\mu\nu}\varepsilon_\nu$. Therefore the vector $\mathbf{k}$ satisfies $\mathbf{k} \cdot \boldsymbol{\varepsilon} = k^\mu \varepsilon_\nu = 0$. The polarization of the wave is orthogonal in spacetime to the direction of propagation of the wave. The vector $\mathbf{k}$ is null, and the polarization vector $\boldsymbol{\varepsilon}$ must be spacelike.

Enforcing the Maxwell equations on this wave solution yields

$$\partial_\mu \partial^\mu A_\alpha = \partial_\mu(\eta^{\mu\nu}ik_\nu A_\alpha) = (i\partial_\mu k^\mu - k_\mu k^\mu)A_\alpha$$
$$= \left(i\partial_\mu \partial^\mu S - \mathbf{k}^2\right)A_\alpha = 0. \tag{5.199}$$

The imaginary part of this equation vanishes if $\partial_\mu \partial^\mu S = 0$, which means the phase function $S(x)$ itself satisfies a Lorentz-invariant wave equation. The real part of $\partial_\mu \partial^\mu A_\alpha = 0$ is $\mathbf{k}^2 = 0$, which means the normal vector $\mathbf{k}$ is a null vector in $\mathbf{M}^4$. The propagation vector $\mathbf{k}$ for a general traveling wave solution of the Maxwell equations in empty space behaves like the spacetime momentum $\mathbf{p}$ of a massless particle, which satisfies $\mathbf{p}^2 = 0$. A light wave can be spread out all over space so we can't say that it has a definite path. However, we can talk about a ray of light as being the curve to which the vector $\mathbf{k}$ is tangent. Using that definition, we can say that, according to Maxwell equations, *light rays travel along the paths of massless particles*.

The most important class of waves for physicists are plane waves, defined by a constant propagation vector $\mathbf{k}$. Since $\partial_\mu \partial^\mu S = \partial_\mu k^\mu = 0$, then S can be written as $S = \mathbf{k} \cdot \mathbf{x} = -k^0 \tau + \vec{k} \cdot \vec{x}$. The timelike component of $\mathbf{k}$ is $k^0 = \omega/c = 2\pi\nu/c$, where ν is the frequency of the wave. The spacelike components $\vec{k}$ give the wavelength $\lambda = 2\pi/|\vec{k}|$ and direction of propagation $\hat{e}_k = \vec{k}/|\vec{k}|$ of the wave in $\mathbf{E}^3$. Thus we arrive at the basic form of a plane electromagnetic wave

$$\bar{A} = A\bar{\varepsilon}\exp{(i\mathbf{k} \cdot \mathbf{x})}, \qquad \mathbf{k}^2 = 0, \qquad \langle \mathbf{k}, \bar{\varepsilon}\rangle = k^\mu \varepsilon_\mu = 0. \qquad (5.200)$$

Since $\mathbf{k}$ is a null vector in spacetime, the frequency and wavelength are related by $\omega/c = |\vec{k}|$, which gives the familiar relation

$$\omega/|\vec{k}| = \lambda v = c \qquad (5.201)$$

between the frequency, wavelength and speed of light. In relativity it is common to use units with $c = 1$ and just write $\omega = |\vec{k}|$.

Motion of charged particles in spacetime

In Chapter 4 we looked at particle motion in the absence of forces, so the equation of motion for a particle with momentum $\mathbf{p}$ is

$$\frac{d\mathbf{p}}{d\lambda} = u^\mu \partial_\mu \mathbf{p} = 0, \qquad (5.202)$$

where λ is the proper time along the world line of the particle. A particle with electric charge propagating in an electromagnetic field will experience a force from the field. The Lorentz force law for a particle with electric charge q traveling with velocity $\vec{\beta} = \vec{v}/c$ is

$$\frac{d\vec{p}}{dt} = \vec{F} = q(\vec{E} + \vec{\beta} \times \vec{B}). \qquad (5.203)$$

The covariant form of this force law is

$$\frac{d\mathbf{p}}{d\lambda} = q\mathbf{F} \cdot \mathbf{u}, \qquad (5.204)$$

which in coordinate basis components is

$$\frac{dp^\mu}{d\lambda} = q\,F^{\mu\nu}u_\nu. \qquad (5.205)$$

The space components reduce to (5.203), while the time component gives the energy relation

$$\frac{dp^\tau}{d\lambda} = \frac{dE}{d\lambda} = q\gamma\vec{\beta} \cdot \vec{E} \qquad (5.206)$$

equating the change in energy of the particle to the work done on the particle by the electric field. The force from a magnetic field is always orthogonal in space to the velocity of the charged particle, and therefore a magnetic field cannot do any work on a particle with electric charge. Note also that if the particle accelerates, then we can no longer use a Lorentz transformation to reach the rest frame of the particle at all times. But the equation of motion (5.204) for the particle is equally valid in any inertial frame.

If there is more than one charged particle with a current **J**, then we need to use the stress tensor to get the change in the momentum flux. The N-particle stress tensor is $\mathbf{T_p} = \mathbf{p} \otimes \mathbf{n}$, where **n** is the particle number current defined in (5.119). If we assume that charged particles are neither created nor destroyed in the system in question, then the particle number current **n** satisfies $\partial_\mu n^\mu = 0$. The divergence of the stress tensor is in that case given by

$$\partial_\nu T_p^{\mu\nu} = n(x)u^\nu \partial_\nu p^\mu = n(x)\frac{dp^\mu}{d\lambda} = qn(x)F^{\mu\nu}u_\nu = F^{\mu\nu}J_\nu, \qquad (5.207)$$

which generalizes the single particle Lorentz force law (5.204).

Suppose we have a particle with charge q propagating in frame S with coordinates (τ, x, y, z), in a presence of a constant electric field $\vec{E} = E\partial_x$. According to the equations of motion in frame S,

$$\frac{du^0}{d\lambda} = \kappa u^x, \qquad \frac{du^x}{d\lambda} = \kappa u^0, \qquad \frac{du^y}{d\lambda} = \frac{du^z}{d\lambda} = 0, \qquad (5.208)$$

where

$$\kappa \equiv qE/m. \qquad (5.209)$$

A velocity field that satisfies this system of equation is

$$\mathbf{u} = (\cosh(\kappa\lambda), \sinh(\kappa\lambda), 0, 0). \qquad (5.210)$$

This is the spacetime velocity field tangent to the world line of a particle undergoing a uniform acceleration κ in the $+x$ direction, previously discussed in Chapter 4 with $\alpha = 1/\kappa$. One possible world line for the particle in frame S is

$$\mathbf{x} = (\frac{1}{\kappa}\sinh(\kappa\lambda), \frac{1}{\kappa}\cosh(\kappa\lambda), 0, 0). \qquad (5.211)$$

Now consider this same system in frame $\tilde{S}$ traveling at velocity $\vec{\beta} = \beta\partial_y$ relative to frame S. As shown previously, the electric field in frame $\tilde{S}$ is $\vec{E}_{\tilde{s}} = \gamma E\partial_{\tilde{x}}$ and the magnetic field $\vec{B}_{\tilde{s}} = \gamma\beta E\partial_{\tilde{z}}$. The equations of motion in frame $\tilde{S}$ are then

$$\frac{d\tilde{u}^0}{d\lambda} = \gamma\kappa\tilde{u}^x, \qquad \frac{d\tilde{u}^x}{d\lambda} = \gamma\kappa\tilde{u}^0 + \gamma\beta\kappa\tilde{u}^y \qquad (5.212)$$

$$\frac{d\tilde{u}^z}{d\lambda} = 0, \qquad \frac{d\tilde{u}^y}{d\lambda} = -\gamma\beta\kappa\tilde{u}^x. \qquad (5.213)$$

However, we don't actually have to solve these equations. Lorentz covariance of the equations of motion guarantees that the solution to the equations of motion in frame $\tilde{S}$ can be obtained by applying a Lorentz transform to the solution in frame S (5.210), to get

$$\tilde{\mathbf{u}} = (\gamma \cosh(\kappa\lambda), \sinh(\kappa\lambda), -\gamma\beta \cosh(\kappa\lambda), 0).$$
(5.214)

The world line according to an observer at rest in frame $\tilde{S}$ is

$$\tilde{\mathbf{x}} = (\frac{\gamma}{\kappa} \sinh(\kappa\lambda), \frac{1}{\kappa} \cosh(\kappa\lambda), -\frac{\gamma\beta}{\kappa} \sinh(\kappa\lambda), 0).$$
(5.215)

This particle path and the one in (5.210) are both curves that belong to a timelike Lorentz-invariant submanifold of Minkowski spacetime, as discussed in Chapter 4. In this case, both world lines belong to the submanifold of Minkowski spacetime for which

$$\mathbf{x}^2 = -\tau^2 + |\vec{x}|^2 = \left(\frac{m}{qE}\right)^2.$$
(5.216)

A Lorentz transformation maps a curve on this submanifold into another such curve, as we saw in this example.

Conservation of energy and momentum

Notice that the particle stress tensor $\mathbf{T}_p$ in (5.207) has a nonzero divergence. Momentum and energy are not conserved for this system. But there is a perfectly good reason behind this. Electromagnetic forces are acting on the particles in question, and work is being done by those forces to change the motion of the particles. Thus the particles are exchanging energy with the field, and we need to take into account the change in field energy $\mathbf{T}_F$ from work done on the particles, by computing

$$\partial_\nu T_F^{\mu\nu} = \frac{1}{4\pi} \partial_\nu \left(F^\mu{}_\alpha F^{\nu\alpha} - \frac{1}{4}\eta^{\mu\nu} F_{\alpha\beta} F^{\alpha\beta} \right)$$

$$= \frac{1}{4\pi} \left(F^\mu{}_\alpha \partial_\nu F^{\nu\alpha} - F^{\alpha\nu} \partial_\nu F^\mu{}_\alpha - \frac{1}{2}\eta^{\mu\nu} F^{\alpha\beta} \partial_\nu F_{\alpha\beta} \right)$$

$$= \frac{1}{4\pi} \left(F^\mu{}_\alpha \partial_\nu F^{\nu\alpha} - \frac{1}{2}\eta^{\mu\nu} F^{\alpha\beta} \left(\partial_\nu F_{\alpha\beta} + \partial_\alpha F_{\beta\nu} + \partial_\beta F_{\nu\alpha} \right) \right)$$ (5.217)

$$= \frac{1}{4\pi} F^\mu{}_\alpha \partial_\nu F^{\nu\alpha} = -F^\mu{}_\nu J^\nu = -\partial_\nu T_p^{\mu\nu},$$
(5.218)

where we used the source-free Maxwell equation $\partial_\nu F_{\alpha\beta} + \partial_\alpha F_{\beta\nu} + \partial_\beta F_{\nu\alpha} = 0$.

The particle and field stress tensors satisfy $\partial_\nu T_p^{\mu\nu} + \partial_\nu T_F^{\mu\nu} = 0$. Therefore it is the sum $\mathbf{T} = \mathbf{T}_p + \mathbf{T}_F$ of the particle and field stress tensors that satisfies the divergence-free condition $\partial_\nu T^{\mu\nu} = 0$.

Electromagnetism in higher dimensions

In most of this book we have treated the number of dimensions of spacetime as some abstract number d, and then used a particular value of d, for example $d = 2, 3$ or 4, depending on the matter at hand. In this section on electromagnetism, we have stuck entirely to four spacetime dimensions. Why is that? The case of $d = 4$ is special because it's the number we observe in nature, but there is also another reason that is explained below.

A theory that obeys the Maxwell equations can be constructed in any number of spacetime dimensions. If we start with a one form potential $\bar{A} = -\phi d\tau + A_1 dx^1 + \cdots + A_D dx^D$, with $D = d - 1$, then the two form $\bar{F} = d\bar{A}$ automatically satisfies the equation $d\bar{F} = 0$. We can use the Minkowski metric in d dimensions to map the two form $\bar{F}$ into the rank $\binom{2}{0}$ tensor $\mathbf{F}$, and use the spacetime volume form in d dimensions $\bar{\epsilon} = d\tau \wedge dx^1 \wedge \ldots \wedge dx^D$ to make the dual form $*\bar{F} = \bar{\epsilon}(\mathbf{F})$, which is a $(d - 2)$ form. The spacetime current vector $\mathbf{J} = \rho_e \partial_\tau + J^i \partial_i$ has the dual form $*\bar{J} = \bar{\epsilon}(\mathbf{J})$, which is a $(d - 1)$ form. The Maxwell equations in arbitrary spacetime dimension d are then

$$d\bar{F} = 0, \qquad d*\bar{F} = \frac{S_{D-1}}{c}*\bar{J}, \qquad (5.219)$$

where $S_{D-1} = 2\pi^{D/2}/\Gamma(D/2)$ is the area of a unit $(D - 1)$ sphere around a point charge in D space dimensions. The function $\Gamma(D/2)$ is the Euler beta function defined by $\Gamma(x) = \int_0^\infty t^{x-1} e^{-t} dt$. For the case of $D = 3$, that is, $d = 4$, we get the usual factor of 4π as the area of a unit two sphere surrounding a point charge in three space dimensions.

What about the electric and magnetic fields? In $\mathbf{M}^d$, in a frame S with coordinates $(\tau, x^1, \ldots, x^D)$, the electric field $\vec{E} = F^{i0} \partial_i$ is a Euclidean vector field. However, the magnetic field can no longer be represented as a vector field. The magnetic field has $D(D - 1)/2$ components $F_{ij} = \partial_i A_j - \partial_j A_i$. These tensor components only make a Euclidean vector in $\mathbf{E}^D$ if $D = 3$. The magnetic field is really a two form field $\bar{B} = d\bar{A}$ in the Euclidean space $\mathbf{E}^D$ defined by the space axes in the frame S. The electric and magnetic fields still mix under a Lorentz transformation, but not using the formula (5.163). The Lorentz transformations of $\vec{E}$ and $\bar{B}$ will be left for the reader as an exercise.

If we're in a particular Lorentz frame S, that means we've chosen a particular way to slice Minkowski spacetime $\mathbf{M}^d$ into spacelike slices of constant time. Each spacelike slice is a copy of Euclidean space $\mathbf{E}^D$. We can apply Stokes's theorem (5.95) on any spacelike slice using the $\vec{E}$ and $\bar{B}$ fields defined above.

The volume form on our spacelike slice is the D form $\bar{\epsilon} = dx^1 \wedge \ldots \wedge dx^D$. The dual field $*\vec{E} = \bar{\epsilon}(\vec{E})$ is a $(D - 1)$ form, and therefore $d*\vec{E}$ is a D form. We can apply Stokes's theorem as

$$\int_V d^*\bar{E} = \oint_{\partial V} {}^*\bar{E},$$

(5.220)

which can also be written as

$$\int_V \vec{\nabla} \cdot \vec{E} \, dV = \oint_{\partial V} \vec{E} \cdot d\vec{S}.$$

(5.221)

In this formula, V is the D-dimensional volume on the chosen spacelike slice of spacetime, and $d\vec{S}$ is the area element normal to the $(D-1)$-dimensional hypersurface that makes up the boundary ∂V of V.

The magnetic field $\bar{B}$ is a two form in our chosen Euclidean slice of spacetime. We can use the Euclidean inverse metric δ^{ij} to raise the component indices of B_{ij} and make the components B^{ij} of a rank $\binom{2}{0}$ tensor $\mathbf{B}$. Contracting that tensor with the volume form $\bar{\epsilon}$ on the spacelike slice gives a $(D-2)$ form $^*\bar{B} = \bar{\epsilon}(\mathbf{B})$. The exterior derivative $d^*\bar{B}$ is a $(D-1)$ form, so to apply Stokes' theorem to the magnetic field, we must integrate over a $(D-1)$-dimensional region of the chosen spacelike slice. Stokes's theorem in any dimension D says

$$\int_W d^*\bar{B} = \oint_{\partial W} {}^*\bar{B},$$

(5.222)

where W is a $(D-1)$-dimensional region of our chosen Euclidean slice of spacetime, and ∂W is the $(D-2)$-dimensional boundary of W. For $D = 3$, the region W is the two-dimensional area A through which the magnetic flux passes, and the boundary ∂A is the one-dimensional line that bounds the area. Only in the case of $D = 3$ (that is, in four spacetime dimensions) can we reduce (5.222) to the usual vector equation

$$\int_A (\vec{\nabla} \times \vec{B}) \cdot dA = \oint_{\partial A} \vec{B} \cdot d\vec{\ell}.$$

(5.223)

There is one other important fact about electromagnetism that makes four spacetime dimensions special. Let's look first at the Lorentz-invariant quantity $\text{Tr}\,\mathbf{T} = T^\mu{}_\mu = \eta_{\mu\nu} T^{\mu\nu}$. In some particular frame S, if we write the stress tensor $\vec{\mathbf{T}}_F$ in terms of fields $\vec{E}$ and $\vec{B}$, then

$$\text{Tr}\,\mathbf{T} = \frac{1}{4\pi}(\frac{d}{4} - 1)\, F_{\mu\nu} F^{\mu\nu} = \frac{1}{4\pi}(\frac{d}{4} - 1)\left(|\vec{E}|^2 - |\vec{B}|^2\right),$$

(5.224)

where $|\bar{B}|^2 = B_{ij} B^{ij}$ and we've used the fact that $\eta_{\mu\nu} \eta^{\mu\nu} = d$.

Only in four spacetime dimensions does $\text{Tr}\,\mathbf{T} = 0$ for any and all fields $\vec{E}$ and $\bar{B}$. What is the significance of this? The trace of the stress tensor is related to the behavior of the fields under a scale transformation $x^\mu \to \lambda x^\mu$. This scale invariance symmetry is part of a larger symmetry, called conformal invariance, of the source-free Maxwell theory. If this Lorentz-invariant quantity Tr T vanishes, then the system is invariant under rescaling and there is no natural length scale L_F in the theory.

To see the behavior of electromagnetism under a rescaling, consider the energy density $\mathcal{E}$ which, if we're using $c = 1$, has units $[\mathcal{E}] \sim M/L^D$. The electric and magnetic fields have units $[E, B] \sim Q/L^{D-1}$ and therefore the field energy density also has units $[\mathcal{E}] \sim Q^2/L^{2(D-1)}$. If we work in units where Planck's constant $\hbar = 1$, then $[M] \sim 1/L$, and the units for electric charge Q in $d = D + 1$ spacetime dimensions

$$[Q^2] \sim L^{d-4} \sim M^{4-d}. \tag{5.225}$$

Therefore in $d \neq 4$ spacetime dimensions, electromagnetism has a natural length scale $L_{em} \sim Q^{2/(4-d)}$. In the special case of $d = 4$, the charge is dimensionless, which means that there is no natural scale for the theory, and electromagnetic fields in free space behave the same no matter at what length scale we examine them. In any other number of dimensions, this is not true – there is a natural length scale L_{em}, or equivalently, a mass or energy scale M_{em} related to the value of the electric charge through $L_{em} = 1/M_{em} \sim Q^{2/(4-d)}$. The physics does not look the same at all length scales.

Electromagnetism is scale-invariant when the trace of the stress tensor vanishes, Tr T $= 0$, which only happens in four spacetime dimensions. This property of scale invariance is shared by certain generalizations of Maxwell theory, called Yang–Mills theories, that are used to describe nuclear forces. However, this scale invariance only applies to the classical theory, and is ultimately spoiled by quantum corrections that introduce a fundamental scale. Scale invariance sometimes can be protected from quantum corrections by further generalizations involving a symmetry called *supersymmetry*, which will be discussed in Chapter 9. Of course, the real world is certainly not scale-invariant.

Magnetic monopoles

According to the Maxwell equations, the source of all electric and magnetic fields is the current $\mathbf{J}_e$, which appears in the equations in the unsymmetrical fashion

$$d\bar{F} = 0, \qquad d^*\bar{F} = \frac{4\pi}{c} {}^*\bar{J}_e. \tag{5.226}$$

If there were such a thing as magnetic charge, then there would be a magnetic current $\mathbf{J}_m$, and we could write the Maxwell equations in a symmetric manner as

$$d\bar{F} = \frac{4\pi}{c} * \bar{J}_m, \qquad d*\bar{F} = \frac{4\pi}{c} * \bar{J}_e. \qquad (5.227)$$

In some frame $\tilde{S}$ with coordinates (τ, x, y, z), we could then write the equations for the fields around a point source of electric charge q and magnetic charge g on the world line $(\tau, 0, 0, 0)$ as

$$\vec{\nabla} \cdot \vec{E} = 4\pi q \, \delta^3(\vec{x}), \qquad \vec{\nabla} \cdot \vec{B} = 4\pi g \, \delta^3(\vec{x}). \qquad (5.228)$$

The field $\vec{B}$ would then be the magnetic field around a *magnetic monopole* with magnetic charge g, and would take the same form as the electric field around an electric point charge with electric charge q.

The problem with this pleasingly symmetric setup is that the Euclidean vector $\vec{B}$ is really a Euclidean two form $\bar{B}$ in disguise. The two form $\bar{B}$ is defined as the exterior derivative $\bar{B} = d\bar{A}$ of the one form potential $\bar{A}$, which means $d\bar{B} = 0$ automatically, so there are no exceptions whatsoever to the equation $\vec{\nabla} \cdot \vec{B} = 0$.

However, things are not so simple when we get to the details. Previously in this section we showed that all one form potentials $\bar{A}$ that differ by the gradient $d\psi$ of a scalar function on spacetime $\psi(x)$ are physically equivalent and lead to the same electric and magnetic fields. This physical equivalence is called gauge invariance. The way to define $\bar{B} = d\bar{A}$ and still get around the restriction imposed by $d\bar{B} = 0$ is to use two different one form potentials $\bar{A}_1$ and $\bar{A}_2$ defined in separate but overlapping regions of spacetime, with the difference between the two potentials given by the gradient $d\psi$ of some function ψ. Gauge invariance means these two different fields are physically equivalent. This physical equivalence can be used as a loophole for defining magnetic monopoles.

For the following example, let's use spherical coordinates (E5.1) for the space-like coordinates in our frame S. Let's define two fields $\bar{A}^+$ and $\bar{A}^-$

$$\bar{A}^+ = g(1 - \cos\theta)d\phi, \qquad \bar{A}^- = -g(1 + \cos\theta)\, d\phi. \qquad (5.229)$$

The field $\bar{A}^+$ is well defined everywhere except for $\theta = \pi$, where there is a singularity called the *Dirac string*. The field $\bar{A}^-$ is well defined everywhere except for $\theta = 0$, where there is also a Dirac string singularity. Apart from the Dirac string, both fields produce the same magnetic field two form $\bar{B} = d\bar{A}^\pm$. Using the metric in spherical coordinates, and the volume form $\bar{\epsilon} = r^2 \sin\theta dr \wedge d\theta \wedge d\phi$, we can write $\vec{B}$ as the vector

$$\vec{B} = \frac{g}{r^2} \partial_r. \qquad (5.230)$$

The two fields $\bar{A}^+$ and $\bar{A}^-$ differ by $\bar{A}^- - \bar{A}^+ = 2gd\phi$, so they are equivalent up to the gradient $d\psi(x)$ of the function $\psi(\phi) = 2g\phi$, which means they are equivalent up to a gauge transformation. There is no way to construct a monopole solution using a single field $\bar{A}$, the only way is to use two gauge fields that differ by a gauge transformation. The physics of magnetic monopoles thus depends on details about gauge invariance, group theory, and topology that won't be covered here.

In spacetime dimensions other than four, the apparent symmetry between the electric and magnetic currents $*\bar{J}_e$ and $*\bar{J}_m$ in (5.227) becomes complicated by the fact that $*\bar{J}_m$ and $*\bar{J}_e$ are differential forms of different degree. In d spacetime dimensions, the Maxwell tensor $*\bar{F} = 1/2\,\bar{\epsilon}(\mathbf{F})$ is a $(d-2)$ form, so $d*\bar{F}$ is a $(d-1)$ form. The electric source term $*\bar{J}_e = \bar{\epsilon}(\mathbf{J}_e)$ must also be a $(d-1)$ form, which means the electric current density $\mathbf{J}_e$ is a spacetime vector. The Faraday tensor $\bar{F}$ is a two form in any number of dimensions because it is the exterior derivative of a one form $\bar{A}$. This means the magnetic source term $*\bar{J}_m$ must be a three form in any number of spacetime dimensions, and so the source current $\mathbf{J}_m$ is a rank $\binom{d-3}{0}$ tensor, which is only a vector if $d=4$. Therefore, the structure of Maxwell theory implies that magnetic monopoles can only be point particles in four spacetime dimensions. In d spacetime dimensions, they are $(d-4)$-dimensional extended objects known as *branes*. For example, in a universe with ten spacetime dimensions, this gives six-dimensional objects called six-branes.

5.6 What about the gravitational field?

Electrostatics started out as a scalar field theory, but electricity and magnetism turned out to be different parts of a unified theory. As we have seen, this unified theory is a relativistic theory described by the Maxwell equations. The story for gravity is somewhat similar, though there are also important differences. In this section we sketch the similarities and the differences.

In electrostatics the $1/r^2$ force law translates into a differential equation for a scalar potential ϕ

$$\nabla^2 \phi_E = \frac{4\pi}{c}\rho(\vec{x}), \qquad (5.231)$$

where $\rho(\vec{x})$ is the electric charge density. As we have seen, in the full relativistic theory described by Maxwell's equations, ρ is identified with the time component of the electric current J^μ and ϕ_E is identified as the time component of the vector potential A_μ. Then Eq. (5.231) is identified as the time component of the Maxwell equation

$$\partial_\nu F^{\mu\nu} = \frac{4\pi}{c} J^\mu(x). \tag{5.232}$$

A more naive relativistic generalization of Eq. (5.231) would be obtained by simply replacing the Laplacian on the left-hand side by the Lorentz-invariant d'Alembertian (or wave operator) giving the equation

$$\left(\nabla^2 - \frac{\partial^2}{\partial\tau^2}\right)\phi_E = \frac{4\pi}{c}\rho(\vec{x}). \tag{5.233}$$

This is mathematically sensible but physically wrong for describing electromagnetism. A field ϕ satisfying an equation of this type is called a Klein–Gordon field. It transforms under Lorentz transformations as a spacetime scalar rather than as the time component of a spacetime vector. Such fields will be discussed further in Chapter 7 in the context of quantum theory.

In static nonrelativistic Newtonian gravity (let's call it *gravitostatics* in analogy with electrostatics) there is also an inverse square force law, and therefore one has an equation for a scalar gravitational potential ϕ_G, namely

$$\nabla^2\phi_G = 4\pi G\mu(\vec{x}), \tag{5.234}$$

where G is Newton's constant, and $\mu(\vec{x})$ represents mass density. This equation is strikingly similar to Eq. (5.231), but there are also important differences. For one thing, the charge density $\rho(\vec{x})$ can be either positive or negative, but the mass density $\mu(\vec{x})$ is never negative.

Another important fact in the gravitational case is that the strength of the source is proportional to its mass. It is a highly nontrivial fact that this mass, called the gravitational mass, is the same as the inertial mass that occurs in Newton's second law. This fact, called the *equivalence principle*, is one of the key ingredients that guided Einstein in his construction of a relativistic theory of gravity. It has been tested to very high precision, beginning with the classic Eötvös experiment and continuing to modern times. On occasion, experimental discrepancies have seemed to appear, which were interpreted as evidence for a new "fifth force" (carried by a Klein–Gordon field), but so far they have always gone away.

How are we going to make Eq. (5.234) relativistic? For matter at rest in relativity, the mass density is the energy density divided by c^2. Moreover we have seen that the energy density is the time–time component T^{00} of the stress tensor. Thus just as $\rho(\vec{x})$ generalizes to the vector $J^\mu(x)$ in the case of electromagnetism, it is natural to suppose that $\mu(\vec{x})$ should generalize to the tensor $T^{\mu\nu}(x)$ in the case of gravity. This reasoning suggests that the correct generalization of Eq. (5.234) should be of the form

$$G^{\mu\nu} = -8\pi G\, T^{\mu\nu}(x), \qquad\qquad (5.235)$$

where $G^{\mu\nu}$ is an expression, called the Einstein tensor, that involves two derivatives of a fundamental field. These are the Einstein field equations, which are the fundamental equations of *general relativity*. It is the gravitational analog of Eq. (5.232).

What is the fundamental field for gravity? In the case of electromagnetism the source is a vector and the fundamental field is a vector. In the case of gravity, we have just seen that the source is a symmetric tensor (the stress tensor), and so the fundamental field should also be a symmetric tensor. In a brilliant stroke of insight, Einstein realized that this symmetric tensor, called $g_{\mu\nu}(x)$, should be identified as a metric tensor that describes the geometry of spacetime. In special relativity there is no gravity, and spacetime geometry is described by the metric $\eta_{\mu\nu}$, which describes flat Minkowski spacetime. The proposal is that the presence of a nonzero stress tensor acts as a source that distorts the spacetime geometry to a curved spacetime described by a metric tensor that satisfies the Einstein field equations. All that remains to complete the story of general relativity is to give the formula for the Einstein tensor. However, it is rather complicated, and we don't want to do that here. Instead, we will settle for a couple of general remarks. (We will say a little more about general relativity in the first section of Chapter 10.)

The first general remark is that whereas the Maxwell equations are linear in the vector potential, the Einstein field equations are nonlinear in the metric tensor. This makes the Einstein field equations much more difficult to analyze, but it also leads to profound new possibilities for physical phenomena, some of which are discussed in Chapter 10. We can understand this difference between the two theories roughly as follows: the electromagnetic field is itself electrically neutral, and therefore it does not act as its own source. Gravitational disturbances, on the other hand, can carry energy and momentum, and thus they do act as gravitational sources. This self-interaction is built into the structure of the Einstein tensor and accounts for its nonlinearity.

The second general remark concerns the relationship between the Einstein field equations and the equation of gravitostatics in Eq. (5.234). In the case of electromagnetism, we could obtain Eq. (5.231) by simply restricting the Maxwell equations to the special case of a static source. In the case of gravity this doesn't work, because restricting the Einstein equations to the case of a static source given by a mass distribution doesn't eliminate the nonlinearity of the equations. To do that one also has to make a weak gravity approximation. This is achieved by writing the metric tensor in the form $g_{\mu\nu} = \eta_{\mu\nu} + h_{\mu\nu}$, and expanding the Einstein field equations to linear order in h, dropping all higher powers of h. This approximation assumes that the spacetime geometry can be described as a small perturbation of flat Minkowski spacetime. Then if one assumes that the only nonzero component of h is $h_{00} = -2\phi_G(\vec{x})$, and substitutes in the linearized Einstein field equations, one obtains Eq. (5.234).

Exercises

5.1 For the purposes of the following exercises, spherical coordinates (τ, r, θ, ϕ) for $\mathbf{M}^4$ are related to Minkowski coordinates by

$$(\tau, x, y, z) = (\tau, r \sin\theta \cos\phi, r \sin\theta \sin\phi, r \cos\theta). \qquad \text{(E5.1)}$$

Null coordinates (u, v, x, y), also known as light cone coordinates, are related to Minkowski coordinates by

$$(\tau, x, y, z) = \left(\frac{1}{2}(u + v), x, y, \frac{1}{2}(u - v)\right). \qquad \text{(E5.2)}$$

Null cylindrical coordinates (u, v, ρ, ϕ) are related to Minkowski coordinates by

$$(\tau, x, y, z) = \left(\frac{1}{2}(u + v), \rho \cos\phi, \rho \sin\phi, \frac{1}{2}(u - v)\right). \qquad \text{(E5.3)}$$

Compute the four-dimensional Lorentz-invariant interval $ds^2 = -d\tau^2 + dx^2 + dy^2 + dz^2$ and the volume element $\bar{\epsilon} = d\tau \wedge dx \wedge dy \wedge dz$ in each of the coordinate systems given above.

5.2 Expand the following geometrical objects in each of the coordinate systems given above:
(a) $df = \partial_\mu f \, dx^\mu$, where $f(x)$ is a scalar function on $\mathbf{M}^4$.
(b) $d\bar{A} = \partial_\mu A_\nu dx^\mu \wedge dx^\nu$, where $\bar{A} = A_\mu dx^\mu$ is a one form on $\mathbf{M}^4$.
(c) $\partial_\mu \partial^\mu f = \partial_\mu (g^{\mu\nu} \partial_\nu f)$, where the components $g^{\mu\nu}$ are the components of the inverse Minkowski metric tensor $\eta^{\mu\nu}$ in the relevant coordinate system.

5.3 Show that (5.27) is consistent with (5.21).

5.4 Compute the exterior derivative $d\bar{\omega}$ of a two form $\bar{\omega} = 1/2\omega_{\mu\nu} \, dx^\mu \wedge dx^\nu$ for the following manifolds: $\mathbf{E}^2$ with coordinates (x, y), $\mathbf{E}^3$ with coordinates (x, y, z), $\mathbf{M}^4$ with coordinates (τ, x, y, z).

5.5 Using Eq. (5.30), show that the Lie derivative $\mathcal{L}_\mathbf{v} \bar{\omega}$ of a p form $\bar{\omega}$ satisfies $d(\mathcal{L}_\mathbf{v} \bar{\omega}) = \mathcal{L}_\mathbf{v}(d\bar{\omega})$.

5.6 Let $\mathbf{g}$ be a symmetric rank $\binom{0}{2}$ tensor field. Take the Lie derivative of the scalar invariant $\mathbf{g}(\mathbf{u}, \mathbf{u}) = g_{\mu\nu} u^\mu u^\nu$ and use the Leibniz rule to arrive at (5.48).

5.7   Let $\mathbf{k}_i$ and $\mathbf{k}_j$ be Killing vectors of $\mathbf{g}$. Show that the Lie derivative $\mathbf{k}_k = [\mathbf{k}_i, \mathbf{k}_j]$ is also a Killing vector of $\mathbf{g}$.

5.8   Compute the Lie derivatives $[\mathbf{j}_i, \mathbf{k}_j]$ of the Killing vectors in (5.56) and (5.58).

5.9   Show that the Jacobian J of a transformation from Euclidean coordinates to some other coordinates satisfies $J = \sqrt{g}$, where g_{ij} are the components of the Euclidean metric in the new coordinates. Compute J for the transformation from Minkowski coordinates (τ, x, y, z) to each of the three coordinate systems given in the first exercise.

5.10   Given a Euclidean two form $\bar{\alpha} = 1/2\, \alpha_{ij}\, dx^i \wedge dx^j$ and vector $\vec{\alpha} = \alpha^i\, \partial_i$, with $\bar{\alpha} = \bar{\epsilon}(\vec{\alpha})$, show that

$$\vec{\alpha} = \frac{1}{2}\, \epsilon^{ijk} \alpha_{jk} \partial_i, \qquad \bar{\alpha} = \frac{1}{2}\, \epsilon_{ijk}\, \alpha^i\, dx^j \wedge dx^k. \qquad \text{(E5.4)}$$

5.11   The tensor components $\epsilon^{\mu\nu\sigma\lambda}$ are obtained from the components $\epsilon_{\mu\nu\sigma\lambda}$ of the volume form in $\mathbf{M}^4$ by raising the indices with the Minkowski inverse metric $\eta^{\mu\nu}$. Show that

$$\epsilon^{\mu\nu\sigma\lambda} \epsilon_{\alpha\beta\sigma\lambda} = -2 \left(\delta^\mu{}_\alpha\, \delta^\nu{}_\beta - \delta^\mu{}_\beta\, \delta^\nu{}_\alpha \right), \qquad \text{(E5.5)}$$

where $\delta^\mu{}_\alpha$ is the Kronecker delta symbol, with $\delta^\mu{}_\alpha = 1$ if $\mu = \alpha$ and $\delta^\mu{}_\alpha = 0$ otherwise.

5.12   Show that the particle number density $\mathbf{n}$, as defined in (5.116), is a spacetime vector in $\mathbf{M}^d$.

5.13   Apply Stokes's theorem to the particle number density $\mathbf{n}$ and charge density $\mathbf{J}$ in (5.119). Under what physical circumstances is it true that $\partial_\mu n^\mu = 0$? Under what physical circumstances is it true that $\partial_\mu J^\mu = 0$?

5.14   Put back all of the factors of c in (5.132) and (5.133). Show that in the nonrelativistic limit $|\vec{v}| \ll c$, $P \ll \rho c^2$, and the two equations reduce to (5.134) and (5.135).

5.15   What is the limit of (5.132) and (5.133) for $\beta = 1 - \epsilon$, with $\epsilon \ll 1$?

5.16   Consider a perfect fluid at a constant temperature T. Let the number density vector $\mathbf{n} = n(x)\mathbf{u}$, where $\mathbf{u}$ is the spacetime velocity of the fluid. The stress tensor of the fluid is given by (5.108), with pressure P and mass density ρ.

        (a) Use conservation of number density $\partial_\mu n^\mu$ and conservation of energy $\partial_\nu T^{0\nu} = 0$ to derive an equation relating the change in mass density $d\rho/d\lambda$ to the change in number density $dn/d\lambda$ for the fluid. Remember that $d/d\lambda = u^\alpha \partial_\alpha$.

(b) The first law of thermodynamics says that the change in energy of a fluid is related to the changes in volume V and entropy S through the relation

$$dE = d(\rho V) = -P dV + T dS. \tag{E5.6}$$

Let the number density in the fluid be $n = N/V$, and the entropy $S = sN$, where s is the amount of entropy per particle in the fluid. Find $ds/d\lambda$ as a function of $d\rho/d\lambda$ and $dn/d\lambda$. Use the previous result from energy conservation to compute the value of $ds/d\lambda$ for a perfect fluid.

5.17   An imperfect fluid that is viscous can have gradients in the velocity that create shear stress. Consider the $\binom{0}{2}$ Minkowski tensor $\nabla \mathbf{u}$ with components $\partial_\mu u_\nu$.

(a) Show that $\partial_\mu u_\nu$ can be written as the sum of its irreducible parts

$$\partial_\mu u_\nu = \omega_{\mu\nu} + \sigma_{\mu\nu} + \frac{1}{d} \eta_{\mu\nu} \theta, \tag{E5.7}$$

where $\theta = \partial_\mu u^\mu$, and

$$\omega_{\mu\nu} = \frac{1}{2} \left(\partial_\mu u_\nu - \partial_\nu u_\mu \right),$$

$$\sigma_{\mu\nu} = \frac{1}{2} \left(\partial_\mu u_\nu + \partial_\nu u_\mu \right) - \frac{1}{d} \eta_{\mu\nu} \theta. \tag{E5.8}$$

(b) Use the projection tensor $\mathbf{h}$ with components $h^{\mu\nu} = \eta^{\mu\nu} + u^\mu u^\nu$, to project out the components of $\omega_{\mu\nu}$ and $\sigma_{\mu\nu}$ in the hyperplane normal to the fluid velocity $\mathbf{u}$, to create new tensors $\hat{\omega}_{\mu\nu}$ and $\hat{\sigma}_{\mu\nu}$ such that $\hat{\omega}_{\mu\nu} u^\nu = \hat{\sigma}_{\mu\nu} u^\nu = 0$. The new tensors $\hat{\omega}_{\mu\nu}$ and $\hat{\sigma}_{\mu\nu}$ are called the rotation and shear, respectively, of the fluid.

(c) Show that the fluid velocity gradient $\partial_\mu u_\nu$ can be written in terms of the fields in the hyperplane normal to the fluid velocity $\mathbf{u}$ as

$$\partial_\mu u_\nu = \hat{\omega}_{\mu\nu} + \hat{\sigma}_{\mu\nu} + \frac{1}{D} \theta h_{\mu\nu} - u_\mu \left(u^\alpha \partial_\alpha u_\nu \right) \tag{E5.9}$$

(d) The stress tensor for a viscous fluid with shear stress, but no rotation, is then

$$\mathbf{T} = \rho \mathbf{u} \otimes \mathbf{u} + (P - \zeta\theta) \mathbf{h} - 2\eta\hat{\sigma}, \tag{E5.10}$$

where η is shear viscosity and ζ the bulk viscosity of the fluid in question. Substitute into $\partial_\nu T^{\mu\nu} = 0$ to derive the Navier–Stokes equation.

5.18   Compute the components of $\partial_\nu T^{\mu\nu}$ for the stress tensor for N pointlike particles (5.117) and show that $\partial_\nu T^{\mu\nu} = 0$ is only true if each particle in the sum propagates as a free particle, that is, with no acceleration.

5.19   The requirement that energy be a positive quantity for all observers is enforced for a continuous system through the *weak energy condition* $T_{\mu\nu} u^\mu u^\nu \geq 0$, where $\mathbf{u}$ is the spacetime velocity of some observer in spacetime. What does this condition reduce to in the rest frame of an observer? If this condition is satisfied in the rest frame, does it follow that it is satisfied in all other inertial frames? If this condition is satisfied in Minkowski coordinates for $\mathbf{M}^d$, does it follow that it is satisfied in all other coordinate systems for $\mathbf{M}^d$?

5.20 Compute the components of the Faraday and Maxwell tensors in the null coordinate system given in (E5.2) in terms of the Minkowski components of $\mathbf{E}$ and $\mathbf{B}$. Write out the vacuum Maxwell equations $d\bar{F} = 0$ and $d\,{}^*\bar{F} = 0$ in these coordinates.

5.21 Suppose an observer in frame S sees an electric field with magnitude E in the x direction and a magnetic field with magnitude B pointing in the y direction. What electric and magnetic fields are observed in frame $\tilde{S}$ traveling at velocity $\vec{\beta} = \beta \partial_z$ relative to frame S? Are there any conditions under which either the electric or magnetic field will vanish in frame $\tilde{S}$?

5.22 Show that the four components of Eq. (5.168) are equivalent to the four components of the Maxwell equations (5.152) with charge and current sources.

5.23 Verify Eq. (5.174).

5.24 Compute the values of the Lorentz-invariant quantities

$$I_1 = \frac{1}{2} F_{\mu\nu} F^{\mu\nu}, \qquad I_2 = \frac{1}{4} F_{\mu\nu} \,{}^*F^{\mu\nu}, \qquad \text{(E5.11)}$$

where ${}^*F^{\mu\nu} = \eta^{\mu\alpha} \eta^{\nu\beta} \,{}^*F_{\alpha\beta}$, in terms of $\vec{E}$ and $\vec{B}$. Show that I_1 is invariant under any coordinate transformation, not just under a Lorentz transformation. Is the same true for I_2?

5.25 Consider a point electric charge at rest in frame S, with an electric field given in spherical coordinates by $\vec{E} = (Q/r^2)\partial_r$.
   (a) Compute the Faraday tensor $\bar{F} = F_{0r}\,dt \wedge dr$ for this field.
   (b) Transform $\bar{F}$ from spherical coordinates to Minkowski coordinates according to (E5.1).
   (c) Compute the components of $\bar{F}$ for an observer in frame $\tilde{S}$ traveling at velocity $\vec{\beta} = \beta \partial_x$ relative to frame S.
   (d) Transform the previous result to spherical coordinates $(\tilde{t}, \tilde{r}, \tilde{\theta}, \tilde{\phi})$ centered at the origin of frame $\tilde{S}$.
   (e) In the rest frame of a point charge, the electric field is isotropic, that is, the same in all directions around the charge. Is the electric field of a point charge isotropic according to an observer who sees the charge moving?

5.26 Using the Lorentz gauge $\partial_\mu A^\mu = 0$, compute the Maxwell equations in Minkowski coordinates and show that the equation $d\,{}^*\bar{F} = {}^*\bar{J}$ reduces to the equation

$$\partial_\alpha \partial^\alpha A_\mu = -\frac{4\pi}{c} J_\mu, \tag{E5.12}$$

where $J_\mu = \eta_{\mu\nu} J^\mu$.

5.27 Find the electric and magnetic fields corresponding to the plane wave solution (5.200). Compute the components of the stress tensor $T^{\mu\nu}$ in Minkowski coordinates (τ, x, y, z) in terms of the components of $\mathbf{k}$ and $\bar{\varepsilon}$.

5.28 Consider a constant magnetic field $\vec{B} = B^z \partial_z$ in frame S with Minkowski coordinates (τ, x, y, z). Solve the equations of motion to get the particle motion in this field. Compute the electric and magnetic fields and equations of motion in the $\tilde{S}$ frame traveling at velocity $\vec{\beta} = \beta^y \partial_y$ relative to frame S. Show that the Lorentz transformation of the solution to the equations of motion obtained in frame S solves the equations of motion as computed in frame $\tilde{S}$.

6

Causality and relativity

Hands-on exercise

Make a two-dimensional coordinate grid on a piece of heavyweight paper that is stiff enough to bend into a circle without folding over. Make a time axis labeled τ and a space axis labeled x, with the origin of the coordinate system in the center of the sheet. Pick a direction on the τ axis to be the direction of increasing τ and label it.

Roll the paper into a cylinder with the x axis running lengthwise and lightly tape it shut. The circular axis should be the τ axis. Using a marking pen, draw two null lines starting the origin, towards the direction of increasing τ, which should be 45° lines in the (τ, x) grid. Draw a timelike line in one direction, starting at the origin. Keep drawing the lines around the cylinder until the end of the paper in both directions. Untape the paper and unroll it. Is there any area on this diagram that cannot be reached from the origin by a future-directed timelike or null world line? Is there any difference between the past and the future light cone of an event in such a spacetime? Are there any two events in this spacetime that cannot be connected by a null or timelike path?

6.1 What is time?

Flat spacetime appears on its face to be very simple. We have space in the form of D coordinates $(x^1, \ldots, x^D)$ and time in the form of the τ coordinate, and we have Lorentz transformations between the two.

In physics before Einstein, time was something external, universal and absolute. The Universe evolves in time, objects move in space as a function of time. But the nature of time is unquestioned and unquestionable within Newtonian physics. The simplicity of time in Newtonian physics comes from the deliberate ignorance about time built into the assumption in Newtonian mechanics that time is universal and absolute.

The nature of time appears to be very simple in special relativity – after all, it's just the τ coordinate in spacetime, right? But this is not the same simple universal time we learn to use in Newtonian physics. As it turns out, the subtle complexity of the notion of time in special relativity is exactly what is necessary to ensure that cause precedes effect in physics. In this section we will look at time in flat spacetime, and in the next section we will look at the principle of causality, and how it is protected by the relativistic notion of time.

Back to Pythagoras

Back in Chapter 1 we looked at the Pythagorean rule for the sums of squares of the sides of a right triangle

$$(\Delta X)^2 + (\Delta Y)^2 = (\Delta L)^2 \tag{6.1}$$

as a rule for measuring the geometry of space. In modern times this ancient rule has evolved into the differential line element for Euclidean space in D dimensions, with coordinates $(x^1, \ldots, x^D)$

$$dl^2 = \delta_{ij}\, dx^i\, dx^j = |d\vec{x}|^2, \tag{6.2}$$

where the coefficients δ_{ij} are the components of the metric tensor for Euclidean space in Euclidean coordinates.

Euclidean space, being flat and infinite, looks the same in every direction. In reflection of this fact, the Euclidean metric is left unchanged by rotation of Euclidean coordinates by some constant angle θ, according to the linear coordinate transformation

$$\tilde{x}^n = R^{\tilde{n}}_m\, x^m, \qquad R^T I R = I, \tag{6.3}$$

where I is the $D \times D$ identity matrix. The matrices R are rotation matrices. In Euclidean space $\mathbf{E}^D$, they belong to the rotation group $SO(D)$. The simplest useful case to understand is $D = 2$, where

$$\begin{pmatrix} \tilde{x} \\ \tilde{y} \end{pmatrix} = \begin{pmatrix} \cos\theta & \sin\theta \\ -\sin\theta & \cos\theta \end{pmatrix} \begin{pmatrix} x \\ y \end{pmatrix}. \tag{6.4}$$

In Euclidean space, it makes no difference what we call the x direction and what we call the y direction. The x and y directions have no physical meaning. Nothing that is physically measurable is changed when we rotate the coordinate axes to make the Euclidean axes face in some other direction.

As has been discussed at length previously in this book, the analog of a Pythagorean rule for spacetime with $d = D + 1$ dimensions with coordinates

$(\tau, x^1, \ldots, x^D)$ is

$$ds^2 = -d\tau^2 + \delta_{ij} \, dx^i dx^j = \eta_{\mu\nu} \, dx^\mu dx^\nu, \tag{6.5}$$

where $\tau = ct$ is the time coordinate t expressed in units of length using the velocity c. The difference between this spacetime line element ds^2 and the line element in (6.2) for Euclidean space is the minus sign in front of the time coordinate. Because of this minus sign, the spacetime line element is invariant under a different kind of rotation. Minkowski coordinates in d spacetime dimensions transform as

$$\tilde{x}^\mu = L^{\tilde{\mu}}_\nu x^\nu, \quad L^T \eta L = \eta, \tag{6.6}$$

where η is the Minkowski metric, with nonzero components $\eta_{00} = -1$, $\eta_{ij} = \delta_{ij}$. This spacetime rotation mixes space and time. The new coordinate frame is related to the old one not by a rotation but by a state of relative motion.

In two spacetime dimensions with Minkowski coordinates (τ, x) this transformation can be written in a manner similar to a Euclidean rotation in two space dimensions but with hyperbolic functions

$$\begin{pmatrix} \tilde{\tau} \\ \tilde{x} \end{pmatrix} = \begin{pmatrix} \cosh \xi & -\sinh \xi \\ -\sinh \xi & \cosh \xi \end{pmatrix} \begin{pmatrix} \tau \\ x \end{pmatrix}. \tag{6.7}$$

This hyperbolic transformation mixes space and time axes, but does not allow them to be interchanged. In the limit $\xi \to \pm\infty$, the transformation degenerates so that the new $\tilde{\tau}$ axis and the new $\tilde{x}$ axis become parallel. As was shown in Chapter 1, the relative velocity between the frames S and $\tilde{S}$ is given by $\beta = v/c = \tanh \xi$, so the degenerate limit $\xi \to \pm\infty$ corresponds to $\beta = \pm1$, or $v = c$. There is no value of β for which the time axis becomes space and the space axis becomes time.

The maximum possible speed in this system is the speed of light. This maximum speed is physically meaningful and so is the distinction that it enforces between time and space. With rotational symmetry in Euclidean space, there is no physical restriction on the relative angle between two coordinate frames, so it can take any value (though it is only defined modulo 2π). The relative "angle" between two frames in spacetime has a physical maximum that represents the speed of light, and this is physically meaningful. In fact as we shall see, we can say that this maximum angle for a spacetime rotation is what makes and keeps all physics physically sensible, by allowing and protecting the causal propagation of information in physical systems in spacetime.

Even though time and space are mixed by a Lorentz boost, the minus sign in the spacetime metric gives us a notion of time that is the same for all observers. The infinitesimal interval ds^2 between any two events in spacetime falls into one

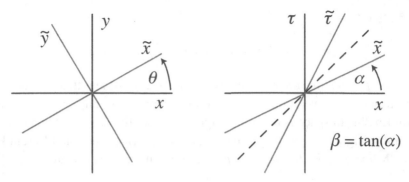

Fig. 6.1. A rotation of the coordinate axes in two space dimensions by some angle θ is shown in the figure to the left. The figure on the right shows a Lorentz boost of the time and space axes n two spacetime dimensions. Unlike a rotation in space, a Lorentz boost has a degenerate limit where the time and space axes meet each other and become parallel. This limit is reached when the boost velocity β is the speed of light, so there is a physical limit on a Lorentz boost that does not have a counterpart for a rotation in space.

of three classes:

$$ds^2 < 0 \quad \text{timelike, with } d\tau^2 > |d\vec{x}|^2$$
$$ds^2 > 0 \quad \text{spacelike, with } d\tau^2 < |d\vec{x}|^2$$
$$ds^2 = 0 \quad \text{null, with } d\tau^2 = |d\vec{x}|^2. \tag{6.8}$$

These three classes of intervals are not only Lorentz-invariant, but also coordinate-invariant. A general coordinate transformation $\tilde{x}(x)$ will change the coordinate components $\eta_{\mu\nu}$ of the Minkowski metric tensor η, but won't change the sign or value of $ds^2 = \eta_{\mu\nu} dx^\mu dx^\nu$, because

$$ds^2 = \eta_{\mu\nu} dx^\mu dx^\nu = \left(\frac{\partial \tilde{x}^\sigma}{\partial x^\mu} \frac{\partial \tilde{x}^\rho}{\partial x^\nu} \right) \tilde{\eta}_{\sigma\rho} \left(\frac{\partial x^\mu}{\partial \tilde{x}^\alpha} \frac{\partial x^\nu}{\partial \tilde{x}^\beta} \right) d\tilde{x}^\alpha d\tilde{x}^\beta$$
$$= \left(\frac{\partial \tilde{x}^\sigma}{\partial x^\mu} \frac{\partial x^\mu}{\partial \tilde{x}^\alpha} \right) \left(\frac{\partial \tilde{x}^\rho}{\partial x^\nu} \frac{\partial x^\nu}{\partial \tilde{x}^\beta} \right) \tilde{\eta}_{\sigma\rho} \, d\tilde{x}^\alpha d\tilde{x}^\beta$$
$$= \delta^\sigma_\alpha \delta^\rho_\beta \tilde{\eta}_{\sigma\rho} d\tilde{x}^\alpha d\tilde{x}^\beta = \tilde{\eta}_{\alpha\beta} d\tilde{x}^\alpha d\tilde{x}^\beta = \tilde{\eta}_{\mu\nu} d\tilde{x}^\mu d\tilde{x}^\nu. \tag{6.9}$$

The invariant interval ds^2 is invariant under all coordinate transformations, not just Lorentz transformations. Therefore a timelike interval for one observer is a timelike interval for all observers in spacetime. So even though we don't have one definition of time that works for all observers in spacetime, or even for all inertial observers in spacetime, a timelike interval between two events is timelike in all frames.

The importance of a timelike interval is that it protects our ability to order events uniquely in time. If two events E_1 and E_2 are separated by a timelike interval, then we can say for certain that one event happens before the other, and this will be true in all inertial frames. Let $\Delta\tau = \tau_2 - \tau_1$ and $\Delta x = x_2 - x_1$ in frame S. The time interval $\Delta\tilde{\tau} = \tilde{\tau}_2 - \tilde{\tau}_1$ in frame $\tilde{S}$ is then

$$\Delta\tilde{\tau} = \cosh\xi\,\Delta\tau - \sinh\xi\,\Delta x. \tag{6.10}$$

If the two events are separated by a timelike interval with $\Delta s^2 < 0$, then $\Delta\tau^2 > \Delta x^2$, therefore $|\Delta\tau| > |\Delta x|$. The value of $\Delta\tilde{\tau}$ passes through zero at $\Delta\tau = \tanh\xi\,\Delta x \leq \Delta x$. If the interval between two events has $\Delta\tau \leq \Delta x$, then that interval is spacelike or null, but not timelike. Therefore if two events E_1 and E_2 are separated by a timelike interval in frame S, if event E_2 occurs at a later time than event E_1 in frame S, then E_2 occurs at a later time than E_1 in frame $\tilde{S}$. This is true for all observers, not just for inertial observers.

If instead the two events are separated by a spacelike interval with $\Delta s^2 > 0$, then $\Delta x^2 > \Delta\tau^2$, and there exists a frame in which the two events occur at the same time. There is no unique time ordering possible between two events separated by a spacelike interval. A difference in time is not a reliable quantity for describing the relationship between two such events.

These two examples show that the geometry of spacetime determines the circumstances under which it is meaningful to say that one event occurs before or after another. In spacetime physics, the ordering of two events in time is a feature of the geometry that can't be universally relied upon. The ordering of two events in spacetime is provisional, in that it only makes sense providing that the interval between the events in question is timelike. Flat spacetime is not just space with time added. The phenomenon of time is not separate from geometry but something that arises from it.

Time along a world line

If a pointlike object is moving through spacetime, then the events along the world line of the object have a unique ordering in time. A one-dimensional curve in spacetime can be described in terms of a parameter λ by $\mathbf{x}(\lambda) = (\tau(\lambda), x^1(\lambda), \ldots, x^D(\lambda))$. The tangent vector to this curve is

$$\mathbf{u} = \frac{d\mathbf{x}}{d\lambda} = \left(\frac{d\tau}{d\lambda}, \frac{dx^1}{d\lambda}, \ldots, \frac{dx^D}{d\lambda}\right). \tag{6.11}$$

If nearby events along this curve are separated by a timelike interval, so that $ds^2 < 0$, then we can define λ by $ds^2 = -d\tau^2 + |d\vec{x}|^2 = -d\lambda^2$, and the parameter λ is the proper time along the curve. This is the time measured in the rest frame of

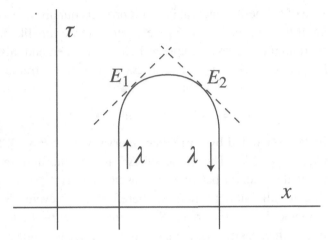

Fig. 6.2. The figure above shows the world line of an object that turns around in time. The object sits at rest with proper time λ increasing towards the future. The object then speeds up and reaches the speed of light at event E_1. At event E_2 the object is again traveling at the speed of light again, but the proper time λ is increasing towards the past. Between events E_1 and E_2, the tangent vector to the curve is spacelike, and $ds^2 = -d\lambda^2 > 0$, so there is no proper time defined along the curve between events E_1 and E_2.

the object whose world line we are following. With this choice for λ, then we see that $\mathbf{u}^2 = \eta_{\mu\nu}u^\mu u^\nu = -1$, and the tangent vector to the world line is a timelike vector.

The direction in coordinate time τ of this spacetime curve is given by $u^0 = d\tau/d\lambda$. If $d\tau/d\lambda > 0$, then the coordinate time and the proper time are both increasing along the curve, and the curve can be called future-directed. If $d\tau/d\lambda < 0$, then the curve is past-directed. People who might want to try to travel in time should now ask: under what conditions can we have a curve that starts out future-directed, and ends up heading towards the past?

If a curve were to change direction in time, heading first towards the future and then back towards the past, the derivative $d\tau/d\lambda$ would have to pass through zero at some $\lambda = \lambda_0$. If $d\tau/d\lambda = u^0 = 0$, then $\mathbf{u}^2 = |\vec{u}|^2 > 0$, so the time component of $\mathbf{u}$ can't change sign without making $\mathbf{u}$ into a spacelike vector. Therefore a path in spacetime with a tangent vector $\mathbf{u}$ that is everywhere timelike can never change its direction in time.

The spacetime momentum $\mathbf{p}$ of an object is related to the tangent vector to its world line by $\mathbf{p} = m\mathbf{u}$, hence $\mathbf{p}^2 = -m^2$. Spacetime geometry thus dictates that a particle with real mass m cannot change direction in time. If $u^0 > 0$ and $m > 0$, then the particle energy $E = p^0 = mu^0 > 0$. Therefore a positive energy particle propagates forward in time. Notice that a negative energy particle propagating

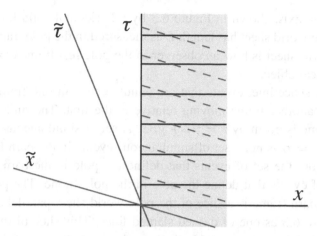

Fig. 6.3. The pole world sheet is shown by the shaded area above. The solid horizontal lines are the pole moving forward in time τ as measured in the frame S
in which the pole is at rest. The dashed slanted lines are slices of constant $\tilde{\tau}$ that
represent the pole moving forward in time as seen in the barn rest frame $\tilde{S}$. The
volume of space occupied by the pole is defined by the solid lines according to an
observer in the pole frame, and the dashed lines according to an observer in the
barn frame.

forward in time, with $u^0 > 0$, is equivalent to a positive energy particle propagating backwards in time, with $u^0 < 0$. This interesting fact becomes important when
quantum mechanics comes into the picture in Chapter 7.

Time as a spacelike hypersurface

An extended object occupies space, and the space occupied by an extended object
can be used to define time in spacetime.

Recall the example of the pole and the barn discussed in Chapter 3. In this
discussion we're going to call the pole rest frame coordinates (τ, x) and call the
barn rest frame coordinates $(\tilde{\tau}, \tilde{x})$, as shown in Figure 6.3. In the barn frame, the
barn has length L, while the pole, moving at speed β towards the barn, has length
L/γ. In the pole rest frame, the pole has length L, while the barn, moving at speed
β towards the pole, has length L/γ. The symmetric discrepancy in the lengths of
the pole and the barn comes about because the pole and barn are defined differently
as objects in each other's rest frame than they are in their own.

Consider the pole in the pole rest frame. Let's model the pole as a one-
dimensional object propagating in spacetime, with a two-dimensional world sheet,
shown by the shaded area in Figure 6.3. In the pole rest frame, the pole as a physical object occupies the space between $x = 0$ and $x = L$ continuously in coordinate
time τ. The pole at each moment in time $\tau = \tau_i$ is a slice of the pole world sheet

parallel to the x axis, shown in Figure 6.3 by a horizontal solid line. Each such slice of the pole world sheet has length L as measured in the pole frame. This slice of the pole world sheet is how an observer in the pole rest frame experiences the pole as a physical object.

However, in spacetime, events that are simultaneous in one frame are not simultaneous in another frame moving relative to the first. The time as measured in the barn frame is given by $\tilde{\tau} = \gamma\tau + \gamma\beta x$. A set of simultaneous events in the pole frame at $\tau = \tau_i$ is not a set of simultaneous events in the barn frame at time $\tilde{\tau} = \tilde{\tau}_i$. Therefore the set of events that define the pole in the barn frame is not the same set of events that define the pole in the pole frame. The pole at $\tilde{\tau} = \tilde{\tau}_i$ as seen in the barn frame is a slice of the pole world sheet parallel to the $\tilde{x}$ axis, shown in Figure 6.3 as one of dashed slanted lines. This slice of the pole world sheet has length L/γ as measured in the barn frame. This slice of the pole world sheet is how an observer in the barn rest frame experiences the pole as a physical object.

If we define a physical object's extent in space by a set of events that all occur at the same time, then we can define an extended object's motion in time as a succession of spacelike surfaces of simultaneity. This view of the motion in time of an extended object gives us still another definition of time, in this case as a particular way to slice spacetime into spacelike surfaces. Such a slicing is called a *foliation* of spacetime.

In special relativity, spacetime is flat, and in a flat spacetime, the topology is a direct product of time $\otimes$ space, or $\mathbf{R} \otimes \Sigma$, where Σ represents D-dimensional space, and $\mathbf{R}$ is the line described by some time coordinate in spacetime. This topology guarantees that we can slice spacetime into spacelike surfaces of constant time.

Consider a D-dimensional submanifold S of $\mathbf{M}^d$ defined by some function $\Sigma(x)$ such that $\Sigma = $ constant. If the tangent space of S is spanned by D spacelike vectors $(\mathbf{v}_1, \ldots, \mathbf{v}_D)$, then S is a spacelike submanifold, or a spacelike hypersurface, of $\mathbf{M}^d$.

The normal vector $\mathbf{n}$ to the surface S is by definition orthogonal to all D spacelike vectors that span the tangent space. There are only D linearly independent spacelike vectors in $\mathbf{M}^d$. A spacetime vector that is orthogonal to D linearly independent spacelike vectors must therefore be timelike. Using this fact, we can define a spacelike hypersurface S of $\mathbf{M}^d$ as a D-dimensional submanifold of $\mathbf{M}^d$ with a normal vector field $\mathbf{n}$ that is timelike everywhere on S.

The normal vector to the spacelike hypersurface defined by $\Sigma = $ constant is a timelike vector associated with the one form $\bar{n} = \pm d\Sigma = \pm\partial_\mu\Sigma\,dx^\mu$, where there is a choice of two signs because S has two sides in spacetime – one pointing to the future, and one to the past. The vector $\mathbf{n}$ has components $n^\mu = \pm\eta^{\mu\nu}\partial_\nu\Sigma$. Because $\eta^{00} = -1$, the minus sign gives the future-directed vector.

If spacetime is foliated by spacelike hypersurfaces, then each surface S_i in the foliation has a normal vector $\mathbf{n}_i$ that is a timelike vector field in spacetime. If each $\mathbf{n}_i$ is also normal to all of the other hypersurfaces in the foliation, then all of the $\mathbf{n}_i$ are parallel, and can be considered as one vector field $\mathbf{n}$. In that case, $\mathbf{n}$ is a *hypersurface orthogonal* timelike vector field, and the integral curves of $\mathbf{n}$ make a smooth congruence of curves representing the propagation of space in time. In this case, the function Σ is called a *time function*.

The simplest example of a time function is $\Sigma = \tau$ in two spacetime dimensions. The future-pointing normal vector to this surface is $\mathbf{n} = \partial_\tau$. The spacelike hypersurfaces $\tau = c_i = \text{constant}$ are straight lines parallel to the x axis. Minkowski spacetime in two dimensions can be thought of as an infinite stack of straight lines with $\Sigma = \tau_i(x) = c_i$. This stack in its entirety is a foliation of $\mathbf{M}^2$.

In the pole in the barn example, the spacelike hypersurfaces $\tau = \text{constant}$ determine the time in the pole frame. Each spacelike hypersurface in the foliation contains the pole at that value of time. The normal vector pointing towards the future of the pole is $\mathbf{n} = \partial_\tau$. This vector field is tangent to a smooth congruence of curves parallel to the τ axis shown as dashed lines in Figure 6.3.

An alternative time function is given by $\Sigma = \tilde{\tau} = \gamma\tau + \gamma\beta x$. This is the time function for the barn frame. The surfaces of constant Σ are lines $\gamma\tau_i(x) = -\gamma\beta x + c_i$ in the (τ, x) coordinate frame. These are the spacelike hypersurfaces that contain the spatial volume of the pole according to an observer at rest relative to the barn. The future-directed normal vector is $\mathbf{n} = \gamma\partial_\tau - \gamma\beta\partial_x = \partial_{\tilde{\tau}}$. This vector field is tangent to a smooth congruence of curves parallel to the $\tilde{\tau}$ axis in Figure 6.3.

Time for a pointlike object moving through Minkowski spacetime is measured along the world line of the object. For an extended object, time is measured along a set of curves whose tangent field is normal to the spacelike hypersurface containing the object. This notion of time is important for understanding the initial value problem in physics, which shall be discussed below.

6.2 Causality and spacetime

The term causality refers to the principle that cause should precede effect. For example, a golf ball should not sail across the green until after the golfer has swung the golf club and hit the ball. Causality is a profoundly important principle of physics, because physics is about examining the physical causes of the physical effects we observe in the world around us. But what is it that protects this principle from being violated in physics?

Einstein trusted the Maxwell equations more than Newtonian mechanics in his quest to reconcile the two most important ideas in classical physics in his time.

When Einstein realized that the Maxwell equations forbade a rest frame for a traveling light wave, he seized upon this fact and used it as a basis for his theory of relativity. Einstein made the right choice when he focused his attention on that particular aspect of Maxwell theory. It is the lack of a rest frame for light that is the basis for the causal structure of flat spacetime that enforces and protects the principle of cause and effect in physics.

Time, space and light

Suppose a person at rest in frame S is bored with going forward in time and wants to turn around and go backwards for a change. A possible world line for this person is shown in Figure 6.2. The time traveler accelerates, achieves the speed of light at E_1, then travels faster than the speed of light until event E_2, where she/he slows down to a rest again, only this time going backwards in time rather than forwards.

What is wrong with this scenario? It was shown above that in order to turn around in time, a curve in spacetime has to become spacelike at the turning point. But what's wrong with an observer's world line becoming spacelike? What's wrong is that the geometry of spacetime may allow this in a theoretical sense, but not in any real sense. The speed of light is a physical barrier that prevents the world line in Figure 6.2 from being realizable.

In Chapter 5 we showed that the propagation vector $\mathbf{k}$ of a light wave is a null vector with $\mathbf{k}^2 = 0$. When we talk about the propagation of light along a path, what we mean by path is the world line to which the propagation vector $\mathbf{k}$ is tangent. Since light cannot be accelerated by any force, these world lines are always straight lines in flat spacetime. This is also true of the path of a massless particle. Since the propagation vectors of light waves and momentum vectors of massless particles are tangent to the same set of paths in spacetime, in what follows they are to be considered equivalent.

Consider the total spacetime momentum $\mathbf{p} = (\gamma m, \gamma m \vec{\beta})$ of an object with mass m. The energy $E = \gamma m$ becomes infinite if $\beta \to 1$ while the mass remains finite. It takes an infinite amount of energy to accelerate a massive object to the speed of light. But this is what we should expect, because objects that normally propagate at the speed of light are massless. An object with zero mass has a finite energy even though it travels at the speed of light. An object with finite mass can never reach the speed of light with a finite energy.

Objects that propagate at the speed of light are not only massless, they are also timeless and spaceless. The invariant interval $ds^2 = 0$ along the world line of a massless object or a flash of light. There is zero proper time and zero proper distance along a null world line.

The speed of light is the barrier between time and space in spacetime, and it is a barrier that no massive object can pass through. Massless objects cannot go through the barrier either, they are constrained to propagate on it. This means that null surfaces – surfaces made of the world lines of massless particles – make up a barrier that timelike world lines cannot pass through.

In the hands-on exercise, you made a universe where the time dimension is circular. In this kind of spacetime, a person traveling towards the future ends up repeating the past over and over again. This revisiting of the past does not happen because the person manages to go faster than the speed of light and turns around in time. Even in this spacetime, it's still not possible to travel faster than the speed of light, though there are other serious problems.

Null hypersurfaces in spacetime

A null line in spacetime represents a barrier created by spacetime geometry that prevents an observer from traveling at a speed faster than the speed of light. A null hypersurface $\mathcal{N}$ is a submanifold of d-dimensional spacetime with a normal vector $\mathbf{n}$ that is null everywhere on $\mathcal{N}$. One example of a null hypersurface was shown in Chapter 5. The wave fronts of an electromagnetic wave with potential $\bar{A} = A_0 \bar{\varepsilon} \exp i S(x)$ are surfaces of constant phase S whose normal vector $\mathbf{k}$ satisfies the null condition $\mathbf{k}^2 = 0$. The wave fronts of an electromagnetic wave are therefore null hypersurfaces in spacetime.

One point to remember about a null hypersurface $\mathcal{N}$ is that the normal vector to $\mathcal{N}$ also belongs to the tangent space of $\mathcal{N}$, by virtue of the fact that $\mathbf{n}^2 = 0$. So $\mathbf{n}$ is orthogonal to itself. A null hypersurface is generated by null lines whose tangent vector field is also the normal vector field to the surface made by the null lines. This is an inevitable consequence of dealing with a metric product that is not positive-definite, and hence allows vectors that have a vanishing metric product with themselves. A metric for spacetime cannot be positive-definite, or it would be a metric for space, and there would be no time.

Because the null tangent vector to a null hypersurface is also the normal vector to the surface, a null hypersurface is technically a $(d - 2)$-dimensional surface. The length of a null hypersurface in the null direction is effectively zero, because there is no proper time or proper distance along a null direction. So instead of having a volume $V \sim L^{d-1}$, a null hypersurface has an area $A \sim L^{d-2}$.

The most important null hypersurface in physics is the light cone $\mathcal{L}$ of an event E_i. If event E_i happens at $\mathbf{x} = \mathbf{x}_i$, then the light cone of event E_i consists of all events in spacetime for which

$$(\mathbf{x} - \mathbf{x}_i)^2 = -(\tau - \tau_i) + |\vec{x} - \vec{x}_i|^2 = 0. \tag{6.12}$$

The future light cone $\mathcal{L}^+$ of E_i consists of events for which $\tau > \tau_i$. The past light cone $\mathcal{L}^-$ of E_i is made up of events with $\tau < \tau_i$.

The light cone as a whole is generated by null lines that pass through the event E_i in all directions. Events on $\mathcal{L}^+$ represent events that can be reached from E_i by traveling at the speed of light. Events on $\mathcal{L}^-$ represent events from which one can reach E_i by traveling at the speed of light.

The causal regions of an event

The *causal future* of event E_i, written $J^+(E_i)$, consists of all events in spacetime that can be reached from E_i by a future-directed null or timelike world line. These are all of the events in spacetime that can be influenced by something that happens at event E_i. This set of events includes the future light cone of E_i plus all of the spacetime events that are inside the future light cone, with a timelike separation from E_i. The future light cone $\mathcal{L}^+(E_i)$ is the boundary of the causal future of event E_i. The causal future of an event is shown in Figure 6.4.

Events that are outside of the causal future of E_i cannot be influenced by anything that happens at E_i. These events have either a spacelike separation from E_i or belong to the *causal past* of event E_i, written $J^-(E_i)$. The causal past of E_i consists of all the events in spacetime that can influence what happens at event E_i, or, in other words, all events in spacetime that reach E_i by a future-directed null or timelike world line. An event that can reach E_i by a future-directed null or timelike path can be reached from E_i by a past-directed null or timelike path. The past light cone $\mathcal{L}^-(E_i)$ is the boundary of the causal past of event E_i.

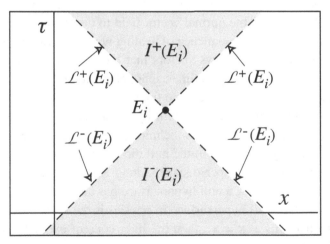

Fig. 6.4. The causal future of event E_i consists of the chronological future $I^+(E_i)$ plus the future light cone $\mathcal{L}^+(E_i)$ that serves as its boundary. The causal past of event E_i consists of the chronological past $I^-(E_i)$ plus the past light cone $\mathcal{L}^-(E_i)$ that serves as its boundary.

The *chronological future* of E_i, written $I^+(E_i)$, consists of the timelike interior of the causal future, minus the null boundary. There is no proper time along a null line, therefore the set of all events that occur at a later proper time than E_i, as measured by an observer whose world line intersects E_i, does not include events that are on the future light cone of E_i. The chronological future of E_i is the set of all events perceived to happen to the future of E_i according to observers who were at event E_i. There is also a *chronological past* of event E_i, written $I^-(E_i)$. The chronological past of E_i consists of the causal past of E_i, minus the past light cone that serves as its boundary.

When Einstein developed special relativity, he didn't set out to find the mathematics underlying the principle of causality in nature. But that's what he accomplished in the end with his two simple postulates. Minkowski geometry defines and protects the principle of cause and effect in physics, by limiting the causal influence of any one event in spacetime to the causal future of the event. The speed of light is the maximum speed for all observers in all frames, and so information cannot propagate outside the barrier made by light. An event cannot influence other events that cannot be reached by traveling at or less than the speed of light. The causal barrier made by light that enforces this principle is the future light cone, the null hypersurface that is the boundary of the causal future of any event.

The causal regions of an object

We've just learned about the causal regions of a single event in spacetime. The volume of a physical object propagating in spacetime consists of an infinite number of single events in spacetime. How do we define the causal future of an extended object in spacetime? Let's go back to our previous example of the pole and the barn and look at the causal future of the pole.

In the pole rest frame, the pole at time $\tau = 0$ consists of the events along the spacelike hypersurface $\tau = 0$ between $x = 0$ and $x = L$. Let's label this set of spacetime events $\mathcal{P}$, short for the points along the pole. The causal future of every event in $\mathcal{P}$ overlaps with the causal future of every other event in $\mathcal{P}$ in the amount of time τ that it takes for light to get from one end of the pole to another. The causal future of the pole at time $\tau = 0$ is therefore the union of all the causal futures of all the events between $(0, 0)$ and $(0, L)$ in the pole frame.

Events in spacetime with a spacelike separation from either end of the pole at $\tau = 0$ are events that can only be reached from an event in the set $\mathcal{P}$ by "breaking the light barrier." Therefore the boundary of the causal future of the pole is generated by null lines leaving the physical boundary of the pole at $\tau = 0$, heading away from the pole. This is shown in Figure 6.5.

We can define the causal past of the spacelike set of events $\mathcal{P}$ in the same way as we defined the causal future, but with time reversed. The chronological future

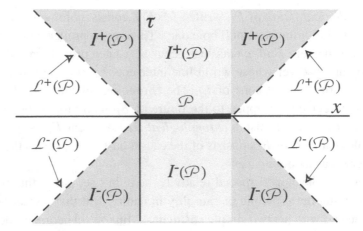

Fig. 6.5. The set of events $\mathcal{P}$ represents the pole at time $\tau = 0$. Any event in the causal future of $\mathcal{P}$ can be reached by a future-directed timelike or null line from $\mathcal{P}$. This includes the chronological future $I^+(\mathcal{P})$ plus the null boundary $\mathcal{L}^+(\mathcal{P})$. Any event in the causal past of $\mathcal{P}$ can be reached from $\mathcal{P}$ by a past-directed timelike or null line. This includes the chronological past $I^-(\mathcal{P})$ plus the null boundary $\mathcal{L}^-(\mathcal{P})$. Notice that the entire pole world sheet is included in the causal past and future of the pole at $\tau = 0$. An object cannot lose causal contact with itself!

and chronological past of $\mathcal{P}$ are defined as they are for an event, by subtracting the null boundary from the causal future and causal past, leaving only the events that can be reached from $\mathcal{P}$ by traveling along a timelike world line at speeds less than the speed of light. These regions of spacetime are shown in Figure 6.4.

What happens to these regions in the barn frame, where the pole is moving at velocity β? A null hypersurface of spacetime is invariant under any change of coordinates, not just those that constitute a Lorentz transformation. The set of events on the boundary of the causal future of the pole in one frame are also on the boundary of the causal future of the pole in any other frame.

However, as noted in the previous section, we have to be careful about how we define the pole in the barn frame. The pole defined as an object in the barn frame is not the Lorentz transformation of the Pole defined as an object in its own rest frame. The pole in the barn frame is the spacelike slice $\tilde{\tau} = 0$ of the pole world sheet. The difference between the causal future of the Lorentz transformation of the slice $\tau = 0$ and the causal future of the slice $\tilde{\tau} = 0$ will be explored as an exercise.

The initial value problem in spacetime

In classical physics, the time evolution of a system is supposed to be unique. In practice, differential equations have constants of integration, and these constants of integration encode the lack of knowledge the equations of motion have about the initial condition of the system being evolved. In order for physics to make sense, a given set of initial data should lead to a unique solution of the equations of motion.

However, how do we know whether our set S of initial data is complete? How can we be sure that some information not belonging to S won't spoil our unique solution and introduce the unfortunate, and in many circumstances, deadly, element of surprise into the time evolution of our system?

The problem is spacetime is infinite and we can't control everything that happens there all the time. The best we can hope for in finding a unique solution to equations of motion is to find the region of spacetime that cannot be influenced by events that do not belong to S.

The causal future of S is the set of all events in spacetime that could in principle be influenced by some event that occurs in S. But events in the causal future of S can still be influenced by other events in spacetime that don't belong in S. The set of events that can only be influenced by events in S consists of events that can only be reached by a future-directed null or timelike world line from an event in S. This set of events is called the *future domain of dependence* of S, written $D^+(S)$.

For example, consider the pole at rest at $\tau = 0$ between $x = 0$ and $x = L$ in the pole frame. This is the set of events we called P above. The causal future of the pole at $\tau = 0$ is bounded by the two null lines $\tau = -x$ and $\tau = x - L$. An observer at rest at $x = -L$ in this frame enters the causal future P at time $\tau = L$, and could introduce new information at that time to spoil the predicted time evolution of the system.

The future domain of dependence of P is bounded by the null lines $\tau = x$ and $\tau = L - x$. These two null lines intersect at $\tau = x = L/2$, and so the future domain of dependence of P is a finite region of spacetime. It doesn't extend infinitely in all directions as does the causal future of P. This is shown in Figure 6.6.

The boundary formed by the two null lines $\tau = x$ and $\tau = L - x$ is called the *future Cauchy horizon* of P, written $H^+(P)$. Events that are outside of the Cauchy horizon of P can be influenced by events that don't belong to the set P.

A spacelike set of events S also has a *past domain of dependence* $D^-(S)$ consisting of all of the events in the causal past of S that can only reach events S by traveling on a null or timelike world line. The null boundary of the past domain of dependence of S is called the *past Cauchy horizon*, written $H^-(P)$. For the pole example, the past Cauchy horizon of P is formed by two null lines $\tau = -x$ and $\tau = x - L$, and the past domain of dependence of P is the region of spacetime between the events $(0, 0)$, $(0, L)$ and $(L/2, -L/2)$.

The union of the past and future domains of dependence of a spacelike set of event S is called the domain of dependence of S. The domain of dependence of S

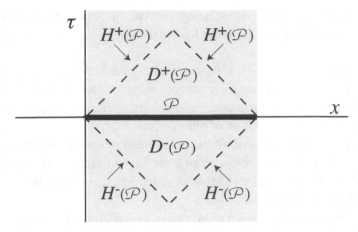

Fig. 6.6. Events in the domain of dependence $D(\mathcal{P}) = D^+(\mathcal{P}) \cup D^-(\mathcal{P})$ are contained within the past and future Cauchy horizons $H^\pm(\mathcal{P})$. Suppose a set of initial
data for the equations of motion for some field in spacetime is specified in the region $\mathcal{P}$. The solution is only constrained by the initial data given in $\mathcal{P}$ inside
the domain of dependence $D(\mathcal{P})$. Outside of the past and future Cauchy horizons $H^\pm(\mathcal{P})$, the time evolution of the field cannot be predicted by data specified
at $\mathcal{P}$.

consists of all events in spacetime that have exclusive influence over or are exclu-
sively influenced by events in $\mathcal{S}$. Events inside of the domain of dependence of $\mathcal{S}$
have a unique causal relationship with events in $\mathcal{S}$ and events outside the domain
of dependence have a non-unique causal relationship with events in $\mathcal{S}$.

If the domain of dependence of the set $\mathcal{S}$ encompasses all of spacetime, then
$\mathcal{S}$ is said to be a *Cauchy surface* for said spacetime. For Minkowski spacetime
in d dimensions, any spacelike hypersurface that includes all of space, that is,
is infinite in all D spacelike directions, is a Cauchy surface. If a set of data is
specified on a Cauchy surface, then the entire past and future of the spacetime is
determined.

Now let's go back to the pole in the barn and recall the time functions $\Sigma_P = \tau$.
The condition $\Sigma_P = 0$ defines the spacelike hypersurface that contains the set of
events $\mathcal{P}$ representing the physical object of the pole at time $\tau = 0$. The surface
$\tau = 0$ as a whole extends from $-\infty < x < \infty$. The domain of dependence of this
infinite surface is indeed all of $\mathbf{M}^2$. Any spacelike surface defined by $\Sigma = constant$
is a Cauchy surface for this spacetime.

In the case of the pole, if we specify that the pole is at rest at $\tau = 0$ and occupies
the space between $x = 0$ and $x = L$, and nothing else at all happens anywhere else
in spacetime at $\tau = 0$, and there are no forces on the pole, so the pole has zero
acceleration, then we've not only predicted the motion of the pole infinitely far

into the future, but we've also postdicted the motion of the pole infinitely far back into the past.

A spacetime that has a Cauchy surface is called a *globally hyperbolic* spacetime. A globally hyperbolic spacetime can be foliated by Cauchy surfaces, and these Cauchy surfaces serve as the global time function for the spacetime.

Causality is a fairly straightforward affair in Minkowski spacetime, which is flat and hence has a line element that can be written

$$ds^2 = -d\tau^2 + |d\vec{x}|^2. \tag{6.13}$$

General relativity is Einstein's theory of gravity. In general relativity, spacetime is curved and can look something like

$$ds^2 = -f(x)d\tau^2 + g(x)(dx^1)^2 + \cdots. \tag{6.14}$$

The time direction is determined by the minus sign in the metric, not by the fact that we call the coordinate by the name τ. If the functions $f(x)$ and $g(x)$ change sign by passing through zero, or $1/0$, at some value of x, then time and space can switch places, so that vectors in the x direction become timelike, and vectors in the τ direction become spacelike. This happens because null lines in curved spacetime can get trapped inside regions of space, or repulsed from them. When time and space switch identities, we end up with phenomena such as black holes or closed timelike loops. But to learn about those phenomena, you have to go beyond the material in this book and learn to do physics in curved spacetime using Einstein's general theory of relativity.

Causality and the wave equation

In Chapter 5 we showed that the equation of motion for the one form potential $A_\mu(x)$ in Lorentz gauge $\partial_\mu A^\mu = 0$ is

$$\partial_\alpha \partial^\alpha A_\mu = \frac{4\pi}{c} J_\mu. \tag{6.15}$$

We saw that for the case $J_\mu = 0$, the solution was a traveling wave that propagates at the speed of light on null hypersurfaces with a null propagation vector **k**. But is the solution for a general current density causal?

The general solution to the equation (6.15) is the sum of a general potential A_μ^f that solves the free space equation $\partial_\mu \partial^\mu A_\mu^f = 0$ and the function $A_\mu^J(x)$ obtained by integrating the source equation (6.15) using the Green function $D(\mathbf{x} - \mathbf{x}_s)$

$$A_\mu^J(x) = \frac{4\pi}{c} \int d^4x_s \, D(\mathbf{x} - \mathbf{x}_s) J_\mu(x_s). \tag{6.16}$$

The Green function $D(\mathbf{x} - \mathbf{x}_s)$ satisfies the wave equation for a single source event at $\mathbf{x}_s$

$$\partial_\alpha \partial^\alpha D(\mathbf{x} - \mathbf{x}_s) = \delta^4(\mathbf{x} - \mathbf{x}_s), \tag{6.17}$$

where $\delta^4(\mathbf{x} - \mathbf{x}_s)$ is the Dirac delta function. The Dirac delta function has a representation as a momentum space integral

$$\delta^4(\mathbf{x} - \mathbf{x}_s) = \left(\frac{1}{2\pi}\right)^4 \int d^4k\, e^{-i\mathbf{k}\cdot(\mathbf{x}-\mathbf{x}_s)}. \tag{6.18}$$

Using (6.17) and (6.18) one can show that

$$D(\mathbf{x} - \mathbf{x}_s) = -\left(\frac{1}{2\pi}\right)^4 \int \frac{d^4k}{k^2} e^{-i\mathbf{k}\cdot(\mathbf{x}-\mathbf{x}_s)}. \tag{6.19}$$

The integral for $D(\mathbf{x} - \mathbf{x}_s)$ is doable as a contour integral, once one decides on how to handle the places where the denominator vanishes, but we won't go over the details here. For a particular prescription, the Green function is

$$D_R(\mathbf{x} - \mathbf{x}_s) = -\frac{1}{2\pi}\theta(\tau - \tau_s)\,\delta((\mathbf{x} - \mathbf{x}_s)^2), \tag{6.20}$$

whereas for another choice it is

$$D_A(\mathbf{x} - \mathbf{x}_s) = \frac{1}{2\pi}\theta(\tau_s - \tau)\,\delta((\mathbf{x} - \mathbf{x}_s)^2). \tag{6.21}$$

The average of these two would be a third possibility. The function $\theta(x)$ is the Heaviside step function

$$\theta(x) = \begin{cases} 0 & x < 0 \\ 1 & x \geq 1. \end{cases} \tag{6.22}$$

The function $D_R(\mathbf{x} - \mathbf{x}_s)$ is called the *retarded Green function* and vanishes everywhere except on the future light cone of the source point $\mathbf{x}_s$. The function $D_A(\mathbf{x} - \mathbf{x}_s)$ is called the *advanced Green function* and vanishes everywhere except on the past light cone of the source point $\mathbf{x}_s$. These two functions show that information about any change in the current is transmitted at the speed of light from the source point $\mathbf{x}_s$ to the field point $\mathbf{x}$.

The retarded Green function is ordinarily the physically preferred choice for implementing causality. A general solution $A_\mu(x)$ can be written as the sum of a free potential $A_\mu^f(x)$ that is specified by boundary conditions in the infinite past, plus the field $A_\mu^J(x)$ coming from the source current. These boundary conditions on the field pick out the retarded Green function so that the full solution for the potential field is

$$A_\mu(x) = A_\mu^f(x) + \frac{4\pi}{c} \int d^4x_s \, D_R(\mathbf{x} - \mathbf{x}_s) J_\mu(x_s). \tag{6.23}$$

When we use the retarded Green function, changes in the source current at the source point $\mathbf{x}_s$ only influence the electromagnetic field along the future light cone of x_s. Thus cause precedes effect. This also shows that there is no action at a distance in electromagnetism. The effect of the current on the field propagates at the speed of light.

What about the advanced Green function, which seems to send information about the present into the past? The wave equation (6.15) is invariant under time reversal $\tau \to -\tau$. The equations of motion don't care which direction along the τ axis we choose to call the future. The most general solution to the equation should, therefore, reflect this same symmetry. The choice of a time direction in the previous case was made by choosing to fix the boundary conditions of the free field portion of the solution $A_\mu^f(x)$ in the infinite past, as an "incoming" field, which picks out the retarded Green function, because we want the field $A_\mu^J(x)$ to vanish in the past.

According to time reversal invariance of the wave equation, we could choose the free field boundary conditions in the infinite future, and then use the advanced Green function to compute the fields in the past. This is not useful for realistic calculations, but it is consistent with the mathematics, just as choosing a past-directed world line is possible in spacetime kinematics, just not realistic.

There is one place, however, where the advanced Green function does become realistic and crucial to our ability to understand Nature as we see it, and that is when quantum mechanics enters the picture, which will be the subject of Chapter 7.

Exercises

6.1 Two observers are at rest in $\mathbf{M}^2$ in frame S with coordinates (τ, x). There is a pole located between $x = 0$ and $x = L$, with one observer standing at each end with a stop-watch and a cell-phone. What is the least amount of coordinate time τ that it takes for the two observers to measure the length L of the pole? Assume that the cell-phone sends messages at the speed of light and the observers take zero time to think and talk.

6.2 A massless particle in $\mathbf{M}^2$ with coordinates (τ, x) travels on the world line $\mathbf{x}(\lambda) = (\lambda, \lambda)$. What is the proper length in terms of λ of a segment of the particle's path connecting the events $(0, 0)$ and (L, L)? What happens to the square of the spacetime momentum of this massless particle if we scale λ by an arbitrary constant α so that $\lambda \to \alpha\lambda$? What happens to the squared spacetime momentum of a massive particle if the proper time $\lambda \to \alpha\lambda$? What happens if $\alpha \to 0$? Does this tell you anything about the nature of time and space along a null world line compared with a timelike world line?

6.3 A particle with mass m and electric charge q in $\mathbf{M}^4$ is accelerated by a constant electric field $\vec{E} = E\partial_x$, as described in Chapter 5. Suppose the charge is placed at rest at $x = 1/\kappa = m/qE$ at time $\tau = 0$. How much energy does it take to maintain the constant electric field long enough to accelerate the particle to speed $\beta = 0.99$? How much energy must be supplied to accelerate the particle to $\beta = 0.999$? How much energy must be supplied to accelerate the particle to $\beta = 0.9999$?

6.4 Consider the function $\Sigma = \tau - \sqrt{1 + x^2}$ in two spacetime dimensions with coordinates (τ, x). Plot the surface $\Sigma = 0$ and find the future-directed vector field $\mathbf{n}$ normal to this surface. What happens to $\mathbf{n}^2$ as $x \to \pm\infty$? Find the set of curves to which $\mathbf{n}$ is tangent. Do these curves ever intersect anywhere else in the spacetime? Given what you have deduced, is Σ a time function for $\mathbf{M}^2$? In other words, do the curves to which $\mathbf{n}$ is tangent constitute a smooth congruence of timelike curves along which we can define a proper time parameter to serve as a reliable time coordinate for $\mathbf{M}^2$?

6.5 Consider $\mathbf{M}^4$ in null coordinates (u, v, y, z), with $u = t + x$ and $v = t - x$.
(a) Compute the Minkowski line element $ds^2 = -d\tau^2 + |d\vec{x}|^2$ in these coordinates and find the components of the metric tensor.
(b) Using null coordinates, find the null vector $\mathbf{k}$ that is tangent to the world line $(u, v, y, z) = (0, \lambda, 0, 0)$ and points towards the future. Transform $\mathbf{k}$ to Minkowski coordinates.
(c) Using null coordinates, find the null vector $\mathbf{n}$ that is normal to the surface $u = 0$ and points towards the future. Transform $\mathbf{n}$ to Minkowski coordinates.
(d) Find a null vector $\mathbf{l}$ that is not orthogonal to $\mathbf{k}$. Transform $\mathbf{l}$ to Minkowski coordinates.

6.6 If event E_1 belongs to the causal future of event E_2, does event E_2 then belong to the causal past of E_1? In other words, is it true that $E_1 \in J^+(E_2) \Rightarrow E_2 \in J^-(E_1)$? If event E_1 belongs to the chronological future of event E_2, does event E_2 then belong to the chronological past of E_1? In other words, is it true that $E_1 \in I^+(E_2) \Rightarrow E_2 \in I^-(E_1)$? The easiest way to solve this problem is to draw the light cones in question.

6.7 Consider the two-dimensional periodic spacetime you created in the hands-on exercise. Let's call this spacetime $\mathbf{M_P}^2$ because we've made the time dimension periodic. Show that for any event $E_i \in \mathbf{M_P}^2$, $J^+(E_i) = J^-(E_i) = \mathbf{M_P}^2$.

6.8   A pole occupies the space between $x = 0$ and $x = 1$ at $\tau = 0$ in its rest frame S with coordinates (τ, x). Let's call this set of events $\mathcal{P}$. A barn occupies the space between $\tilde{x} = 2$ and $\tilde{x} = 3$ at $\tilde{\tau} = 0$ in its own rest frame with coordinates $(\tilde{\tau}, \tilde{x})$. Let's call this set of events $\mathcal{B}$. The pole moves at speed β in the $\tilde{x}$ direction relative to the barn.

     (a) Draw the causal futures of the pole $J^+(\mathcal{P})$ and the barn $J^+(\mathcal{B})$ in the pole frame S.

     (b) Draw the causal futures of the pole $J^+(\mathcal{P})$ and the barn $J^+(\mathcal{B})$ in the barn frame $\tilde{S}$.

6.9   Find the domain of dependence $D(\mathcal{S})$ if $\mathcal{S} = H^-(\mathcal{P})$ for the pole as shown in Figure 6.6. Is it true or not true that $D(\mathcal{S}) = D(\mathcal{P})$ for *any* set $\mathcal{S}$ of events with null or spacelike separation that smoothly connect the events $(0, 0)$ and $(0, L)$ in Figure 6.6?

6.10   Suppose the Jewel of Spacetime diamond has been stolen at $\tau = 1$ from the museum located at $x = 1$ in two-dimensional Minkowski spacetime. Detective Cauchy of the Initial Value Police inspects the Cauchy surface Σ at $\tau = 0$ and discovers that the only world line intersecting Σ between $x = 0$ and $x = 2$ belongs to the renowned jewel thief Yucant Katchmi. At Katchmi's trial, Cauchy testifies that the evidence found on the Cauchy surface at $\tau = 0$ proves that only Katchmi could have been in the museum at $\tau = 1$, and therefore he is guilty beyond any shadow of a doubt. However, Katchmi's lawyer reveals proof that a point had been removed from spacetime at $(0, 0)$, so the Cauchy surface Σ fails in being the unique predictor of events that take place in the museum at $\tau = 1$. Does the jury convict Katchmi, or set him free because Cauchy's evidence was incomplete?

Part II
Advanced Topics

7

When quantum mechanics and relativity collide

The twentieth century gave birth to three major conceptual revolutions in human understanding of the fundamental physical principles that characterize how the Universe works. They are relativity (both special and general), quantum mechanics, and string theory. The first two arose in the early part of the century and are supported by overwhelming experimental and observational evidence. The latter (string theory) arose much later – around 1970 – and is much more speculative than relativity or quantum mechanics, though most practitioners are completely convinced that it is also an essential ingredient of the story. String theory is certainly every bit as mind-bending in its implications as relativity and quantum mechanics.

The primary purpose of this book, as its title indicates, is to explain and explore the special theory of relativity. However, this also gives us license to explore its interplay with other profound ideas. By confronting quantum theory with relativity physicists were led to uncover remarkable concepts and insights. A thorough treatment of relativistic quantum mechanics is the subject of quantum field theory and is beyond the scope of this book. This chapter will present an overview of the basics. (Why should all the fun be reserved for the specialists?) Our brief treatment will serve as a general introduction for students who have not studied these matters before. Those who have may find a few surprises.

This chapter begins by surveying the origins and properties of nonrelativistic quantum mechanics. Having done that, the stage is set to address the issues that arise when a reconciliation with relativity is sought. We explain why simply replacing the nonrelativistic Schrödinger equation with a relativistic analog is not sufficient to take account of all the phenomena that are associated with the merger of relativity and quantum mechanics. In particular, the mathematical description of particle creation and annihilation, which is an inevitable feature of a relativistic quantum theory, requires quantum field theory.

7.1 Yet another surprise about light

Maxwell's theory, which was so successful in describing electrodynamics, did not fit neatly into the fabric of physics as it was understood at the time. We have discussed at great length how it conflicts with Galilean relativity and how influential that was in Einstein's development of the special relativity. But there was a second problem, which concerned its marriage with thermodynamics, a subject in which Maxwell's contributions were also very important. The issue has to do with the energy carried by electromagnetic radiation in thermal equilibrium with ordinary matter. The standard theory of the time unambiguously gave an infinite answer for the total energy in the radiation field. This conflicted both with common sense and with observations.

Maxwell, Jeans, and various other clever people agonized about this for a couple of decades before Max Planck came up with the right answer. This launched the quantum revolution.

The conflict between classical electrodynamics and thermodynamics

Electromagnetic radiation that is in thermal equilibrium with its surroundings, at a temperature T, is said to have a blackbody spectrum. Imagine a hot oven maintained at temperature T with a small hole through which the radiation can escape. Then one can ask about the spectral distribution $I(v, T)$ of the emitted radiation as a function of its frequency v. The rate at which energy escapes (per unit area) in a frequency interval of width dv, centered at frequency v, is given by $I(v, T)\, dv$.

In classical thermodynamics energy is shared among all the degrees of freedom of the system in such a way that the average energy carried by each degree of freedom is $k_B T$, where k_B is Boltzmann's constant. In the case of an oscillation mode, for example, the average kinetic energy is $k_B T/2$ and the average potential energy is also $k_B T/2$ for a total of $k_B T$. This average value arises from a continuous distribution of possible energies given by the Maxwell distribution that characterizes thermal equilibrium, according to which the probability of finding energy E is

$$p(E) = \frac{1}{Z} \exp(-\frac{E}{k_B T}). \tag{7.1}$$

The coefficient Z is adjusted to ensure that the total probability is 1.

According to a carefully analyzed and well-understood line of reasoning, known to Maxwell and his colleagues,

$$I(v, T) = \frac{8\pi v^2}{c^2} < E(T) >, \tag{7.2}$$

where $< E(T) >$ is the average energy carried by a degree of freedom in the radiation field at temperature T. Applying the standard rule, $< E(T) > = k_B T$, gives a formula for $I(\nu, T)$ called Rayleigh's law. This is the prediction of classical physics. In particular, classical theory predicts that $< E(T) >$ should be independent of the frequency. Experimentally, Rayleigh's law agrees with observations at low frequencies, but deviates from it at high frequencies, where the spectrum cuts off very sharply. This sharp cutoff is important, because it ensures that the total energy (obtained by integrating over all frequencies) is finite. It cannot be understood within the framework of classical physics.

The quantum understanding of light begins

At the turn of the twentieth century, Max Planck proposed a resolution to the problem of the blackbody spectrum. He observed first that excellent empirical agreement with observations is achieved by replacing the classical formula $< E(T) > = k_B T$, with the formula

$$< E(\nu, T) > = \frac{h\nu}{\exp(h\nu/k_B T) - 1}, \tag{7.3}$$

where h is an empirically determined fundamental constant, which later became known as Planck's constant. Note that this formula reduces to $k_B T$ for $h\nu \ll k_B T$, but cuts off exponentially fast when $h\nu \gg k_B T$.

Given this successful empirical formula, Planck's next step was to give it a physical interpretation. He soon realized that such an understanding could be achieved by postulating that light consists of discrete particles or "quanta," whose energy is proportional to their frequency. Specifically, for a quantum of light (later called a *photon*), Planck's proposal was that

$$E = h\nu. \tag{7.4}$$

The constant of proportionality h, Planck's constant, has a value that is determined experimentally. The 2002 official value is $6.62606876 \times 10^{-31}$ J s with an uncertainty of 78 parts per billion.

It is often convenient to replace the frequency ν by the angular frequency ω. Since there are 2π radians in a cycle, $\omega = 2\pi\nu$. Then Planck's relation can be replaced by the formula

$$E = \hbar\omega, \tag{7.5}$$

where, by definition, $\hbar = h/(2\pi)$. Replacing joules by millions of electronvolts (MeV), which also is often convenient, the current value of $\hbar$ is

$6.58211889 \times 10^{-22}$ MeV s. This is known with greater accuracy (39 parts per billion), because the value of an electron-volt is defined with more precision than that of a joule.

It is natural to wonder how Planck's formula relates to the famous relativity formula $E = mc^2$. Recall that this is the formula for the energy of a massive particle at rest. For a particle with momentum p, we have seen that it generalizes to $E = \sqrt{m^2c^4 + p^2c^2}$. Now suppose we wish to apply this formula to a photon. A photon, being a particle of light, travels with the speed of light. According to the formulas of relativity, specifically $m = E\sqrt{1 - v^2/c^2}$, the only way this can be achieved with a finite amount of energy is if the rest mass m of a photon is zero. In that case we have $E = pc$, where p represents the magnitude of the momentum three-vector. Comparing this relation with Planck's formula, we deduce that a photon carries momentum whose magnitude is given by

$$p = h\nu/c = \hbar\omega/c. \tag{7.6}$$

One consequence of this is that a light beam carries momentum. When the light is absorbed by an object that the light is incident on, there is a pressure due to the impinging light. This is called *radiation pressure*. This has been confirmed experimentally.

In classical Maxwell theory, light propagates in waves. This means that in empty space the components of the electric and magnetic fields satisfy the wave equation

$$\left(\frac{\partial^2}{\partial t^2} - c^2 \frac{\partial^2}{\partial x^2} \right) \phi(x, t) = 0. \tag{7.7}$$

This equation has been written for one spatial dimension. In higher dimensions the second x derivative is replaced by a sum of second derivatives, that is, the Laplacian. A solution to this equation is provided by a wave propagating in the x direction with wave number k and angular frequency ω,

$$\phi(x, t) \sim e^{i(kx - \omega t)} \tag{7.8}$$

provided that $(\omega)^2 = (kc)^2$ or $\omega = kc$. We see that the differential operator $\partial/\partial t$ acting on the wave solution brings down a factor of $-i\omega$. Combining this with the Planck relation suggests the identification

$$E = \hbar\omega \sim i\hbar \frac{\partial}{\partial t}. \tag{7.9}$$

Similarly, one is led to the identification

$$p = \hbar k \sim -i\hbar \frac{\partial}{\partial x}. \tag{7.10}$$

Representing energy, momentum, and other quantities by differential operators, in this way, turns out to be a very useful viewpoint in quantum mechanics. It is the approach that was pioneered by Heisenberg.

Perhaps this is a good place to point out one of the profound implications of quantum mechanics. We have argued that momentum should be associated to the differential operator given in Eq. (7.10). This implies that acting on a function of position $f(x)$ the operators $x^i p_j$ and $p_j x^i$ give different answers when $i = j$. In the first case the derivative in p_j does not act on the factor x^i, whereas in the second case it does. The difference for an arbitrary function f is summarized by the commutation relation

$$[x^i, p_j] = i\hbar \, \delta^i_j. \tag{7.11}$$

This equation is the mathematical origin of the famous Heisenberg uncertainty relation, which says that position and momentum cannot be simultaneously determined with arbitrary precision. Rather, denoting the uncertainties in the determination of x and p by Δx and Δp, one finds that $\Delta x \, \Delta p \geq h$.

7.2 The Schrödinger equation is not covariant

Even though the founders of quantum mechanics were fully cognizant of relativity, they started by formulating quantum theory for nonrelativistic systems. This was very sensible, because many of the atomic and molecular systems that they wanted to understand have typical electron velocities on the order of 1 percent the speed of light and sometimes even less than that. Also, as we will discuss, the relativistic generalization raises a host of profound issues. While some of them were understood early on, it took decades to sort out others. So the fundamental Schrödinger equation, which is described below, was constructed with the aim of understanding quantum theory in the nonrelativistic approximation. For this it is spectacularly successful.

From classical mechanics to the Schrödinger equation

We have shown above how energy and momentum can be represented by differential operators in quantum mechanics. This allows one to take a kinematic relation between energy and momentum and associate to it a partial differential equation. In particular, in the case of light we saw that the relation $E^2 - p^2 c^2 = 0$ is associated to the wave equation. In a similar manner, one can associate other partial differential equations to other kinematical relationships. The case we wish to focus on here is the Schrödinger equation, which is the quantum mechanical wave equation for a nonrelativistic particle.

The idea is as follows. Consider a nonrelativistic particle of mass m propagating under the influence of an external potential $V(x)$, so that its total energy is given by the sum of its kinetic energy $p^2/(2m)$ and its potential energy $V(x)$

$$E = \frac{p^2}{2m} + V(x). \tag{7.12}$$

Then the Schrödinger equation is the corresponding differential equation obtained by substituting differential operators for E and p in the manner indicated above. This yields the equation

$$i\hbar \frac{\partial}{\partial t} \psi(x,t) = \left(-\frac{\hbar^2}{2m} \frac{\partial^2}{\partial x^2} + V(x) \right) \psi(x,t). \tag{7.13}$$

This equation has the general structure

$$i\hbar \frac{\partial}{\partial t} \psi(x,t) = H\psi(x,t), \tag{7.14}$$

where H is the Hamiltonian. In the context at hand, the Hamiltonian is given by the differential operator

$$H = -\frac{\hbar^2}{2m} \frac{\partial^2}{\partial x^2} + V(x). \tag{7.15}$$

There are various possible generalizations in which H can contain additional terms.

Equation (7.14) is the time-dependent form of the Schrödinger equation. If the Hamiltonian is time-independent, then energy is conserved. In this case it makes sense to consider configurations of definite energy E. To analyze them, consider wave functions of the following special form

$$\psi(x,t) = e^{-iEt/\hbar} \psi_E(x). \tag{7.16}$$

Substitution in Eq. (7.14) gives the time-independent form of the Schrödinger equation

$$H\psi_E(x) = E\psi_E(x). \tag{7.17}$$

In this equation E is a real number, whereas H is a differential operator. Thus this equation can be interpreted as an eigenvalue equation in which E is an eigenvalue of the Hamiltonian operator and $\psi_E(x)$ is the associated eigenvector.

Associating relations among kinematical quantities, such as E and p, to differential equations is just mathematics. The new physics, which make quantum mechanics such a profound advance, concerns the interpretation of the solutions of the Schrödinger wave equation. The basic idea is that the wave function of a particle $\psi(x, t)$ encodes the probability for finding the particle at the position x at the time t. More precisely, if one has a normalized solution, for which $\int |\psi(x, t)|^2 dx = 1$, then $|\psi(x, t)|^2$ is a probability density. In other words, $|\psi(x, t)|^2 dx$ is the probability of finding the particle in the interval between x and $x + dx$ at time t. For N-particle systems the wave function is a function of the positions of all N particles $\psi(x_1, \ldots, x_N, t)$, and $|\psi|^2$ is a joint probability density for finding the entire collection of particles in specified locations at time t.

Consider the case of a bound state in which the particle described by the wave function does not have enough energy to escape the influence of the potential. If the zero point of energy is associated to the value of the potential at infinity, then the energy E of a bound state is negative. Similarly, for a particle that can escape to infinity, the energy is positive. For a bound state one should require that the total probability is 1, namely

$$\int |\psi_E(x)|^2 dx = 1. \tag{7.18}$$

Therefore the wave function $\psi_E(x)$ must fall sufficiently rapidly at spatial infinity to ensure convergence of the integral. Such a wave function is said to be *normalizable*. Now for the punch line: The eigenvalue equation Eq. (7.17) only has normalizable bound state solutions for discrete values of the energy E. This is the mathematical explanation of the discreteness of energy levels that is observed in atomic and molecular systems.

Prior to the discovery of quantum mechanics the stability of atoms was mysterious. Consider, for example, an electron orbiting a proton under the influence of its electrical attraction as happens in the case of a hydrogen atom. According to classical Maxwell theory, an accelerating charge emits radiation and thereby loses energy. Since the orbiting electron is in accelerated motion, one therefore would expect it to radiate and its orbit to decay by spiraling inwards. However, this is not what happens, and the Schrödinger equation allows one to understand why.[1]

What Bohr proposed, and the Schrödinger equation implements, is that the electron can only occupy certain discrete energy states. One says that the allowed energy levels are quantized. In the leading approximation, the potential energy of the electron orbiting a proton in a hydrogen atom is given by the coulomb attraction $V = -e^2/r$. Solving the eigenvalue problem in this case, one finds that the discrete binding energies are given by

$$E_n = -\frac{e^4 m}{2\hbar^2 n^2}, \tag{7.19}$$

---

[1] Actually, Niels Bohr had figured out the basic idea before the discovery of the Schrödinger equation.

where $-e$ is the electric charge of an electron, and n is a positive integer. When the hydrogen atom is in its ground state, given by $n = 1$, it is absolutely stable. However, when it is in an excited state $n > 1$, it can decay to a lower energy state m by emitting a photon of energy $\Delta E = E_n - E_m$. These energy differences beautifully account for the spectrum of frequencies of emitted radiation (using $\Delta E = h\nu$) that were observed before the discovery of quantum mechanics, and had been regarded as mysterious.

This interpretation also implies that $E_1 \approx 13.6$ eV is the minimum energy required to ionize a hydrogen atom that is initially in its ground state. For example, if one shines light of frequency ν on the atoms, the individual photons will carry energy $h\nu$. If the frequency is high enough, so that this energy exceeds 13.6 eV, then absorption of a single photon by the electron will result in ionization of the hydrogen atom. Other atoms have other characteristic ionization energies.

This phenomenon of a frequency threshold for ionization of an atom is called the photoelectric effect. It deserves special mention in a book about relativity, because it is the result that earned Einstein the Nobel prize. Apparently, the committee was too timid to award it for relativity. It is also noteworthy, because it is one of several important contributions that Einstein made to the development of quantum theory. Despite this fact, Einstein was very troubled by quantum theory. He found the probability interpretation philosophically disturbing.[2]

The fact that the wave function may be complex gives the possibility of interesting interference effects when the wave function is given by the sum of several distinct contributions. The classic example of this is when a coherent beam of particles is incident on a screen with two parallel slits and then projected on a second screen. People were familiar with the fact that passage of coherent light through a double slit results in an interference pattern on a screen behind the slits. The fact, implied by quantum mechanics, that the same thing can happen for electrons or neutrons came as quite a revelation. Interference and diffraction effects were soon observed in scattering from crystals.

Time and space in quantum mechanics

The Schrödinger equation is obviously nonrelativistic, since it is based on nonrelativistic kinematics in the way that we have indicated. An important consequence of this fact is that it involves a first-order time derivative and second-order spatial derivatives. This does obvious violence to any notion of space and time transforming into one another under linear Lorentz transformations. Time is treated

---

[2] One could argue that his greatly reduced productivity in his later years was a consequence of his reluctance to accept quantum mechanics. His efforts were mostly devoted to constructing a "unified field theory." Certainly all modern work along those lines is based on quantum theory.

as absolute, and the wave function gives probability amplitudes throughout all of space at a given instant of time. We can contrast this with the wave equation for the electromagnetic field, which involves second-order time derivatives and is Lorentz-invariant. Quite clearly, a relativistic wave equation for the electron should have space and time derivatives of the same order. Either the time derivative should be second-order (as for the photon) or the spatial derivatives should be first-order. As we will explain later in this chapter, the latter turns out to be the right answer.

When is the Schrödinger equation a good approximation?

The Schrödinger equation is very useful when relativistic effects are small. In the case of the hydrogen atom, one can estimate the average speed of an electron in a particular energy state, such as the ground state. One finds that is of order αc, where $\alpha = e^2/(\hbar c) \approx 1/137.036$. The quantity α, which is a dimensionless number that characterizes the strength of electromagnetic interactions, is called the fine structure constant. It is fortunate that it is much smaller than 1. If this were not the case nonrelativistic quantum mechanics would not be so useful – and the world would be a different place. Relativistic corrections to energies computed using the Schrödinger equation are of order $\alpha^2 \approx 10^{-4}$. Moreover, the leading relativistic corrections can be taken into account by adding new terms into the Hamiltonian. Thus one can get quite far in atomic physics without introducing a full-fledged relativistic formulation of quantum theory.

The harmonic oscillator

An especially simple and important example of a quantum mechanical system is the harmonic oscillator. It is a very useful building block for analyzing more complicated problems. Basically this is the quantum version of an ideal spring in one dimension. When the spring is stretched by an amount x the magnitude of the restoring force is kx, where k is the spring constant. This is the derivative of the potential energy function $V(x) = kx^2/2$. Thus the Hamiltonian for the harmonic oscillator is

$$H = -\frac{\hbar^2}{2m}\frac{d^2}{dx^2} + \frac{1}{2}kx^2. \tag{7.20}$$

The shape of the potential energy function confines the particle to the vicinity of the origin, preventing it from escaping to infinity. As a result, all the solutions of the Schrödinger equation in this case describe bound states, and the spectrum of energy eigenvalues of the Hamiltonian is discrete. This is to be contrasted with problems, such as the hydrogen atom, where the electron can escape to infinity if it acquires enough energy. In such problems it is conventional to measure energies relative to the ionization energy, and therefore bound state energies are negative. In the case of the harmonic oscillator by contrast, the ionization energy is infinite, so a different choice of zero point for the energy is needed. The bottom of the potential well is a natural choice.

One can find normalizable solutions of differential equation $H\psi(x) = E\psi(x)$ directly. They turn out to be Hermite polynomials times exponential factors. By a crude analysis of the leading large x behavior of the equation one easily sees that the exponential factors are $\exp(-1/2(x/x_0)^2)$, where $x_0 = \hbar/\sqrt{mk}$. Thus the wave function is very small except within a distance of order x_0 of the origin.

There is a nice algebraic way of analyzing the harmonic oscillator in terms of raising and lowering operators a and $a^\dagger$. These are defined by

$$a = \frac{m\omega x + ip}{\sqrt{2m\hbar\omega}} \quad \text{and} \quad a^\dagger = \frac{m\omega x - ip}{\sqrt{2m\hbar\omega}}, \tag{7.21}$$

where the spring frequency ω is given by $\omega = \sqrt{k/m}$. The operators a and $a^\dagger$ are dimensionless, and they are complex conjugates of one another. (Since they are operators, it is better to say that they are Hermitian conjugates of one another.) They have the commutation relations

$$[a, a^\dagger] = 1, \tag{7.22}$$

as an immediate consequence of $[p, x] = -i\hbar$. In terms of the raising and lowering operators, the Hamiltonian takes the simple form

$$H = (a^\dagger a + \frac{1}{2})\hbar\omega. \tag{7.23}$$

The harmonic oscillator has an infinite number of bound states each of which corresponds to a linearly independent wave function $\psi_n(x)$. These can be regarded as defining basis vectors in an infinite-dimensional vector space (that is, a Hilbert space). The ground state is annihilated by the lowering operator $a\psi_0 = 0$. Therefore, acting on ψ_0 with H, we see that its energy is $E_0 = \hbar\omega/2$. This energy is referred to as the *zero-point energy*. Moreover, since the commutation relation implies that $(a^\dagger a)a^\dagger = a^\dagger(a^\dagger a + 1)$, we see that each time $a^\dagger$ acts on a state, the eigenvalue of $a^\dagger a$ increases by 1. Therefore its energy (the eigenvalue of H) increases by $\hbar\omega$. The nth state, which is obtained by acting on the ground state n times with $a^\dagger$ (in other words, $\psi_n \sim (a^\dagger)^n \psi_0$) has energy

$$E_n = (n + \frac{1}{2})\hbar\omega. \tag{7.24}$$

We now note a curious coincidence. A harmonic oscillator with the natural frequency of ω has energy levels that are equally spaced, and the spacing is $\hbar\omega$. This is the same amount of energy that Planck discovered is carried by a photon of angular frequency ω. Why should that be so? Later we will see that when the electromagnetic field is treated quantum-mechanically, it is an operator that is, in a certain sense that we will explain, built out of harmonic oscillators. This picture will make the connection.

7.3 Some new ideas from the Klein–Gordon equation

There is an obvious first guess for a relativistic generalization of the Schrödinger equation, called the Klein–Gordon equation, which we describe in this section. It was considered very early in the history of quantum mechanics and quickly rejected. The reason that it was rejected is that it predicts values for relativistic corrections to the hydrogen atom energy levels that conflict with observation. The equation that gives the right values – the Dirac equation – will be described later in this chapter. Even though the Klein–Gordon equation is not appropriate for electrons, it is appropriate for certain other particles, and it illustrates some of the issues that arise in relativistic quantum mechanics. Therefore it seems reasonable to discuss it first.

The Klein–Gordon equation

Let us try to construct a Lorentz-invariant wave equation for a massive particle. Based on what we have said previously, there is a rather obvious thing to try. Namely, we should start with the relativistic relation between energy and momentum, $E^2 = p^2c^2 + m^2c^4$ and convert this into a partial differential equation using the prescription given in Eqs. (7.9) and (7.10). This yields the partial differential equation

$$-\hbar^2 \frac{\partial^2}{\partial t^2} \phi(x, t) = \left(-c^2\hbar^2 \frac{\partial^2}{\partial x^2} + m^2c^4 \right) \phi(x, t). \tag{7.25}$$

This equation, known as the Klein–Gordon equation, certainly is invariant under Lorentz transformations. In other words, it is a relativistic equation. Note that it reduces to the ordinary wave equation when $m = 0$. As before, if there are multiple spatial dimensions the second derivative with respect to x should be replaced by the Laplacian. Note that in writing the Klein–Gordon equation, we have not included a potential energy term. Thus, as it stands, it is only appropriate for the description of a free (noninteracting) relativistic particle. One could discuss the addition of a potential energy term, but that is not the direction we wish to pursue here. There are interesting lessons to be learned even without doing that.

A plane wave solution

Like the other free wave equations we have considered, the Klein–Gordon equation admits plane-wave solutions. In fact, we constructed the equation just so that this would be the case. To check this, consider the trial solution $\phi \sim e^{i(\omega t - kx)}$. Clearly, this satisfies the Klein–Gordon equation provided that

$$\hbar^2 \omega^2 = c^2 \hbar^2 k^2 + m^2 c^4, \tag{7.26}$$

which corresponds to the kinematic relation $E^2 = p^2 c^2 + m^2 c^4$.

In the nonrelativistic limit $p \ll mc$, one has $E = mc^2 + p^2/(2m) + \ldots$. Aside from the rest mass, which is just an overall additive constant, this is the usual nonrelativistic relation, just as one would deduce by the corresponding analysis of the free $(V = 0)$ Schrödinger equation. However, there is a second solution, $E = -mc^2 - p^2/(2m) + \ldots$, which does not have a counterpart in the Schrödinger equation. The essential feature that is responsible for the existence of this second solution is that the Klein–Gordon equation is second-order in time derivatives, whereas the Schrödinger equation is only first-order in time derivatives. More generally, without making any non-relativistic expansions, we can observe that Eq. (7.26) has two classes of solutions: those for which the frequency ω is positive and those for which the frequency ω is negative. Equivalently, using Planck's relation $E = \hbar\omega$, we can speak of positive-energy and negative-energy solutions. What are we to make of the negative-energy solutions?

Particles and antiparticles

One might consider discarding the negative-energy solutions of the Klein–Gordon equation on the grounds that they are unphysical. However, one doesn't really have the right to throw away solutions that are not wanted. One either has to give them an acceptable interpretation or discard the theory. So let's think about the possible meaning of a negative-energy solution. It looks like an ordinary positive-energy solution except that it is time-reversed, t is replaced by $-t$. What does it mean for a particle to propagate backwards in time? This question is a recasting of the earlier question: what does it mean to have negative energy?

The answer to these questions, which turns out to be an inevitable consequence of a consistent relativistic quantum theory, is that a particle propagating backwards in time should be reinterpreted as an antiparticle propagating forward in time. In other words, every particle has an associated antiparticle, and the negative-energy solutions describe the antiparticles. Which is which, is a matter of convention. One usually says that the electron is a particle, the antiparticle of which is the positron. But one could equally well say that the positron is a particle, the antiparticle of which is the electron. The latter phraseology is unnatural only because electrons are so much more ubiquitous.

Suppose, for example, that we interpret a positive-energy solution of the Klein–Gordon equation as giving a description of a π^+ meson, which has spin zero. Then the negative-energy solutions describe the antiparticle of the π^+ meson, which has the opposite electrical charge and is a π^- meson. Another way of saying this is that if ϕ is a possible wave function of a π^+, then the complex conjugate ϕ^* is a possible wave function of a π^-. There are also neutral π mesons, which are their own antiparticles. How can we describe them? The answer is to require that the wave function of a π^0 is real, so that it is its own complex conjugate. This implies that the wave function contains a superposition of positive and negative frequency components.

7.4 The Dirac equation and the origin of spin

In an important breakthrough, Dirac discovered a relativistic wave equation for the electron that is first-order in space and time derivatives. The key idea that makes this possible is the use of matrices. Thus the wave operator that appears in the Dirac equation is actually a matrix of operators, and the wave function that they act on is actually a set of wave functions. This vector of wave functions is called a *spinor*.

We will discuss the case of two dimensions (one space and one time) first. This example allows us to describe the essential ideas with less mathematical complication. Moreover, the two-dimensional Dirac theory has important applications in string theory, so it is useful in its own right. The massless and massive cases will be examined separately. The generalization to four dimensions (three space and one time) will be considered afterwards. With the two-dimensional example already under our belt, it will be pretty obvious how to proceed.

The Dirac equation that is appropriate to the description of electrons is massive and four-dimensional. This is the most important case, since all of atomic, molecular, and condensed matter physics is (to good approximation) described by the behavior of electrons in a certain specified environment. Since electrons are so much lighter than atomic nuclei, the motion of the nuclei in response to the motion of the electrons is a small effect.

The massless Dirac equation in two dimensions

How can we replace the two-dimensional (one space dimension and one time dimension) Klein–Gordon equation

$$-\hbar^2 \frac{\partial^2}{\partial t^2} \phi(x, t) = \left(-c^2 \hbar^2 \frac{\partial^2}{\partial x^2} + m^2 c^4 \right) \phi(x, t) \qquad (7.27)$$

by an equation that is first-order in derivatives? When the mass m is zero, the solution is quite easy, so let us consider that case first. For $m = 0$ the wave operator factorizes, and the equation can be rewritten in the form

$$\left(i\frac{\partial}{\partial t} + ic\frac{\partial}{\partial x}\right)\left(i\frac{\partial}{\partial t} - ic\frac{\partial}{\partial x}\right)\phi(x, t) = 0. \tag{7.28}$$

Let us therefore consider as a new wave equation the first-order differential equation obtained by keeping one of the factors only:

$$\left(i\frac{\partial}{\partial t} + ic\frac{\partial}{\partial x}\right)\psi(x, t) = 0. \tag{7.29}$$

This is the operator implementation of the kinematic relation $E - cp = 0$, which is sensible for a massless particle in one spatial dimension. The plane wave solutions are evidently of the form $\psi \sim e^{i(\omega t - kx)}$, where $\omega = ck$.

Notice that $\psi \sim e^{i(\omega t + kx)}$, which would have been a solution of the original second-order differential equation, is not a solution of our first-order equation. By passing to the first-order equation, the space of solutions has been roughly cut in half. The surfaces of constant phase in the wave function $e^{i(\omega t - kx)}$ have $ct - x$ constant. This describes waves (or wave functions of particles) that are moving in the direction of increasing x with the speed of light. This is a Lorentz-invariant statement. After all, as we learned long ago, the speed of light is the same in all inertial frames. Such a particle in one dimension is sometimes called a "right-mover." There are no left-moving solutions, since they are solutions that have been eliminated in passing to the first-order equation.

Even though the left-moving solution has been eliminated, there is still a second possibility. Above, we assumed that ω and k are both positive. The second possibility, which also solves the same first-order equation, is that they are both negative, still with $\omega = ck$. In this case we have a negative-energy solution describing the antiparticle. The surfaces of constant phase for the negative-energy solutions also have $ct - x$ constant. Therefore they also describe right-movers. As a result it is possible in this case for the particle to be its own antiparticle. In that case we would require (as we discussed earlier for the π^0) that the wave function is real. This corresponds to a superposition of the positive frequency and negative frequency solutions.

The massive Dirac equation in two dimensions

The assumption that $m = 0$ played an essential role in the preceding discussion, so let us think about what to do when this is not the case. Physically, it is clear that a clean separation between left-movers and right-movers is not possible in the massive case. A particle that is moving to the right with a speed less than that of light in one inertial frame will appear to be moving to the left in another suitably boosted inertial frame. Thus, the notions of left-moving and right-moving become observer-dependent in the massive case.

What we can do in the massive case is to take the original Klein–Gordon equation and recast it as a pair of coupled first-order partial differential equations:

$$\left(i\hbar \frac{\partial}{\partial t} + ic\hbar \frac{\partial}{\partial x} \right) \psi_1(x, t) = mc^2 \psi_2(x, t) \tag{7.30}$$

and

$$\left(i\hbar \frac{\partial}{\partial t} - ic\hbar \frac{\partial}{\partial x} \right) \psi_2(x, t) = mc^2 \psi_1(x, t). \tag{7.31}$$

By combining these equations, it is easy to show that both ψ_1 and ψ_2 satisfy the original second-order equation.

Note that in the limit $m \to 0$, the wave function ψ_1 describes right-movers and the wave function ψ_2 describes left-movers. However, as we explained, there is no sharp distinction between them when $m \neq 0$. Because both equations are required, the minimum number of degrees of freedom is twice as great in the massive case as in the massless one. This is a general feature of the Dirac equation whenever the number of spatial dimensions is odd. The right-movers and their higher-dimensional analogs are called Weyl spinors. They can exist separately from their mirror-reversed partners only for massless particles.

The pair of equations for ψ_1 and ψ_2 can be rewritten as a single matrix equation for a two-component spinor Ψ. (Its components are just ψ_1 and ψ_2.) The equation, which is the two-dimensional Dirac equation, takes the form

$$\left(i\hbar \frac{\partial}{\partial t} + i\sigma_3 c\hbar \frac{\partial}{\partial x} \right) \Psi(x, t) = mc^2 \sigma_1 \Psi(x, t). \tag{7.32}$$

Here σ_1 and σ_3 are two of the three Pauli matrices

$$\sigma_1 = \begin{pmatrix} 0 & 1 \\ 1 & 0 \end{pmatrix} \quad \sigma_2 = \begin{pmatrix} 0 & -i \\ i & 0 \end{pmatrix} \quad \sigma_3 = \begin{pmatrix} 1 & 0 \\ 0 & -1 \end{pmatrix}. \tag{7.33}$$

The two-dimensional Dirac equation can be recast in a form that makes its Lorentz invariance manifest. To do this let us introduce $x^0 = ct$ and $x^1 = x$ and multiply through by the matrix σ_1 (from the left) to obtain the equivalent equation

$$\left(i\hbar\sigma_1 \frac{\partial}{\partial x^0} + \hbar\sigma_2 \frac{\partial}{\partial x^1} \right) \Psi = mc\, \Psi. \tag{7.34}$$

This can then be rewritten in the form

$$i\left(\gamma^0\frac{\partial}{\partial x^0} + \gamma^1\frac{\partial}{\partial x^1}\right)\Psi = (mc/\hbar)\Psi, \tag{7.35}$$

where $\gamma^0 = \sigma_1$ and $\gamma^1 = -i\sigma_2$. The idea is that when γ^μ multiplies spinors, it effectively transforms as a two-component vector under Lorentz transformations in just such a way that the inner product $\gamma^\mu\partial_\mu$ is Lorentz-invariant. We won't give a complete proof of this here, since Lorentz invariance was built into the construction from the beginning. However, as one piece of evidence, let us note the anticommutation relations of the Dirac matrices

$$\{\gamma^\mu, \gamma^\nu\} = -2\,\eta^{\mu\nu}, \tag{7.36}$$

where $\eta^{\mu\nu}$ is the two-dimensional Lorentz metric.

The massless Dirac equation in four dimensions

Let us approach the construction of the four-dimensional Dirac equation the same way we did the two-dimensional one – namely, by considering the massless case first. This is worth while, both because it is simpler and because it is physically relevant to neutrinos.[3]

Let us begin with the four-dimensional Klein–Gordon equation, written in the form

$$\left(\partial_0^2 - \sum_{i=1}^{3}\partial_i^2 + (mc/\hbar)^2\right)\phi(x) = 0, \tag{7.37}$$

where we have introduced the shorthand notation

$$\partial_\mu = \frac{\partial}{\partial x^\mu} \qquad \mu = 0, 1, 2, 3. \tag{7.38}$$

Also, the argument of the field ϕ is now denoted by the spacetime coordinate x. Now we set $m = 0$ and try to factorize the remaining differential operator into a product of first-order differential operators, just as we did in Eq. (7.28).

The key to factorizing the massless wave equation in four dimensions is to use matrices, rather as we did for the massive case in two dimensions. The factorized equation is given by

$$(\partial_0 + \sigma_i\partial_i)(\partial_0 - \sigma_j\partial_j)\psi(x) = 0, \tag{7.39}$$

---

[3] There is compelling recent experimental evidence that neutrinos are not exactly massless. Still, neglecting the effects of their masses is an excellent approximation for most purposes.

where now the wave function ψ is a two-component spinor. The repeated indices i and j are summed from 1 to 3, of course. The key to understanding this factorization are the Pauli matrix anticommutation relations

$$\{\sigma_i, \sigma_j\} = 2\delta_{ij}. \tag{7.40}$$

This implies that $(\sigma_i \partial_i)^2 = \partial_i \partial_i$. Let us also record the commutation relations of the Pauli matrices, which are

$$[\sigma_i, \sigma_j] = 2i\epsilon_{ijk}\sigma_k. \tag{7.41}$$

The next step is to throw away one of the factors (it doesn't matter which one) to obtain the desired first-order equation

$$(\partial_0 + \sigma_i \partial_i)\psi(x) = 0. \tag{7.42}$$

This is the massless Dirac equation in four dimensions. Its solutions are two-component spinors, which are sometimes called Weyl spinors.

Let us pause to consider the physical significance of this equation. Using our rule for relating energy and momentum to differential operators it corresponds to the relation $E = cp_i \sigma_i$, but what is the physical interpretation of σ_i? The proposal is that a particle described by the Dirac equation has an intrinsic angular momentum that is given by the three-vector $S_i = \hbar \sigma_i/2$. (This will be explored further later in this chapter.) Since the eigenvalues of each of the matrices σ_i are ± 1, the eigenvalues of the matrices S_i are $\pm \hbar/2$. A particle with this amount of intrinsic angular momentum is said to have spin 1/2. Substituting for the Pauli matrices leaves the relation

$$E = 2cp_i S_i/\hbar. \tag{7.43}$$

Since the magnitude of the three-vector p_i is E/c, this relation tells us that the projection of the spin vector along the direction of motion is $+\hbar/2$.

The projection of the spin along the direction of motion is called the *helicity* of the particle. Thus the massless Dirac equation given above describes a particle with helicity equal to $+1/2$. This is the four-dimensional counterpart of being a right-mover in two dimensions. Had we kept the other factor in choosing the first-order equation (reversing the sign of the $\sigma_i \partial_i$ term), we would have obtained an equation that describes a helicity $-1/2$ particle.

The distinction between the choice of the two possible first-order equations in four dimensions is quite analogous to that in two dimensions. However, there is an interesting difference between the two cases. So far in the four-dimensional analysis we have been discussing the positive-energy solutions, but as before there are also negative-energy solutions that describe antiparticles. In the case of two dimensions we saw that the antiparticle of a right-mover is also a right-mover. Mathematically, the reason for this was that the complex conjugate ϕ^* satisfies the same (right-moving) differential equation as ϕ.

To determine the helicity of the antiparticle of a helicity $+1/2$ massless particle in four dimensions we must examine the complex conjugate of Eq. (7.42). This equation is

$$(\partial_0 + \sigma_1\partial_1 - \sigma_2\partial_2 + \sigma_3\partial_3)\psi^*(x) = 0, \qquad (7.44)$$

where we have used the fact that σ_1 and σ_3 are real, whereas σ_2 is imaginary. Using the anticommutation relations of the Pauli matrices, this can be recast in the equivalent form

$$\sigma_2(\partial_0 - \sigma_1\partial_1 - \sigma_2\partial_2 - \sigma_3\partial_3)\sigma_2\psi^*(x) = 0. \qquad (7.45)$$

Then the left-hand factor of σ_2 can be dropped since it is invertible. From this we conclude that $\tilde{\psi}(x) = \sigma_2\psi^*(x)$ satisfies the negative-helicity wave equation.

$$(\partial_0 - \sigma_i\partial_i)\tilde{\psi}(x) = 0. \qquad (7.46)$$

Thus, in four dimensions, massless particles and their antiparticles have opposite helicity.

To summarize, we have shown that the negative-energy antiparticle solution in four dimensions has the opposite helicity from that of the original particle solution. This is to be contrasted with the two-dimensional result that the antiparticle of a right-mover is also a right-mover. In the application to neutrinos, they are conventionally assigned positive (or left-handed) helicity, whereas the antineutrinos have negative (or right-handed) helicity. The deep truth is that right-handed neutrinos and left-handed antineutrinos do not exist (or else they interact so weakly that they have not yet been discovered). A theory with this kind of an asymmetry is not symmetrical under spatial inversion ($\vec{x} \to -\vec{x}$), a transformation that reverses the sign of the momentum but not the spin. Such a theory is said to have *parity violation*.

The massive Dirac equation in four dimensions

We can now realize the massive four-dimensional Klein–Gordon equation in Eq. (7.37) in terms of a pair of coupled equations, just as we did in the two-dimensional case:

$$i(\partial_0 + \sigma_i\partial_i)\psi_1(x) = (mc/\hbar)\psi_2(x) \qquad (7.47)$$
$$i(\partial_0 - \sigma_i\partial_i)\psi_2(x) = (mc/\hbar)\psi_1(x). \qquad (7.48)$$

Note that ψ_1 and ψ_2 are both two-component spinors. In the limit $m \to 0$, ψ_1 describes a left-handed spinor and ψ_2 describes a right-handed one. This pair of equations implies that all four components of ψ_1 and ψ_2 satisfy the massive Klein–Gordon equation.

Let us now pass to a four-component spinor notation. We define a four-component spinor Ψ to have ψ_1 as its first two components and ψ_2 as its last two components. In this notation we can replace the pair of coupled two-component equations by a single four-component equation of the form

$$i(\partial_0 + \alpha_i \partial_i)\Psi(x) = (mc/\hbar)\beta\Psi(x), \tag{7.49}$$

where

$$\alpha_i = \begin{pmatrix} \sigma_i & 0 \\ 0 & -\sigma_i \end{pmatrix} \qquad \beta = \begin{pmatrix} 0 & 1 \\ 1 & 0 \end{pmatrix}. \tag{7.50}$$

These are understood to be 4×4 matrices written in a notation where each entry represents a 2×2 block. For use later, let us also define

$$\Sigma_i = \begin{pmatrix} \sigma_i & 0 \\ 0 & \sigma_i \end{pmatrix}. \tag{7.51}$$

As a last step we can multiply through by β and recast Eq. (7.49) in a manifestly Lorentz-invariant form

$$i\gamma^\mu \partial_\mu \Psi(x) = (mc/\hbar)\Psi(x), \tag{7.52}$$

where $\gamma^0 = \beta$ and $\gamma^i = \beta\alpha_i$. The Lorentz invariance is manifest in the sense that $\gamma^\mu \partial_\mu$ is a scalar operator if we know that γ^μ can be treated as a four-vector. We have not proved here that this is so. However as supporting evidence we note that these four matrices satisfy covariant anticommutation relations (the Dirac algebra)

$$\{\gamma^\mu, \gamma^\nu\} = -2\eta^{\mu\nu}, \tag{7.53}$$

where $\eta^{\mu\nu}$ is the four-dimensional Lorentz metric.

We have presented the Dirac equation that describes a free (noninteracting) electron. The generalization to motion in an external electromagnetic field is required for applications to atomic physics. This extension is given by a simple rule: the derivative ∂_μ should be replaced by the "covariant" derivative $\partial_\mu - ieA_\mu$, where A_μ is the vector potential of Maxwell theory. By this simple prescription one obtains the equation

$$i\gamma^\mu(\partial_\mu - ieA_\mu)\Psi(x) = (mc/\hbar)\Psi(x). \tag{7.54}$$

As Dirac so eloquently put it, this equation in principle solves all of chemistry. The step from principle to practice is a big one, however.

Spin 1/2 particles

The Dirac equation can be written in a form that looks like a special case of the time-dependent Schrödinger equation in Eq. (7.14). Equation (7.49) is almost in this form already. Using it, one reads off the Hamiltonian

$$H = -i\hbar c\alpha_i \partial_i + \beta mc^2. \tag{7.55}$$

The free Dirac equation is Lorentz-invariant, which means that it takes the same form in any inertial reference frame. This implies, in particular, that it is invariant under spatial rotations. We wish to examine the rotational symmetry here, since this will lead to an explanation of how the equation incorporates spin. Before doing so, we need to make a short digression about symmetry in quantum mechanics, since that is what we will be utilizing.

In quantum mechanics various kinematical quantities (such as momentum, angular momentum, etc.) are represented by differential operators in the way that we have indicated. Our purpose here is to derive the condition that ensures that a kinematic quantity is a constant of the motion, that is, time-independent. Quite generally, suppose that a is a kinematic quantity that is represented by an operator A, which has no explicit time-dependence. The theorem states that a is a time-independent constant of the motion if the associated operator A commutes with the Hamiltonian: $[A, H] = 0$.

To derive this condition we use the time-dependent Schrödinger equation for a wave function $\psi(x, t)$ that is an eigenfunction of the operator A. This requires choosing an inertial frame and introducing separate space and time coordinates. Then we have

$$H\psi(x, t) = i\hbar \frac{\partial \psi(x, t)}{\partial t} \quad \text{and} \quad A\psi(x, t) = a(t)\psi(x, t). \tag{7.56}$$

Using these equations, we first compute $AH\psi$

$$AH\psi(x, t) = Ai\hbar \frac{\partial \psi(x, t)}{\partial t} = i\hbar \frac{\partial}{\partial t}(A\psi(x, t)) = i\hbar \frac{\partial}{\partial t}(a(t)\psi(x, t)), \tag{7.57}$$

and then we compute $HA\psi$

$$HA\psi(x, t) = Ha(t)\psi(x, t) = a(t)H\psi(x, t) = a(t)i\hbar \frac{\partial \psi(x, t)}{\partial t}. \tag{7.58}$$

Comparing these we see that $a(t)$ is t-independent if and only if $[A, H]\psi(x, t)$ vanishes. Since this is supposed to hold for any such wave function, the general condition is $[A, H] = 0$.

With these general remarks out of the way, we can now turn to the problem at hand: the rotational invariance of the free Dirac equation. We need to find the angular momentum operator J_i that commutes with the Hamiltonian in Eq. (7.55);

J is the total angular momentum operator, which will turn out to be the sum of two pieces

$$J_i = L_i + S_i, \tag{7.59}$$

where L_i is the orbital angular momentum operator and S_i is the intrinsic (or spin) angular momentum operator. Let us examine each of these in turn.

The orbital angular momentum is given by the standard classical physics formula $\vec{L} = \vec{x} \times \vec{p}$, or in terms of components

$$L_i = \epsilon_{ijk} x_j p_k = -i\hbar \epsilon_{ijk} x_j \partial_k, \tag{7.60}$$

where we have made the usual quantum operator replacement $p_k \to -i\hbar \partial_k$. Now we compute the commutator with the Hamiltonian

$$[\epsilon_{ijk} x_j p_k, \alpha_l p_l c + \beta mc^2] = \epsilon_{ijk} \alpha_l [x_j, p_l] p_k c = i\hbar \epsilon_{ijk} \alpha_j p_k c. \tag{7.61}$$

Clearly, this is nonzero.

If the Dirac Hamiltonian is rotationally invariant, there must be a compensating contribution S_i to the angular momentum that satisfies

$$[S_i, \alpha_l p_l c + \beta mc^2] = -i\hbar \epsilon_{ijk} \alpha_j p_k c. \tag{7.62}$$

This equation is easily seen to be solved by the spin vector

$$S_i = \frac{1}{2} \hbar \Sigma_i, \tag{7.63}$$

where Σ_i was defined in (7.51). The proof uses

$$[\Sigma_i, \alpha_j] = 2i\epsilon_{ijk} \alpha_k, \quad [\Sigma_i, \beta] = 0, \tag{7.64}$$

which is an immediate consequence of Eq. (7.41) and the definitions of the various matrices. The eigenvalues of each Σ_i are ± 1, and hence the eigenvalues of each S_i are $\pm\hbar/2$. This is indicative of a spin $= 1/2$ particle. Chapter 8 will explore in greater depth the significance of half-integer spin and the relevance of the groups $SU(2)$ and $SL(2, C)$.

Let us conclude this section by considering the special case of a particle at rest. The wave function in this case satisfies $p_i \Psi = 0$, which implies that it is independent of the position coordinates. Let us now introduce the following basis of four-component spinors:

$$u_1 = \begin{pmatrix} 1 \\ 0 \\ 1 \\ 0 \end{pmatrix}, \quad u_2 = \begin{pmatrix} 0 \\ 1 \\ 0 \\ 1 \end{pmatrix}, \quad u_3 = \begin{pmatrix} 1 \\ 0 \\ -1 \\ 0 \end{pmatrix}, \quad u_4 = \begin{pmatrix} 0 \\ 1 \\ 0 \\ -1 \end{pmatrix}. \tag{7.65}$$

To determine the energy E_n of each of these states, we compute $Hu_n = \beta mc^2 u_n = E_n u_n$. Since $\beta u_n = u_n$ for $n = 1, 2$, we deduce that u_1 and u_2 describe states of energy mc^2, as is appropriate for a particle of mass m that is at rest. In the cases $n = 3, 4$, $\beta u_n = -u_n$, and therefore these two states have energy $-mc^2$. As we have explained, negative-energy states such as these should be interpreted as describing antiparticles of positive energy. Finally, let us compute the component of the angular momentum along the x_3 axis for each of these states. (Any axis would do, but x_3 is the standard choice.) This requires computing $S_3 u_n = s_n u_n$. One finds that $s_1 = s_3 = \hbar/2$ and $s_2 = s_4 = -\hbar/2$. To summarize, u_1 describes a spin-up electron, u_2 describes a spin-down electron, u_3 describes a spin-up positron, and u_4 describes a spin-down positron.

7.5 Relativity demands a new approach

Nonrelativistic quantum mechanics successfully accounts for a vast range of physical phenomena for which the nonrelativistic approximation is justified. However, the seemingly innocent extension to relativistic equations, such as the Klein–Gordon or Dirac equation, opens up a host of new issues. The reason for this is not due to any fancy properties of partial differential equations. Rather, it can be traced to a simple algebraic fact: in nonrelativistic physics the dispersion relation that relates the energy and momentum of a free particle is $E = p^2/(2m)$, which is unambiguously positive. In the relativistic setting, this is replaced by $E^2 = p^2 c^2 + m^2 c^4$. Being quadratic, this equation has two solutions, one with $E \geq mc^2$ and another with $E \leq -mc^2$.

In classical physics, one can get away with ignoring the existence of the negative-energy solution, but in quantum physics this is no longer possible. We have already indicated that negative-energy solutions should be associated to antiparticles. In this section we wish to describe more carefully why this is so, and to explore some of the consequences of this interpretation. We will see that there is more to this than the existence of antiparticles. Rather, there is the possibility of particle creation and annihilation. Dealing with this requires a whole new approach, which is called *quantum field theory*.

The Dirac sea and the prediction of antimatter

Dirac recognized the problem of negative-energy solutions, and he proposed an interesting solution, which led him to predict the existence of antiparticles. As no such particles were known at the time, this was certainly a bold step. A few years later, Carl Anderson discovered the antiparticle that Dirac had predicted, the positron, in cosmic rays.

The pure Dirac equation, as we have presented it, describes free noninteracting electrons. The interesting predictions, as well as the problems, arise when one introduces interactions. In Dirac's case that meant including the interaction of electrons with an external electromagnetic field. This is achieved by a rule that was mentioned earlier: replacing ∂_μ by $\partial_\mu - ieA_\mu$, where A_μ is the vector potential of Maxwell theory. Doing this, Dirac was able to compute relativistic corrections to atomic spectra very successfully, so clearly this was a big step in the right direction.

When the electron interacts with an electromagnetic field it can make transitions between different energy states at rates that are computable by the rules of quantum mechanics. The energy difference is carried away as radiation. The rates for transitions between different energy levels of hydrogen and other atoms are computed to great accuracy in this way. The problem that Dirac recognized is that the same reasoning also implies that through interactions with the electromagnetic field transitions from positive energies ($E \geq mc^2$) to negative energies ($E \leq -mc^2$) should also occur at a rapid rate. In fact, adding up the contributions of all the negative energy states, the total computed rate is infinite! This is manifest nonsense, so Dirac made a bold hypothesis to circumvent the problem.

Dirac's idea for dealing with the negative-energy states utilized another important fact of quantum mechanics that we have not yet mentioned: *the Pauli exclusion principle*. Pauli discovered that electrons have the property that in a multielectron system – such as a heavy atom – only one electron can occupy each quantum state. (More precisely, there can be one spin-up electron and one spin-down electron in each quantum state.) This rule plays a crucial role in understanding the periodic table.

Dirac proposed that if all the negative-energy states were ordinarily filled, for the configuration that we call the vacuum, then the Pauli exclusion principle would prevent the possibility of transitions from positive-energy to negative-energy states. This bizarre-sounding suggestion has some interesting consequences. For example, suppose that one of the negative-energy states (with energy $-E_0$, say) is unoccupied. Such a state would have a total energy that is greater than that of the vacuum by an amount $+E_0$. Similarly, it has an electric charge that is greater than that of the vacuum by an amount $+e$. Thus such a "hole" in the negative-energy sea of electrons, would behave for all intents and purposes like a positive-energy particle with the same mass as an electron, and with the opposite electric charge. Thus Dirac identified the antielectron (or positron) as a hole in the otherwise-filled sea of negative-energy states.

Now suppose that one has a system consisting of an electron and a positron, that is, a hole. In this case a transition of the electron into the negative-energy sea is possible, but it must go into the single unoccupied state, since all others are forbidden by the Pauli exclusion principle. The rate for such a transition is computable and finite. When the transition occurs, what is the final state? As usual, the electron making a transition emits radiation, but the rest of the story is very different from an ordinary atomic transition. After the transition there no longer is an electron and there no longer is a hole. Physically, what has happened is that an electron and a positron have been annihilated giving off radiation that carries off all of their energy, which is at least $2mc^2$.

Relativistic quantum mechanics is not a complete theory

This picture of a filled sea of negative-energy electrons led to enormously successful predictions, most notably the predicted existence of the positron. But it is really strange. Shouldn't there be a symmetry between the treatment of matter and antimatter? So why not start with a filled sea of positrons instead and regard an electron as a hole in that sea? Whichever choice one makes, what is it that cancels the charge and energy of the sea so as to make it unobservable? After all, we know that the vacuum does not have a large charge density or a large (negative) energy density. These are good questions, but here is an even better one: spin zero particles, like the π^+, also have antiparticles with the same mass and opposite electric charge. However, spin 0 particles do not satisfy the Pauli exclusion principle. So what prevents a π^+ from making a transition into a negative-energy state? A negative-energy sea of pions would not do the job, so there must be a different answer in this case. Whatever it is might work for electrons, too.

Particle creation and annihilation

To give a satisfactory answer to the problem we have just posed, one needs to go beyond the formalism of relativistic quantum mechanics that we have described so far. It has been developed as a generalization of nonrelativistic quantum mechanics in which one takes account of the change in the dispersion relation required by relativity. The deeper truth that we have just learned, however, is that a complete relativistic quantum theory has to account for the possibility of particle creation and annihilation. The formalism of wave equations for wave functions is based on a fixed number of particles of definite type. It can account for possible changes in their energy and momentum, but it is not designed to describe their annihilation or creation. Even when one describes an atomic transition using the Schrödinger equation, one is failing to account for an important fact: the energy is carried off as electromagnetic radiation, but this is described classically. Yet, as we have already learned, this radiation is also quantized. We should really be speaking of the emission of a photon, which is a quantum of the electromagnetic field.

The lesson to be drawn from the preceding discussion, is that the implications of quantum mechanics are much deeper in the relativistic context. The distinction between radiation and particles washes out. Radiation, which can be treated classically in terms of waves, is actually composed of particles. On the other hand, massive particles – which are described classically by particle mechanics – at the quantum level have probability amplitudes that propagate as waves. Both aspects, particles and waves, are different faces of a common type of object: the quantum field. Both electrons and photons, which at the classical level seem so different, are associated to quantum fields. To be sure the quantum fields of electrons and photons have important differences, but they also have striking similarities. These fields and their interactions can account for the creation and annihilation of particles in a consistent coherent manner. Once one reformulates the theory of the interaction of electromagnetism and electrons in terms of quantum fields – a theory called *quantum electrodynamics* – one can dispense with Dirac's negative-energy sea.

Particles and fields

In this subsection we will sketch briefly the relationship between particles and fields in quantum field theory. The basic issue that needs to be confronted is how to describe the quantum mechanics of fields, such as the electromagnetic field or the electron field. We will first focus on the quantum description of free fields. In the next section we will discuss their interactions.

A field exists throughout space, and the field at each point of space should be regarded as a separate degree of freedom. Therefore a field has an infinite number of degrees of freedom. This is to be contrasted with a nonrelativistic point particle, for example, that has three degrees of freedom – x, y, and z – in three dimensions.

Instead of thinking of the field at each point in space, it is sometimes more useful to represent the field (via a Fourier transform) as a superposition of components of all possible momenta. In this alternative, but equivalent, picture each of the momentum components represents a separate degree of freedom. We will learn that each of these momentum components is quite similar to a harmonic oscillator. Thus a quantum field behaves a lot like an infinite superposition of harmonic oscillators. This is what we will describe. However, to understand how to apply quantum mechanics to such systems, we need a more general characterization of quantum mechanics than we have provided so far.

The required general characterization of quantum mechanics is the following: for any given physical system the relevant degrees of freedom can be split into coordinates (denoted q_i) and conjugate momenta (denoted p_i). As discussed in Appendix 1, the standard way of achieving this is to characterize the system by a

Lagrangian $L(q, \dot{q})$ constructed out of the coordinates q_i and their time derivatives. The conjugate momenta are then defined by

$$p_i = \frac{\partial L}{\partial \dot{q}_i}. \tag{7.66}$$

The Hamiltonian is defined to be given by

$$H = \sum_i p_i \dot{q}_i - L. \tag{7.67}$$

In writing the Hamiltonian it is customary to eliminate $\dot{q}_i$ in favor of p_i by solving Eq. (7.66). Then quantization is achieved by incorporating the canonical commutation relations

$$[q_i, p_j] = i\hbar\delta_{ij}. \tag{7.68}$$

The prescription in the preceding paragraph reduces to the description of quantum mechanics given earlier if the q_i are the position coordinates of particles. However, it is suitable for generalization to situations where the coordinates q_i are allowed to represent other degrees of freedom. One subtlety we will encounter in the application to fields is that the labels i and j will be continuous rather than discrete. In this case the Kronecker delta needs to be replaced by a Dirac delta function.

In order to be concrete, we will focus on Klein–Gordon fields, which are associated to spin 0 particles. In fact, the fields that appear in quantum electrodynamics are the electron field, which is massive and has spin $1/2$ and the photon (or electromagnetic) field, which is massless and has spin 1. The massive complex Klein–Gordon field can be regarded as a simplified model of the electron field, even though there are important differences. Similarly, the massless real Klein–Gordon field can be regarded as a simplified model of the electromagnetic field.

In this discussion it is very convenient to set $\hbar = c = 1$, so that the equations are not too cluttered. If one is careful, they can be restored by dimensional analysis. Doing this, the Klein–Gordon equation becomes

$$\ddot{\phi} - \partial_i\partial_i\phi + m^2\phi = 0, \tag{7.69}$$

where a dot represents a time derivative. The field $\phi(t, \vec{x})$ is complex when we want to describe a charged field and real when we want to describe a neutral one. In the complex case, this equation is the Lagrange equation of motion obtained from the Lagrangian

$$L = \int d^3x \left(|\dot{\phi}|^2 - |\partial_i\phi|^2 - m^2|\phi|^2 \right). \tag{7.70}$$

Here the "coordinate" degrees of freedom are the field ϕ, treated as an independent coordinate for every point $\vec{x}$ in space. The conjugate momenta are obtained by differentiating the Lagrangian with respect to $\dot{\phi}$. (Since the labels x are continuous, these are actually functional derivatives.) The result for the conjugate momenta $\pi(t, \vec{x})$ is simply

$$\pi(t, \vec{x}) = \dot{\phi}^*(t, \vec{x}). \tag{7.71}$$

Now we apply the prescription for canonical quantization, turning these fields into quantum mechanical operators satisfying the equal time commutation relations

$$[\phi(t, \vec{x}), \pi(t, \vec{y})] = i\delta^3(\vec{x} - \vec{y}) \tag{7.72}$$

and

$$[\phi(t, \vec{x}), \phi(t, \vec{y})] = [\pi(t, \vec{x}), \pi(t, \vec{y})] = 0. \tag{7.73}$$

Note that the right-hand side of Eq. (7.72) contains a three-dimensional Dirac delta function, which was defined in Chapter 5. It also contains a factor of $\hbar$, which has been set equal to 1.

Let us now represent the field in momentum space. We use the symbol $\vec{k}$ for the three-momentum here. It is an ordinary triple of numbers and *not* a quantum mechanical operator. Since we are dealing with a particle of mass m, the energy (in units with $c = 1$) is given by

$$E_k = \sqrt{m^2 + k^2}. \tag{7.74}$$

Whenever we write a square-root, we choose the positive branch. Thus this energy is positive. When the other branch is desired, a minus sign is explicitly displayed. The most general solution of the Klein–Gordon equation (which is now an operator equation!) can be written as superposition of terms of arbitrary three-momentum. For each value of the momentum, there is a positive-energy and a negative-energy contribution.

$$\phi(t, \vec{x}) = \int d\mu_k \left(a_-(\vec{k})e^{i(\vec{k}\cdot\vec{x} - E_k t)} + a_+^\dagger(\vec{k})e^{-i(\vec{k}\cdot\vec{x} - E_k t)}\right). \tag{7.75}$$

Let us also record the complex conjugate equation

$$\phi^*(t, \vec{x}) = \int d\mu_k \left(a_-^\dagger(\vec{k})e^{-i(\vec{k}\cdot\vec{x} - E_k t)} + a_+(\vec{k})e^{i(\vec{k}\cdot\vec{x} - E_k t)}\right). \tag{7.76}$$

A convenient choice for the integration measure is

$$d\mu_k = \frac{d^3k}{(2\pi)^3 E_k}, \tag{7.77}$$

which is Lorentz-invariant.

Equations (7.75) and (7.76) require some explanation. First note that the exponential factors ensure that the Klein–Gordon equation is satisfied. Next note that the solution is completely general, because we have included both positive-energy ($e^{-iE_k t}$) and negative-energy ($e^{iE_k t}$) contributions. Also, the angular frequencies are given by the energies, consistent with the rule $E = \hbar \omega$, since we have set $\hbar = 1$.

Since ϕ and ϕ^* are quantum operators, the coefficients $a_{\pm}(\vec{k})$ and their conjugates must also be quantum operators. In fact, one can compute their commutation relations, which are determined by Eqs. (7.72) and (7.73). The result turns out to be

$$[a_+(\vec{k}), a_+^\dagger(\vec{k}')] = [a_-(\vec{k}), a_-^\dagger(\vec{k}')] = E_k (2\pi)^3 \delta^3(\vec{k} - \vec{k}'), \qquad (7.78)$$

while all other commutators involving $a_{\pm}$ and $a_{\pm}^\dagger$ vanish.

The commutation relations given above are those of harmonic oscillators with continuous labels $\vec{k}$. To see how far this interpretation can be pushed, let us examine the Hamiltonian. Using the general prescription given in Eq. (7.67) one obtains

$$H = \frac{1}{2} \int d^3 x \left(|\pi(t, \vec{x})|^2 + |\partial_i \phi(t, \vec{x})|^2 + m^2 |\phi(t, \vec{x})|^2 \right). \qquad (7.79)$$

Substituting the integral representations of ϕ and π and carrying out some integrals, one obtains

$$H = \int d\mu_k E_k \left(a_+^\dagger(\vec{k}) a_+(\vec{k}) + a_-^\dagger(\vec{k}) a_-(\vec{k}) \right). \qquad (7.80)$$

Strictly speaking, we have dropped an infinite zero-point energy in the last step, since each oscillator really gives $a^\dagger a + 1/2$, as we learned in the study of the harmonic oscillator. This overall constant contribution to the energy is unobservable, so this is okay. It would be nicer if it were not required to drop an infinite constant, however. This is the case in supersymmetric theories, where the symmetry ensures that different contributions to zero-point energies cancel.

These equations have a very nice interpretation: $a_+^\dagger(\vec{k})$ and $a_-^\dagger(\vec{k})$ are creation operators producing a positively or negatively charged particle with momentum $\vec{k}$ and energy E_k. Similarly, $a_+(\vec{k})$ and $a_-(\vec{k})$ are annihilation operators annihilating a positively or negatively charged particle with momentum $\vec{k}$ and energy E_k. This interpretation is consistent with the commutation relations

$$[H, a_{\pm}^\dagger(\vec{k})] = E_k a_{\pm}^\dagger(\vec{k}) \qquad (7.81)$$

and

$$[H, a_{\pm}(\vec{k})] = -E_k a_{\pm}(\vec{k}). \qquad (7.82)$$

Thus, if Ψ_0 represents the vacuum with zero energy ($H\Psi_0 = 0$), then we can describe an N-particle state by acting with the creation operators $a^\dagger$ N times:

$$\Psi_N(\vec{k}_1, \ldots \vec{k}_N) \sim a^\dagger(\vec{k}_1) \ldots a^\dagger(\vec{k}_N)\Psi_0. \tag{7.83}$$

We have not displayed the $\pm$ indices that labels the charges in this formula. The total energy of this state is just $E = \sum_I E_{k_I}$.

Let us contrast the role of the $a^\dagger$ and a operators here with those in the earlier description of the quantum-mechanical harmonic oscillator. The harmonic oscillator is a one-particle system with a spectrum of different energy levels. The harmonic oscillator raising and lowering operators act on the wave functions so as to relate the various states of different energy. The amount of energy added or subtracted by each step is $\hbar\omega$. When the operators act in this manner it is customary to speak of *first quantization*. This is quite different from what we have found here for quantum fields. In this context, often called *second quantization*, the analogs of the raising and lowering operators create and destroy particles, and are therefore called creation and annihilation operators. The quantum system they describe is one in which the number of particles can be different for different quantum states.

The quantization of the charged Klein–Gordon field is somewhat similar to that of the electron field. There are creation and annihilation operators for both electrons and positrons that enter pretty much in the way we have indicated. There are two notable differences, however. First, because of the spin degree of freedom, in each case there is a separate operator for a spin up electron or positron and for a spin down electron or positron. The second difference is that electrons and positrons are fermions and satisfy the Pauli exclusion principle, which we described earlier. This is built into the mathematics by replacing the canonical commutation relations with canonical *anticommutation* relations.

A simplified prototype for the electromagnetic (or photon) field is a real massless Klein–Gordon field ϕ. In this case Eq. (7.75) is replaced by

$$\phi(t, \vec{x}) = \int d\mu_k \left(a(\vec{k})e^{i(\vec{k}\cdot\vec{x} - E_k t)} + a^\dagger(\vec{k})e^{-i(\vec{k}\cdot\vec{x} - E_k t)} \right), \tag{7.84}$$

where the second term is the complex conjugate of the first one, so that the field is real. In this case the mass is zero so that $E_k = |\vec{k}|$. Putting back $\hbar$ and c, we recover Planck's relation for the energy of a photon $E = \hbar\omega = \hbar k c$. This field creates or destroys an uncharged particle and therefore a and $a^\dagger$ do not carry a charge subscript. However, the photon does have spin, so a spin label is required in the electromagnetic case.

Microscopic causality

A fundamental notion in special relativity, discussed in Chapter 6, is causality. There are two aspects to this: (1) no signal can travel faster than the speed of light, and (2) cause should precede effect. The first of these is built into the structure of the formulas given above, as we now explain.

Consider, for simplicity, a theory of a single real scalar field $\phi(x)$. This field is a physical observable in the sense that it is an operator with matrix elements that can be measured. The question we wish to address is: when can a field at a spacetime point x influence one at a spacetime point y? To be consistent with special relativity, it is necessary that two points with a spacelike separation have no influence upon one another. The basic criterion for whether there is an influence in quantum mechanics, as we have learned, is whether or not they are commuting operators. Thus what we need is

$$[\phi(x), \phi(y)] = 0 \quad \text{for} \quad \eta_{\mu\nu}(x - y)^{\mu}(x - y)^{\nu} > 0, \tag{7.85}$$

a condition known as microscopic causality. We have previously presented the formula for the commutator in the special case of equal times. In that case we had a result that vanishes unless the spatial coordinates coincide. We now wish to generalize that result to the case of unequal times.

All the ingredients to compute the commutator in the general case are at hand. The expansion in Eq. (7.84) is an explicit formula that can be used in conjunction with

$$[a(\vec{k}), a^{\dagger}(\vec{k}')] = E_k(2\pi)^3 \delta(\vec{k} - \vec{k}'), \tag{7.86}$$

to compute the commutator. After a couple of steps the result can be written in the form

$$[\phi(x), \phi(y)] = iG(x, y) = \int \frac{d^4k}{(2\pi)^3} \delta(k^2 + m^2)\epsilon(k_0)e^{-ik\cdot(x-y)}. \tag{7.87}$$

The crucial fact is that this expression is manifestly Lorentz-invariant. Therefore when the separation of the spacetime points x and y is spacelike, it is possible to make a Lorentz transformation to a new frame that makes the two times the same, with distinct spatial coordinates, without changing the result. Then the expression becomes an equal time commutator, which we already know vanishes.

If one defines advanced and retarded Green functions by $G_A(x, y) = \theta(x^0 - y^0)G(x, y)$ and $G_R(x, y) = -\theta(y^0 - x^0)G(x, y)$, then in the limit $m \to 0$ these give the Green functions D_A and D_R, which were defined in Chapter 6. No linear combination of these Green functions is exactly what one wants to describe particle propagation in quantum field theory, which has the peculiar feature that antiparticles behave like particles propagating backwards in time. In these circumstances the correct boundary conditions that incorporate causality are given by a more subtle prescription due to Feynman, which we will not describe here.

7.6 Feynman diagrams and virtual particles

In the preceding section we described free quantum fields. To get an interesting physical theory, one needs to allow the fields to interact. In quantum electrodynamics (QED), the example we will focus on, the Lagrangian has an additional interaction term that is cubic in the fields (2 powers of the electron field and 1 power of the photon field). We will not go into the mathematical details. Rather, we will give a pictorial description of the interactions implied by such a Lagrangian in terms of Feynman diagrams. These diagrams have precise mathematical meanings, but we will only discuss them qualitatively.

The interaction term has a coefficient that measures the strength by which the electromagnetic field couples to an electron. Not surprisingly, this is just the charge $-e$ of the electron. This gets squared when one computes a probability or a rate. The interaction is weak because, as we discussed earlier, the fine-structure constant $\alpha = e^2/(\hbar c) \approx 1/137$ is small. (We will continue working in units with $\hbar = c = 1$. In these units one simply has $\alpha = e^2$.) This justifies the use of perturbation theory in which quantities of physical interest are computed as power series expansions in α. The first few terms generally give an excellent approximation to the exact answer.

What is a virtual particle?

Let us represent the cubic interaction of QED schematically as $\psi^\dagger \psi A$. Each of the fields has expansions in terms of creation and annihilation operators of the type discussed in the preceding section. Thus A can create or destroy a photon, ψ can create an electron or destroy a positron, and $\psi^\dagger$ can create a positron or destroy an electron. Choosing one of the two options for each of the three fields, the cubic vertex can describe several different reactions $e^- + \gamma \leftrightarrow e^-$, $e^+ + \gamma \leftrightarrow e^+$, and $e^- + e^+ \leftrightarrow \gamma$. Here e^- denotes an electron, e^+ denotes a positron and γ denotes a photon. In each case particles 1 and 2 merge to form particle 3, or particle 3 splits to give particles 1 and 2. Energy and momentum are conserved by the interaction:

$$\vec{k}_1 + \vec{k}_2 = \vec{k}_3 \quad \text{and} \quad E_1 + E_2 = E_3. \tag{7.88}$$

These interactions also satisfy conservation of electric charge, of course.

The reactions described above are not physical processes by themselves. Rather they are building blocks for physical processes. A free electron or positron cannot emit or absorb a real photon as a physical process. For the process to be physical the initial and final particles also have to satisfy the mass-shell condition. This means $E = |\vec{k}|$ for a photon and $E = \sqrt{k^2 + m^2}$ for an electron or positron. (m represents the common mass of the electron and positron.) These equations cannot be satisfied together with the energy and momentum conservation equations.

The Feynman diagrams that represent various contributions to reaction amplitudes in perturbation theory are built up out of repeated cubic interactions of the type described above. In the diagrams, we use solid lines to represent electrons and positrons. Photons are represented by wavy lines. An interaction vertex involves a merger of two solid lines and a wavy line. In principle, there are an infinite number of Feynman diagrams that contribute to any physical process. However, each vertex results in an additional factor of α in the rate, and so one is mostly interested in diagrams that have a small number of vertices.

In any Feynman diagram there are two kinds of lines: internal ones and external ones. The external lines represent the initial and final particles in the reaction of interest. They are required to be physical, and therefore they must satisfy the mass-shell condition. The internal lines, on the other hand, need only be present for a very short period of time, and therefore they are allowed to violate the mass-shell condition. Such unphysical (or "off shell") intermediate particles are called *virtual particles*.

Feynman diagrams for e^+e^- elastic scattering

Let us describe the Feynman diagrams that are needed to compute the scattering amplitude for the reaction $e^- + e^+ \to e^- + e^+$, that is, for elastic electron–positron scattering. (The adjective "elastic" means that the initial and final particles are the same.) At the leading order in the perturbation expansion there are only two Feynman diagrams, which are shown in Figure 7.1. In drawing these figures we use the convention that time increases upwards, so that the initial particles are at the bottom of the drawing and the final ones are at the top.

In Figure 7.1(a) the electron and positron combine to form a virtual photon, which subsequently decays to give the final electron and positron. In Figure 7.1(b) a virtual photon is exchanged between the electron and the positron. If the left-hand vertex is at the earlier time, then the electron emits the virtual photon, which is subsequently absorbed by the positron. Similarly, if the right-hand vertex is at the earlier time, then the positron emits the virtual photon, which is subsequently absorbed by the electron. Figure 7.1(b) is supposed to represent both time orderings for the two vertices, because the mathematical formulas simplify when the two contributions are combined. This could have been anticipated, since we know

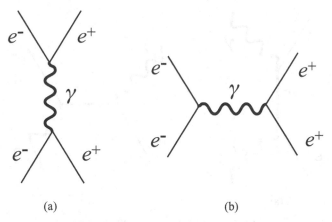

Fig. 7.1. The above Feynman diagrams show electron–positron scattering, with time increasing from bottom to top. In (a) an electron and positron annihilate into a timelike virtual photon, which later decays into another electron and positron. In (b) a spacelike virtual photon is exchanged by the electron and positron. Whether the photon interacts with the electron or positron first depends on the choice of frame.

that when the vertices have a spacelike separation the time-ordering depends on the choice of inertial frame, but the theory is supposed to be Lorentz-invariant.

Both diagrams give a contribution to the total amplitude for the reaction. Calling the amplitude T, we have $T = T_a + T_b$, where the subscripts refer to the two diagrams. The reaction rate (or scattering cross-section) is proportional to $|T|^2$. Therefore there can be interference effects between the various contributions.

Imagine that the electron and positron do not scatter head on, but rather are aimed so as to miss one another by a certain distance, called the impact parameter. We know from classical reasoning that their trajectories will be deflected as a result of the coulomb attraction between them. How is this encoded in the amplitudes we are discussing? Remarkably, this classical coulomb force, appropriately generalized to the relativistic setting, is encoded in diagram (b). (The contribution of (a) is negligible for a significant impact parameter.) The classical force is represented by the exchange of the virtual photon! This remarkable fact takes some of the mystery out of the classical description of the scattering process, which seems to involve action at a distance. Here we see that there is physical object, albeit virtual, that is exchanged between them. Mathematically, the way this works is as follows. If the initial electron has momentum $\vec{k}_i$ and the final one has momentum $\vec{k}_f$, then the momentum transfer $\vec{q} = \vec{k}_i - \vec{k}_f$ is the momentum carried by the virtual photon. According to the Feynman diagram rules, the virtual photon contributes a propagator factor $1/q^2$ to the amplitude. This is the momentum space description of a $1/r$ coulomb potential, related to it by a Fourier transform.

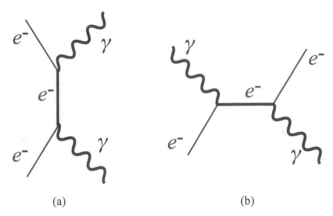

(a)   (b)

Fig. 7.2. These are two leading order Feynman diagrams for Compton scattering. In (a), an electron absorbs a photon and then emits a photon. As we learned in Chapter 4, the emission and absorption are only consistent with spacetime momentum conservation if the intermediate electron is off shell, that is, virtual. In (b), the virtual electron momentum is spacelike, so whether the emission or absorption occurs first depends on the choice of frame.

There is a diagram similar to Figure 7.1(b) that is used to describe the scattering of nucleons (neutrons or protons). In that case a virtual pi meson (or pion) replaces the virtual photon. An important difference in that case is that the pion is massive. As a result the propagator factor is modified to $1/(q^2 + m_\pi^2)$. In position space this corresponds to a Yukawa potential of the form

$$V(r) \sim \frac{1}{r} \exp\left(-r/r_0\right), \tag{7.89}$$

where the range of the force (restoring $\hbar$ and c) is $r_0 = \hbar/(m_\pi c)$. This is a good approximation to the measured range of nuclear forces. Perfect agreement should not be expected, because (unlike QED) the nuclear problem involves lots of subtleties and caveats. Historically, Yukawa predicted the existence of the pi meson, with approximately the right mass, to account for the known range of the nuclear force.

Electron–photon elastic scattering (a reaction called Compton scattering) can be analyzed in a similar manner. Again, there are two Feynman diagrams at the leading order in the perturbation expansion, as shown in Figure 7.2. Note that in Figure 7.2(b) there is a virtual electron line. More precisely, if the left-hand vertex is at the earlier time, a virtual electron propagates from that vertex to the right-hand vertex. On the other hand, if the right-hand vertex is at the earlier time, a virtual positron propagates from there to the left-hand vertex. When their separation is spacelike, the time ordering of the two vertices again depends on the choice of an

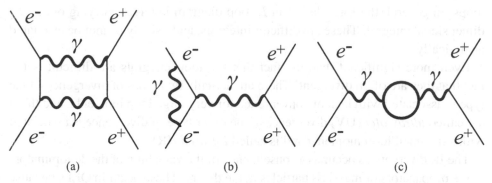

Fig. 7.3. These three fourth-order diagrams are some of the possible terms in the Feynman diagram expansion of the radiative corrections to electron–positron scattering.

inertial frame. This supports Feynman's proposal that a positron can be regarded as an electron that is propagating backwards in time!

Radiative corrections

We have discussed the two Feynman diagrams that are required to compute the amplitude for electron–positron elastic scattering at the leading order in perturbation theory. The leading terms in the perturbation expansion are usually referred to as the *tree approximation*. All the higher-order terms beyond the tree approximation are referred to as *radiative corrections*. At the next order (that is, the leading radiative corrections) there are many more diagrams, some of which are shown in Figure 7.3. Each of these diagrams has four vertices, so that their contributions to the amplitude T are of order α^2. The leading effect of these contributions for $|T|^2$ enters through the interference cross terms with the tree approximation terms T_a and T_b. These interference terms are of order α^3 and thus represent corrections that are smaller than the leading terms by 1 power of α, and thus should give corrections that are roughly of order 1 percent.

Recall that energy and momentum are conserved at each of the interaction vertices. Therefore, if the energy and momentum of each of the external lines are specified, the energy and momentum of the internal lines of any tree diagram are completely determined. However, in the case of the radiative corrections this is no longer the case. These diagrams always contain loops, and these loops can have an arbitrary amount of energy and momentum flowing around them. The Feynman diagram prescription involves carrying out a four-dimensional energy-momentum integral over all possible values of the loop energy and momentum. The leading radiative corrections have one loop and higher radiative corrections have more loops. In general, the formula for an L-loop diagram involves carrying out a $4L$-dimensional integral. These are difficult integrals, and usually cannot be evaluated analytically.

Even more significant than the fact that the loop integrals are difficult is the fact that they are often divergent! There are actually two types of divergences. One type is associated with lack of convergence for very large loop momenta, and these are called *ultraviolet* (UV) divergences. The other type of divergence is associated with very small loop momenta and is called *infrared* (IR).

The IR divergences occur as a consequence of the vanishing of the denominators of the propagators of massless particles in the theory. These occur in QED, because the photon is massless. The IR divergences can be understood physically as being associated with the fact that there is a large rate for emitting very low energy (or *soft*) photons in any reaction. If one makes the reasonable requirement that photons of energy below some very small energy threshold cannot be detected, and therefore should not be counted, then the IR divergence is eliminated.

The UV divergences are a more serious problem. They represent the fact that at very high energies (or, equivalently, very short distances) the theory is incomplete. Such an incomplete theory is only useful if there is an unambiguous prescription for extracting finite results for low energy processes, which are not sensitive to the details of the new physics that is required at high energies. Remarkably, this is sometimes possible.

In the late 1940s Feynman, Schwinger, and Tomonaga independently figured out how to deal with the UV divergences in the case of QED. Their proposal was to absorb the divergences in redefinitions of the mass and charge of the electron as well as the normalization of the fields. This procedure, which is called *renormalization*, ultimately results in unambiguous finite predictions for observable quantities. It has been enormously successful in accounting for a vast amount of high-precision experimental data. For example, theory and experiment agree for the anomalous magnetic moment of the muon to about 1 part in 10^8. To achieve this accuracy requires computing diagrams with up to four loops! The consistency of the renormalization program does not work for every conceivable quantum field theory. Rather, it only works for a special class of theories, which are called *renormalizable*.

Further developments in quantum field theory

Following the success of QED it was natural to attempt to construct quantum field theories to describe the other known fundamental forces, namely the strong and weak nuclear forces and gravitation. There were many obstacles to be overcome before achieving this, however. One was the need to identify what the correct degrees of freedom are to describe these theories. Another was the need to understand the various sorts of symmetries that these theories should possess. Then, one needed to know what the mathematical possibilities are for constructing consistent renormalizable quantum field theories with the desired symmetries and degrees of freedom. Finally, one needed to understand how to extract the physical implications of the theory in order to confront experiment.

In the mid 1950s Yang and Mills constructed a type of quantum field theory that generalizes quantum electrodynamics to a theory that is based on a collection of spin-one fields instead of a single photon field. These fields are associated to symmetry structures of a type described in Chapter 8 – symmetries described by a *nonabelian compact Lie group*. (The group that occurs in the case of QED is Abelian.) The Lie group in Yang–Mills theory describes symmetries that act independently at every point in spacetime. This type of symmetry is called a *local symmetry* or a *gauge symmetry*. It is naturally associated with spin-one fields of the type proposed by Yang and Mills. In Yang–Mills theory the spin-one gauge fields carry the same kind of charge that they couple to, and therefore these spin-one fields interact with one another. This is to be contrasted to the situation in QED, where the electromagnetic field only interacts with charged particles, but the photon itself is electrically neutral.

For a long time it was not known whether or not Yang–Mills theories are renormalizable, and therefore whether it makes sense to use them to construct a physical theory. Renormalizability of Yang–Mills theories was established in the early 1970s by 't Hooft. Following that, a crucial property of these theories, called *asymptotic freedom*, was discovered by Gross, Wilczek, and Politzer. Roughly, asymptotic freedom means that Yang–Mills charges are screened at short distances, in such a way that the effective strength of the interaction decreases logarithmically with decreasing distance. Since the Heisenberg Uncertainty Principle relates short distances to large momenta, one can equivalently say that the interaction strength decreases for large momenta. This is the opposite behavior of QED, where charges get screened at long distances (or small momenta). The source of the difference is the fact that the Yang–Mills fields interact with one another.

These theoretical developments set the stage for the construction of a remarkable theory. At the time one also had a phenomenologically successful quark model of the strongly interacting particles (or hadrons) and lots of experimental information about the various interactions. Putting together these ingredients into a Yang–Mills theory, the so-called *standard model* of elementary particles quickly fell into place. Important contributors to this milestone achievement, in addition to those mentioned above, included Gell-Mann, Weinberg, Glashow, and Salam. The standard model incorporates electromagnetism, the weak nuclear force, and the strong nuclear force within the framework of a renormalizable quantum field theory of the Yang–Mills type. It has been fantastically successful in accounting for a wealth of experimental results over the past 30 years. In fact, experimental physicists have sometimes felt frustrated by not finding anything wrong with it. (Experimentalists generally get a greater sense of achievement from demonstrating that the latest theory isn't quite right than from confirming that it works perfectly.)

Despite the enormous success of the standard model, there are many good reasons to believe that it is not the final theory. We will only mention one of them now: it does not contain gravitation. The inclusion of gravity does not seem to be possible within the framework of renormalizable quantum field theory. We will return to this issue and related matters in Chapter 10.

Exercises

7.1   Let us complete the derivation of Planck's formula Eq. (7.3) for the spectrum of energy in the radiation field. To do this, use Eq. (7.1) for the probability of finding an amount of energy E, but combine this with the fact that at frequency v the allowed energies take the discrete values $E_n = nhv$, where n is a nonnegative integer, corresponding to the presence of n photons. Show that the average number of photons of frequency v at temperature T is given by

$$< n > = \frac{1}{\exp{(hv/k_B T)} - 1} \qquad (E7.1)$$

and explain why Eq. (7.3) follows from this result.

7.2   Suppose a light bulb emits $10\,W$ of light at a wavelength of $5000\,Å$. How many photons are emitted per second?

7.3   For the light bulb in the preceding problem, what is the radiation pressure at a distance of $1\,m$?

7.4   Suppose one wants to probe an atomic nucleus with an electron beam to achieve a spatial resolution 10^{-15} cm. (The size of a proton or neutron is about 10^{-13} cm.) In view of the Heisenberg Uncertainty Principle, what is the minimum beam energy that is required?

7.5   The combination $e^4 m/\hbar^2$, which appears in Eq. (7.19), can be recast as $\alpha^2 mc^2$, where $\alpha = e^2/(\hbar c) \approx 1/137$ is called the fine structure constant. Use this fact and the known rest-mass energy of the electron to compute the hydrogen atom ground state energy E_1.

7.6   What is the energy of the emitted photon when a hydrogen atom makes a transition from an $n = 2$ state to an $n = 1$ state?

7.7 The ground state wave function $\psi_0(x)$ of the harmonic oscillator is a solution of the equation $a\psi_0 = 0$, where a is the lowering operator defined in Eq. (7.21). Find $\psi_0(x)$ by solving this equation. For what coefficient is the solution normalized?

7.8 We have pointed out that the wave function $\psi_n \sim (a^\dagger)^n \psi_0$ is a harmonic oscillator energy eigenstate with $E = (n + \frac{1}{2})\hbar\omega$. For what numerical coefficient is this a normalized wave function?

7.9 Show that $|\lambda\rangle \equiv \exp(\lambda a^\dagger)\psi_0$ is an eigenstate of the lowering operator a. What is its eigenvalue?

7.10 Show that $w^{a^\dagger a}|\lambda\rangle = |w\lambda\rangle$.

7.11 Verify Eqs. (7.36), (7.40), and (7.41).

7.12 Verify Eq. (7.53).

7.13 Verify that Eq. (7.62) is solved by Eq. (7.63).

7.14 Prove that the measure in Eq. (7.77) is Lorentz-invariant.

7.15 Consider a momentum space pion propagator of the form $D(q) = 1/(q^2 + m_\pi^2)$. Compute its three-dimensional Fourier transform, $\int e^{iq \cdot x} D(q)d^3q$, and show that this gives the Yukawa potential described in Eq. (7.89).

8

Group theory and relativity

In previous chapters we have encountered frequently the concept of symmetry of physical laws and symmetry of physical systems under mathematical or physical transformations. The primary examples of symmetry transformations that we have encountered were spatial rotations and Lorentz transformations. The concept of symmetry is very important in physics, not only in the context of transformations of space and time, but also in many others. As it happens, there is a beautiful branch of mathematics, called group theory, which is ideally suited to the description of symmetry. The application of group theory to physics, especially in the context of quantum mechanics, was worked out by Wigner, Weyl, and others in the 1920s and 1930s. By now it is part of the standard toolkit of an educated physicist. This chapter will introduce basic group theory with an emphasis on the application to the descriptions of rotational symmetry and Lorentz symmetry, which are the examples of greatest relevance here.

8.1 What is a group?

Rotations and Lorentz transformations are examples of transformations that one can imagine applying to physical objects, collections of objects, systems of equations (the Maxwell equations, for example), or even the entire Universe. When they are symmetries of the system in question, as in the case of the Maxwell equations, the transformed system is identical to the original one. However, it is also possible to transform a system that does not possess the symmetry, in which case the transformed system looks different from the original one. The set of all transformations that do describe symmetries form a group (which we will define later) that is a characteristic property of the system in question. The more symmetry the system possesses, the larger the symmetry group, and the more powerful are the constraints implied by it.

Let us begin by explaining the notion of a transformation group. For this purpose, suppose that T_1 and T_2 denote two transformations (such as rotations, Lorentz transformations, translations, etc.). Now consider first applying the transformation T_1 and then (to the resulting system) applying the transformation T_2. The combined effect is a new composite transformation, which is denoted $T_2 T_1$. Note the convention that the transformation that is applied first is written to the right. The sequential application of the transformations is represented mathematically by juxtaposition or "multiplication." Of course, this is not ordinary multiplication, because T_1 and T_2 are not numbers. Rather, they are some sort of operators, and this is an operator multiplication. In fact, it often happens that the multiplication is noncommutative, which means that $T_2 T_1 \neq T_1 T_2$.

If one starts with an arbitrary set of transformations and then enlarges the set by including the transformations that undo them (inverse transformations) as well as all transformations that can be obtained by multiplying these transformations any number of times, eventually one reaches a closed algebraic system that is a *group*. Group theory, which is a very mature branch of mathematics, is helpful for analyzing the implications of symmetry. It is useful for classical systems, which is the main concern in this book, but it really comes into its own in the context of quantum mechanics (which was introduced in Chapter 7).

Simple examples of groups

Symmetrical objects or systems are familiar from everyday experience. For example, a perfectly round ball has spherical symmetry. This means that it is invariant under rotations about any axis that passes through the center of the ball. These rotations form a group, called the group of rotations in three dimensions, denoted by the symbol $SO(3)$. (We will elaborate on this notation later.) An even simpler example is the rotational symmetry of a circle. Rotations in the plane of the circle, that leave the center fixed, are symmetry transformations that form the group of rotations in two dimensions and are denoted $SO(2)$. This group was discussed in some detail in Chapter 1.

The notion of rotational symmetry generalizes to an arbitrary (positive integer) number of dimensions. So, as you may have guessed, the group of rotations in D dimensions is denoted $SO(D)$. This group describes the symmetry of a sphere. For example, the unit sphere in D dimensions is described in Euclidean coordinates by the equation $(x^1)^2 + (x^2)^2 + \cdots + (x^D)^2 = 1$. Such a sphere has $D - 1$ dimensions. The fact that it is described by one equation means that it has one fewer dimensions than the space in which it resides. The mathematical notation for such a sphere is S^{D-1}.

The spheres discussed in the preceding paragraphs have additional symmetries beyond rotations. These are reflections. Since the reflections can be carried out in an infinite number of different planes that pass through the center of the sphere,[1] there are an infinite number of them. However, reflections about two different planes are related by a rotation. Altogether, the complete symmetry of a sphere is given by a group that contains reflections as well as rotations. In the case of D dimensions, this group is denoted $O(D)$. Clearly, this group contains the group $SO(D)$ as a subset, which is obtained by discarding the reflections. Whenever one group is a subset of another group, it is called a *subgroup*.

As another example, let us consider the symmetries of the Euclidean plane, described by Cartesian coordinates x and y (or x^1 and x^2). This plane is mapped into itself by rotations by an arbitrary angle about any point in the plane. Also, it is mapped to itself by a rigid shift (or translation) by an arbitrary amount in the x or y directions, or by a combination of both. The set of all possible rotations and translations of the plane form a group, called the Euclidean group in two dimensions. It can be denoted $E(2)$. There is a straightforward generalization to a space of D Euclidean dimensions, for which the symmetry group is the Euclidean group $E(D)$.

The examples provided so far are all infinite groups. This means that there are an infinite number of different symmetry transformations. For example, the rotation angle is a continuous parameter that can take an arbitrary value (between 0 and 2π). There are also examples of systems for which the number of symmetry transformations is finite. For example, consider a regular polygon in the plane with n sides. Such an object is invariant under rotations about the center by an angle that is an integer multiple of $2\pi/n$. Altogether, there are n inequivalent possibilities. So this symmetry group, which is sometimes denoted Z_n and sometimes denoted A_n, has n distinct elements. If one also adjoins the reflection symmetries one obtains a group with $2n$ elements denoted D_n.

The four properties that define a group

Now that we have explained the concept of a group, we can give the formal definition. First of all, a group is a *set*. More specifically, at least in physical applications, it is a set of transformations, which often correspond to symmetries of a physical system. This set can be either finite or infinite, as we illustrated above. When four additional properties, described below, are satisfied the set becomes a group. The group is called finite or infinite according to whether the set is finite or infinite.

---

[1] We imagine here that the planes have $D - 1$ dimensions. A reflection in a plane maps a vector to a reflected vector by reversing the sign of the part of the vector that is perpendicular to the plane.

Let us list the four basic defining properties of a group first, and then following that we will explore them in greater detail.

- A group has a binary composition rule (often called group multiplication) that assigns an element of the group to each ordered pair of elements of the group.
- Group multiplication is required to be associative.
- A group contains a unique identity element.
- Every element of a group has a unique inverse.

The first property (or axiom) that turns a set into a group is that there is a rule for combining any two elements of the group that gives a third element of the group. Thus, if a and b are any two elements of a group G, then the group product ab is also an element of G. Here we have represented the group composition rule as multiplication. This is usually a convenient convention, though sometimes it is more natural to represent it by addition. This property is often referred to as *closure*. This simply means the set G is closed under group multiplication. Mathematically, group multiplication can be viewed as a map from $G \times G$ to G.

The group composition rule is not required to be commutative. This means that we do not require that ab and ba are the same elements of G. The special class of groups for which $ab = ba$ for every pair of elements are called *Abelian groups*. Those for which this is not the case are called *nonabelian groups*. (The word derives the name of the mathematician Abel.) The addition notation for group composition is often used for abelian groups, that is, $a + b = b + a$.

The second defining property of a group is *associativity*. This is the requirement that for any three elements a, b, c of a group G, $(ab)c = a(bc)$. The parentheses serve to indicate the order in which the group multiplications are carried out. Thus, to be very explicit, $(ab)c$ means that the multiplication ab should be carried out first, and the resulting element of the group should then multiply c from the left. A consequence of associativity is that the parentheses are unnecessary. One can simply write abc. The meaning is unambiguous, because either of the two possible interpretations gives the same result.

Multiplication of square $n \times n$ matrices is also associative and need not be commutative. This fact at least raises the possibility that the multiplication table for a group G can be realized by appropriately chosen matrices. Certainly, if we had a multiplication rule that was not associative, we would know that this was not possible. The representation of groups by sets of matrices is an important issue to which we will return later. As an aside, let us remark that nonassociative mathematical systems are sometimes considered by mathematicians (and even by physicists), but they are certainly not groups.

The third defining property of a group is the existence of an identity element. This means that every group G contains a unique element, often called e, with the property that $ea = ae = a$ for every element $a \in G$. Note that e is both a left identity and a right identity. Another possible notation for the identity element is 1, but using the more abstract symbol e serves to emphasize that we are not dealing with ordinary numbers. In an additive notation for group composition, it is natural to represent the identity element by 0, though a more abstract symbol could again be used. Thus, for an additive group, $0 + a = a + 0 = a$. In terms of transformations of a physical system, the identity transformation is the trivial one that leaves the system untouched.

The final defining property of a group is the existence of an inverse. Specifically, for every element $a \in G$, we require the existence of another element of G, denoted a^{-1}, with the property that $aa^{-1} = a^{-1}a = e$. In terms of transformations, a^{-1} is the transformation that undoes the transformation a. Note that the inverse of a^{-1} is a itself, since a satisfies the defining equations for such an inverse. For an Abelian group that is described by an additive notation, the natural notation for the inverse of a is $-a$.

We have now described all four of the defining properties of a group. Any set that has a composition rule with these properties is a group. A group might satisfy additional properties that turns it into an example of a mathematical category with additional structure. However, in such a case, it is still a group.

Group representations

In the preceding we have given the abstract definition of a group. As physicists, we are mainly concerned with specific physical realizations of groups, where they act as transformations of specific physical systems. Such systems usually are described in some coordinate basis, and so the effect of the transformation is to make a linear change of the coordinates. This can be described as multiplication by a matrix. Thus, it is natural from a physical point of view to represent group elements by matrices. This also turns out to be useful for understanding the mathematical structure of groups.

By definition, a group representation is an explicit realization of the group multiplication rule in terms of matrices made out of ordinary numbers (real or complex). Let us consider square matrices with n rows and n columns. The integer n is called the dimension of the group representation. In the special case of a one-dimensional representation, the group elements are represented by ordinary numbers, which can be viewed as matrices with just one row and one column. Given a group G, suppose that we have a rule for assigning a matrix to every element of the group. Specifically, to each element $a \in G$ we assign an $n \times n$ matrix $D(a)$. More explicitly, $D(a)$ consists of n^2 numbers $D_{ij}(a)$, where $i, j = 1, 2, \ldots, n$ label the rows and columns.

The matrices $D(a)$ are not completely arbitrary. Rather, they are supposed to be chosen in such a way as to capture the group multiplication rule. This means that multiplying the matrices should correspond to multiplying the group elements. Expressed as an equation, the rule is

$$D(a)D(b) = D(ab). \tag{8.1}$$

Displaying the row and column labels, this equation takes the form

$$\sum_{k=1}^{n} D_{ik}(a)D_{kj}(b) = D_{ij}(ab). \tag{8.2}$$

Such a rule for assigning matrices to group elements is called a group representation. More specifically, when the matrices have n rows and columns, it is called an n-dimensional representation.

There are a couple of simple facts about group representations that should be noted. First of all, since $D(e)D(a) = D(ea) = D(a)$, the identity element is represented by the unit matrix I:

$$D_{ij}(e) = \delta_{ij} = I_{ij} \quad i, j = 1, 2, \ldots, n. \tag{8.3}$$

Similarly, since $D(a^{-1})D(a) = D(a^{-1}a) = D(e) = I$, we deduce that the inverse element is represented by the inverse matrix

$$D(a^{-1}) = [D(a)]^{-1}. \tag{8.4}$$

Also, since multiplication of finite-dimensional matrices is associative, the associativity rule is essential for the existence of a matrix representation.

Representations are allowed to be infinite-dimensional. For such representations the matrices have an infinite number of rows and columns. Although they do exist, there is no point in considering them if the group itself is finite. The reason is that all the essential structure of the group is captured by a certain collection of finite-dimensional representations. However, for certain types of infinite groups, they can be important. Then associativity is a nontrivial requirement.

The defining equation $D(a)D(b) = D(ab)$ maps one mathematical system (a group) into another one (a collection of matrices) so as to preserve its defining structure (the group composition rule). This is an example of what mathematicians call a *homomorphism*. It should be emphasized that there need not be a one-to-one correspondence between the group elements and the matrices. $D(a) = D(b)$ need not imply that $a = b$. For some group representations this is true and for others it is not. When there is a one-to-one correspondence, the group representation is called *faithful*. In this case, the homomorphism is also an *isomorphism*.

There are families of representations that, in a sense to be defined, are *equivalent*. So, in classifying representations of a group, one would only want to count this family once. Put differently, we could choose any convenient representative of this family. To be explicit, let $a \to D(a)$ be an N-dimensional representation of a group G. Then another N-dimensional representation $a \to \tilde{D}(a)$ is considered equivalent to the first one if and only if there exists a nonsingular (that is, invertible) $N \times N$ matrix A such that $\tilde{D}(a) = A^{-1}D(a)A$ for all $a \in G$. This definition makes sense because

$$\tilde{D}(a)\tilde{D}(b) = A^{-1}D(a)AA^{-1}D(b)A$$
$$= A^{-1}D(a)D(b)A = A^{-1}D(ab)A = \tilde{D}(ab). \tag{8.5}$$

A representation D is called *unitary* if all the representation matrices $D(a)$ are unitary matrices.[2] In general, group representations need not be unitary. However, for all finite groups, and for large classes of infinite ones, it turns out that every representation is equivalent to a unitary one.

Suppose we are given two representations $D^{(1)}$ and $D^{(2)}$ of a group G, having dimensions N_1 and N_2, respectively. Then we can form a new representation D of dimension $N_1 + N_2$ as a direct sum as follows.

$$D(a) = \begin{pmatrix} D^{(1)}(a) & 0 \\ 0 & D^{(2)}(a) \end{pmatrix}. \tag{8.6}$$

Here we have written an $(N_1 + N_2) \times (N_1 + N_2)$ matrix in block form. These matrices can be verified easily to satisfy the definition of a representation. A representation of this type is called *reducible*. Clearly, it does not add knowledge about the group to what we already had from knowing $D^{(1)}$ and $D^{(2)}$. So the important representations are those that are *irreducible*. By definition, they are not of this block form, nor are they equivalent to a representation of this block form.

To conclude this brief introduction to group representations, the representations that are usually of greatest interest are the unitary irreducible representations. For finite groups and a large class of infinite ones that is pretty much the whole story. However, for certain types of infinite groups (noncompact Lie groups) the story is more complicated. Rotation groups are of the simpler (compact) type, but the Lorentz group is of the more complicated (noncompact) type.

---

[2] Let us recall the definitions. The adjoint of a matrix M, denoted $M^\dagger$, is the transpose and complex conjugate, that is, $(M^\dagger)_{ij} = (M_{ji})^*$. A unitary matrix is one for which the adjoint is the inverse, that is, $MM^\dagger = M^\dagger M = I$.

8.2 Finite and infinite groups

Permutation groups

The groups of greatest relevance to this book are groups containing rotations and Lorentz transformations. However, finite groups are also important – and the general theory is somewhat simpler. They arise, for example, in characterizing the symmetries of crystals. An important class of finite groups are groups of permutations. In fact, every finite group can be realized as a subgroup of a permutation group. The group of permutations of N objects is usually called the *symmetric group S_N*. The idea is that given N objects, in a definite order, we can consider changing (or permuting) the order of the objects. Each such reordering is a group operation. The number of distinct elements in the group S_N is the same as the number of distinct orderings, which is $N!$.

There are various ways of describing a specific permutation, and we won't explore every one of them here. One possibility is to write the integers 1 through N in some order inside square brackets as follows $[p_1 p_2 \ldots p_N]$. The idea is that the permutation p takes the first object and puts it in the p_1th place, it puts the second object in the p_2th place, and so forth. It is essential that all the p_is are distinct, so that each of the N integers occurs exactly once. The identity element in this notation is $[12 \ldots N]$. The inverse permutation p^{-1} simply undoes the original one. So the object in the p_1th place is moved to the first place, and so forth.

A special case of a permutation is one that simply interchanges two of the objects and leaves the other $N - 2$ unaffected. Such simple interchanges generate the entire group. This means that by making a suitable sequence of interchanges one can build up an arbitrary permutation. The way one does this is not unique, of course. However, we can divide the group of permutations into two sets, called the even permutations and the odd permutations. Even permutations are built up by an even number of interchanges and odd permutations are built up by an odd number of interchanges. Despite the non-uniqueness pointed out a moment ago, this is a meaningful distinction. Even permutations can only be built up from an even number of interchanges, and odd ones can only be built up by an odd number of interchanges. The identity element is an even permutation, of course.

The set of even permutations consists of exactly half the elements of the group (for $N > 1$), and the odd permutations are the other half. For example, S_3 has six elements of which three are even and three are odd. The even permutations form a subgroup, called the *alternating group*. This is obvious because the product of two even elements can also be described as a product of an even number of interchanges.

The two simplest irreducible representations of the symmetric group S_N play an important role in quantum mechanics. The two representations in question are both one-dimensional. In fact, they are the only one-dimensional representations, and all other irreducible representations of S_N have more than one dimension. The first one is the trivial representation, which assigns the number 1 to every group element. The second one is a bit more interesting. It assigns the number 1 to every even permutation and the number -1 to every odd permutation. This is a representation because even and odd permutations satisfy the rules: odd $\times$ odd $=$ even, odd $\times$ even $=$ odd, and even $\times$ even $=$ even.

In quantum theory there are two classes of particles called bosons and fermions. (This is discussed in much greater detail in Chapter 9.) A system of N identical bosons or fermions is described by a wave function $\Psi(x_1, x_2, \ldots, x_N)$, where x_i represents all the coordinates that are required to describe the ith particle. As usual in quantum mechanics, $|\Psi|^2$ is a probability density. One of the remarkable facts in quantum mechanics is that identical particles are indistinguishable. All that it is meaningful to ask is how the wave function transforms under a permutation of the N identical particles. Nature is kind, and only the two simple representations described above are utilized. Applying a permutation p, the wave function transforms as follows:

$$\Psi(x_{p1}, x_{p2}, \ldots, x_{pN}) = D(p)\Psi(x_1, x_2, \ldots, x_N). \tag{8.7}$$

In the case of bosons, D is the trivial representation, so that $D(p) = 1$ in all cases. This means that the wave function has total symmetry. This simple but profound truth is at the heart of many remarkable phenomena such as superconductivity, lasers, and Bose condensation.

In the case of fermions, the other one-dimensional representation is utilized. This means that the wave function has total antisymmetry. One consequence of this is that the wave function vanishes when the coordinates of two of the fermions approach one another. This is the origin of the Pauli exclusion principle for electrons, which is responsible for the stability of ordinary matter.

Unitary and orthogonal groups

In the previous section we discussed a class of finite groups. Let us now turn to some examples of infinite groups. The groups we will discuss here are examples of *Lie groups*, named for the mathematician Sophus Lie.

In the discussion of group representations we explained that the multiplication rule for a group can be represented by associating a square matrix to each group element and matrix multiplication to group multiplication. Here we will turn the story around and consider groups that are defined in terms of matrices and their multiplication in the first place. For such a group we get one faithful representation for free – namely, the defining representation, which simply assigns each matrix to itself. There will also be other representations, of course.

Let us begin by defining the Lie group $O(N)$. This is the group formed by all $N \times N$ real orthogonal matrices. By definition, an orthogonal matrix A, with matrix elements denoted A_{ij}, is one that satisfies the equation

$$\sum_{k=1}^{N} A_{ik} A_{jk} = \delta_{ij}. \tag{8.8}$$

In matrix notation this takes the more compact form $AA^T = I$. Here I denotes the unit matrix ($I_{ij} = \delta_{ij}$), and A^T denotes the transpose matrix ($(A^T)_{kj} = A_{jk}$).

Let us prove that orthogonal matrices form a group. First we note that the associativity axiom is satisfied automatically, since it is always valid for multiplication of finite-dimensional matrices made from numbers (real or complex). Then we note that the $N \times N$ matrix I is an orthogonal matrix, and it is obviously the identity element of the group. Next we must show that the product of two orthogonal matrices is itself an orthogonal matrix. This requires checking whether, for any two $N \times N$ orthogonal matrices A and B, it is true that $(AB)(AB)^T = I$. This is verified easily using the identity $(AB)^T = B^T A^T$ and associativity to rearrange the parentheses. The final requirement that needs to be checked is that every orthogonal matrix has an inverse, and that this inverse is itself an orthogonal matrix. To prove this we note that the inverse of an orthogonal matrix A is A^T. Therefore we need to check whether $A^T (A^T)^T = I$. This is true, because $(A^T)^T = A$.

As we have said, any element A of the Lie group $O(N)$ satisfies the equation $AA^T = I$. Let us take the determinant of each side of this equation. To do this we recall the following facts: $\det I = 1$, $\det (AB) = \det A \det B$, and $\det A^T = \det A$. Applying these facts to the equation $AA^T = I$, one deduces that $(\det A)^2 = 1$. Hence $\det A = \pm 1$. The orthogonal matrices that satisfy $\det A = 1$ form a subgroup of $O(N)$, which is denoted $SO(N)$. The letter S stands for "special," meaning that the determinant is unity, or the matrix is *unimodular*.

The physical interpretation is as follows: orthogonal matrices with $\det A = 1$ can be interpreted as rotations of an N-dimensional Cartesian space. (The N-dimensional rotation group was discussed previously in Chapter 5.) Those with $\det A = -1$ can be interpreted as reflections of the same N-dimensional Cartesian space.

This construction has a straightforward generalization to $N \times N$ complex matrices. As mentioned in an earlier footnote, a complex $N \times N$ matrix A is *unitary* if and only if it satisfies the equation $AA^\dagger = 1$. Here $A^\dagger$ is called the *adjoint* (or Hermitian conjugate) of A. It is the complex conjugate of the transpose. In terms of components $(A^\dagger)_{ij} = (A_{ji})^*$. The claim is that the set of all such matrices form a group called $U(N)$. The proof that the axioms of a group are satisfied works the same way as before, and we leave it as an exercise. Note that, as a special case, the matrix A is allowed to be real. For such a matrix, adjoint and transpose are the same thing. Therefore $O(N)$ is a subgroup of $U(N)$.

Let us identify an important subgroup of $U(N)$ by the same method we used for $O(N)$. We form the determinant of both sides of the equation $AA^\dagger = 1$. By the same reasoning as before, this leads to the equation $|\det A|^2 = 1$. Therefore $|\det A| = 1$, and $\det A$ is a phase. As before, the unimodular ($\det A = 1$) unitary matrices form a subgroup denoted $SU(N)$.

8.3 Rotations form a group

Rotations in two dimensions were discussed in Chapter 1, where the group properties were introduced (without mentioning groups). Specifically, we discussed the 2×2 matrix $R(\theta)$ describing a rotation by an angle θ. It was shown that $R(0)$ is the identity matrix I, and that $R(\theta)R(-\theta) = I$, so that $R(-\theta)$ is the inverse of $R(\theta)$. In an exercise you showed the group closure rule $R(\theta_1)R(\theta_2) = R(\theta_1 + \theta_2)$. In this section we wish to generalize these facts to rotations in an arbitrary number of dimensions with special emphasis on the particular case of three dimensions.

As we discussed in the beginning of this chapter, one way of thinking about rotation groups is as symmetries of spheres. Since an n-dimensional sphere S^n is naturally embedded in $n + 1$ Euclidean dimensions (by the equation $\sum_i (x^i)^2 = R^2$), the rotations of S^n can be thought of as acting in $n + 1$ dimensions. The group of symmetries is $O(n + 1)$ if we include reflections as well as rotations, or $SO(n + 1)$ if we restrict to rotations only. Recall that the $(n + 1) \times (n + 1)$ matrices that represent the group elements have determinant -1 for reflections and $+1$ for rotations. The group $SO(2)$, which describes the rotations of a circle, is relatively trivial since it is Abelian. In this section, we wish to explore rotations in three dimensions, given by the group $SO(3)$, which is decidedly less trivial. Needless to say, it is the example of greatest physical interest. We will explore this Lie group by focusing on infinitesimal transformations – ones that differ infinitesimally from the identity. Following that, we will explore a closely related group, $SU(2)$, which is required to describe rotations of fermions in quantum mechanics. Finally, we will make some brief remarks about rotations in higher dimensions.

Infinitesimal rotations and the SO(3) algebra

In Section 1.4 we considered infinitesimal rotations in two dimensions. The type of analysis we described there is very powerful, and it will be generalized here to the case of rotations in three dimensions. The basic conceptual fact is that in the case of a continuous group the structure of the group in an infinitesimal neighborhood of the identity element contains all the essential information required to understand arbitrary finite group transformations. The only caveat is that different groups can be identical in the neighborhood of the identity. When this happens, the two groups are distinguished by their global topology. As we will see, the two groups have the same Lie algebra. A pair of groups that are related in this way, which we will describe in detail, are $SO(3)$ and $SU(2)$.

Recall that rotation matrices are orthogonal matrices with determinants of $+1$, and an orthogonal matrix is one that satisfies $AA^T = A^T A = I$. Here I denotes the unit matrix. Since we wish to focus on rotations in three dimensions, we will be interested in the case that A is a 3×3 matrix. For example, a rotation by an angle θ in the $x-y$ plane is given by the matrix

$$R_z(\theta) = \begin{pmatrix} \cos\theta & -\sin\theta & 0 \\ \sin\theta & \cos\theta & 0 \\ 0 & 0 & 1 \end{pmatrix}. \tag{8.9}$$

Now suppose that the rotation angle θ is very small. Expanding to first order we have

$$R_z(\theta) = \begin{pmatrix} 1 & -\theta & 0 \\ \theta & 1 & 0 \\ 0 & 0 & 1 \end{pmatrix} + O(\theta^2) = I + \theta \tilde{L}_z + O(\theta^2), \tag{8.10}$$

where

$$\tilde{L}_z = \begin{pmatrix} 0 & -1 & 0 \\ 1 & 0 & 0 \\ 0 & 0 & 0 \end{pmatrix}. \tag{8.11}$$

This matrix encodes infinitesimal rotations about the z axis. It is referred to as an infinitesimal generator, because finite transformations can be built up by repeated action of infinitesimal ones.

Frequently in the physics literature, one includes an extra factor of i in the definition of the infinitesimal generators in order that the matrices are Hermitian. Of course, if one does this, compensating factors of $-i$ have to be added to their coefficients so that the formulas are unchanged. Therefore we define $\tilde{L}_z = iL_z$, and it follows that $R_z(\theta) = I + i\theta L_z + \cdots$ and

$$L_z = -i \begin{pmatrix} 0 & -1 & 0 \\ 1 & 0 & 0 \\ 0 & 0 & 0 \end{pmatrix}. \tag{8.12}$$

A rotation by an angle θ about a specified axis can be obtained by rotating N times by an angle θ/N. Infinitesimal rotations enter in the limit $N \to \infty$. This limit gives an exponential of the infinitesimal generator, that is,

$$e^{\theta \tilde{L}_z} = e^{i\theta L_z} = \begin{pmatrix} \cos\theta & -\sin\theta & 0 \\ \sin\theta & \cos\theta & 0 \\ 0 & 0 & 1 \end{pmatrix} = R_z(\theta). \tag{8.13}$$

To see that this is correct, note that it has the correct small θ expansion, and it has the group property $e^{i\theta_1 L_z} e^{i\theta_2 L_z} = e^{i(\theta_1 + \theta_2)L_z}$. Also, it is an orthogonal matrix, because

$$(e^{i\theta L_z})^T = e^{i\theta L_z^T} = e^{-i\theta L_z}, \tag{8.14}$$

which is the inverse of the original matrix. In a completely analogous way we can derive infinitesimal generators for rotations about the x and y axes:

$$L_x = -i \begin{pmatrix} 0 & 0 & 0 \\ 0 & 0 & -1 \\ 0 & 1 & 0 \end{pmatrix}, \tag{8.15}$$

$$L_y = -i \begin{pmatrix} 0 & 0 & 1 \\ 0 & 0 & 0 \\ -1 & 0 & 0 \end{pmatrix}. \tag{8.16}$$

The infinitesimal generators of a Lie group form an algebraic system called a Lie algebra. The closure of the group under multiplication translates into closure of the vector space spanned by the infinitesimal generators under commutation. Explicitly, for the example at hand, one has the commutation relation

$$[L_x, L_y] = L_x L_y - L_y L_x = iL_z, \tag{8.17}$$

as is easily verified. Two similar formulas are obtained by cyclically permuting the indices x, y, z.

Representation by differential operators

Let us consider what the preceding analysis implies for the Cartesian coordinates x, y, z. For a rotation by an infinitesimal angle θ about the z axis, we have

$$\begin{pmatrix} x' \\ y' \\ z' \end{pmatrix} = \begin{pmatrix} 1 & -\theta & 0 \\ \theta & 1 & 0 \\ 0 & 0 & 1 \end{pmatrix} \begin{pmatrix} x \\ y \\ z \end{pmatrix} + O(\theta^2) = \begin{pmatrix} x - \theta y \\ y + \theta x \\ z \end{pmatrix} + O(\theta^2) \tag{8.18}$$

This can again be written in the form

$$\begin{pmatrix} x' \\ y' \\ z' \end{pmatrix} = (1 + i\theta L_z + O(\theta^2)) \begin{pmatrix} x \\ y \\ z \end{pmatrix} \tag{8.19}$$

but now with L_z expressed as a differential operator

$$L_z = -i\left(x\frac{\partial}{\partial y} - y\frac{\partial}{\partial x}\right). \tag{8.20}$$

Similar expressions for L_x and L_y are obtained by cyclically permuting the coordinates x, y, z. Formulas of this type are familiar to students of quantum mechanics as operators that (aside from a factor of $\hbar$) represent orbital angular momentum.

The infinitesimal generators, when represented as differential operators, satisfy the same commutation relations as before. Representations of infinitesimal generators by differential operators play an important role in the general theory of Lie groups and Lie algebras.

The relationship between SO(3) and SU(2)

The three-dimensional rotation group $SO(3)$ is related closely to the group $SU(2)$, which consists of all two-dimensional unitary matrices with unit determinants. In fact, there is a two-to-one relationship between the two groups, with two distinct elements of $SU(2)$ corresponding to each element of $SO(3)$. One says that $SU(2)$ is a *covering group* of $SO(3)$. There are a variety ways of understanding this relationship, and we will mention a couple of them here. One crucial fact is that the two groups are identical in the vicinity of the identity element, and therefore they have the same Lie algebra.

Let us begin by giving a couple of explicit descriptions of the matrices that form the group $SU(2)$. One possible parametrization is

$$U = \begin{pmatrix} a & b \\ -b^* & a^* \end{pmatrix} \tag{8.21}$$

subject to the restriction that $\det U = |a|^2 + |b|^2 = 1$. As the reader can verify, this specific parametrization ensures that the matrices are unitary ($U^\dagger U = 1$). If one separates real and imaginary parts by writing $a = a_1 + ia_2$ and $b = b_1 + ib_2$, then the determinant condition takes the form $(a_1)^2 + (a_2)^2 + (b_1)^2 + (b_2)^2 = 1$, which is the equation for a three-dimensional sphere S^3. Therefore, one says that the *group manifold* of $SU(2)$ is S^3.

Let us now consider $SU(2)$ matrices that differ from the identity matrix by an infinitesimal amount. Such matrices can be written in terms of three real infinitesimal parameters $\alpha_x, \alpha_y, \alpha_z$ as follows

$$U = \begin{pmatrix} 1 + i\alpha_z & i\alpha_x + \alpha_y \\ i\alpha_x - \alpha_y & 1 - i\alpha_z \end{pmatrix} + O(\alpha^2) \qquad (8.22)$$

This corresponds to $a_1 = 1$, $a_2 = \alpha_z$, $b_1 = \alpha_y$, and $b_2 = \alpha_x$, which satisfies the determinant condition to order $(\alpha)^2$. This expression can be recast in terms of a basis of three 2×2 matrices $\vec{\sigma}$ called Pauli matrices:

$$U = 1 + i\vec{\sigma} \cdot \vec{\alpha} + O(\alpha^2), \qquad (8.23)$$

where (as in Chapter 7)

$$\sigma_x = \begin{pmatrix} 0 & 1 \\ 1 & 0 \end{pmatrix} \quad \sigma_y = \begin{pmatrix} 0 & -i \\ i & 0 \end{pmatrix} \quad \sigma_z = \begin{pmatrix} 1 & 0 \\ 0 & -1 \end{pmatrix}. \qquad (8.24)$$

The Pauli matrices are infinitesimal generators of $SU(2)$. Let us therefore examine their commutation relations. One finds that

$$[\sigma_x, \sigma_y] = 2i\sigma_z, \qquad (8.25)$$

and two more equations given by cyclic permutation of indices. Aside from the factor of 2, these are the same commutation relations that we found in the preceding subsections. Thus the three matrices $\vec{\sigma}/2$ have the same commutation relations as the three operators $\vec{L}$ that generate $SO(3)$. This demonstrates that the two Lie groups have the same Lie algebra.

Let us now consider finite $SU(2)$ transformations. We described one way of parameterizing them above, but let us now introduce another one. Consider a matrix of the following form

$$U = \exp(i\vec{\sigma} \cdot \vec{\alpha}), \qquad (8.26)$$

where now the vector $\vec{\alpha}$ represents three finite real numbers. This agrees with the previous expression to first order in α, but it is exactly unitary. To show this, note that the Pauli matrices are Hermitian, and therefore the adjoint of U is equal to its inverse.

The relationship between $SU(2)$ and $SO(3)$ can be obtained by considering the traceless Hermitian matrix

$$X = \vec{\sigma} \cdot \vec{x} = \begin{pmatrix} z & x - iy \\ x + iy & -z \end{pmatrix}. \qquad (8.27)$$

If one performs a unitary transformation, by an element of the group $SU(2)$ one obtains a transformed matrix

$$X' = U^{-1}XU = \vec{\sigma} \cdot \vec{x}'. \qquad (8.28)$$

Note that X' is also Hermitian and traceless, and therefore $\vec{x}'$ consists of three real numbers that are related to $\vec{x}$ by a real linear transformation. Furthermore this linear transformation is a rotation. To see this, we note that $-\det X = x^2 + y^2 + z^2$, and $\det X' = \det X$. Therefore the vectors $\vec{x}'$ and $\vec{x}$ have the same length. In this way one associates a unique rotation to each element of the group $SU(2)$.

To understand this better, let us specialize to the case of transformations generated by σ_z, which will correspond to rotations about the z axis. One has

$$U = \exp(i\alpha\sigma_z) = \begin{pmatrix} e^{i\alpha} & 0 \\ 0 & e^{-i\alpha} \end{pmatrix}. \tag{8.29}$$

Moreover, for this matrix one finds that

$$X' = U^{-1}XU = \vec{\sigma} \cdot \vec{x}' \tag{8.30}$$

corresponds to a rotation about the z axis by an angle 2α. In particular, the matrix $U = -I$, corresponding to $\alpha = \pi$, corresponds to a rotation by an angle of 2π, which is the identity element of $SO(3)$.

We can now understand the two-to-one relationship between the two groups. Given an $SU(2)$ matrix U, there is a another distinct $SU(2)$ matrix $-U$, which corresponds to the same $SO(3)$ rotation. A fancier way of saying this is that the two $SU(2)$ matrices I and $-I$ form an invariant subgroup of $SU(2)$, and $SO(3)$ is isomorphic to $SU(2)$ modulo this invariant subgroup. Another way of saying the same thing is that the map from $SU(2)$ to $SO(3)$ described above is a two-to-one homomorphism.

Spinor representations

Because of the two-to-one relationship between the two groups, the three-dimensional matrices that define $SO(3)$ provide a three-dimensional representation of $SU(2)$, whereas the two-dimensional matrices of $SU(2)$ do not provide a two-dimensional representation of $SO(3)$. More generally, it turns out that $SU(2)$ has irreducible representations of every positive dimension, whereas only the ones of odd dimension are also representations of $SO(3)$. The even-dimensional representations of $SU(2)$ are sometimes called *spinor representations*.

It is customary to label the irreducible representations by their "spin" j. The spin is related to the dimension n of the representation by $n = 2j + 1$. Thus the irreducible representations of $SO(3)$ are labeled by $j = 0, 1, 2, \ldots$, and the spinor representations of $SU(2)$, which are not representations of $SO(3)$, have $j = 1/2, 3/2, 5/2, \ldots$.

Rotations in higher dimensions

Rotations in D dimensions are described by the Lie group $SO(D)$. This can be analyzed in much the same way that we analyzed $SO(3)$. However, to describe this generalization it is convenient to make some minor notational changes. First of all, we label the Cartesian coordinates by $x^1, x^2, \ldots, x^D$, rather than using distinct letters x, y, z. Secondly, in D dimensions specific rotations should be described by reference to the two-dimensional plane that is rotated. In three dimensions such a rotation can be characterized by the third direction, which does not transform, and can be identified as the axis of rotation. In D dimensions, the generalization of this statement is that there are $(D - 2)$ orthogonal directions that do not transform under a specific rotation. Thus, what we have been calling a rotation about the z axis, we would now characterize as a rotation in the x^1–x^2 plane.

Let us now consider the generalization of the infinitesimal generators of rotations obtained for $SO(3)$ in Section 8.3. Specifically we wish to generalize the formula in Eq. (8.12) for L_z, which we would now call L_{12}. Altogether, there are $D(D - 1)/2$ infinitesimal generators $L_{ij} = -L_{ji}$. This number is the dimension of Lie group $SO(D)$. In other words, it is the number of real parameters that is required to characterize an arbitrary rotation. The infinitesimal generators L_{ij} describe infinitesimal rotations of two of the D coordinates (x^i and x^j). Therefore they are given by $D \times D$ matrices in which the only nonzero entries occur in the ij and ji positions. There is only one possible generalization of Eq. (8.12) with these properties, namely

$$(L_{ij})_{kl} = i(\delta_{ik}\delta_{jl} - \delta_{il}\delta_{jk}).\tag{8.31}$$

The Lie algebra of $SO(D)$ is obtained by computing the commutators of these matrices, giving formulas that generalize (8.17), for example. The result of this computation is

$$[L_{ij}, L_{kl}] = i(\delta_{ik}L_{jl} - \delta_{il}L_{jk} + \delta_{jl}L_{ik} - \delta_{jk}L_{il}).\tag{8.32}$$

The infinitesimal generators of $SO(D)$ can also be represented by differential operators that act on the coordinates, generalizing (8.20). The formula is derived in the same way as before, and one finds

$$L_{ij} = -i\left(x^i\frac{\partial}{\partial x^j} - x^j\frac{\partial}{\partial x^i}\right).\tag{8.33}$$

One can verify that these differential operators satisfy commutation relations given above.

It is natural to ask whether the existence of a covering group with spinor representations also generalizes from three dimensions to higher dimensions. Not surprisingly, the answer is yes. The covering group of $SO(D)$ is called $Spin(D)$. In particular, in the case of three dimensions, $Spin(3) = 4SU(2)$. In general, there is not another equivalent name for $Spin(D)$, though it does turn out that $Spin(5)$ is isomorphic to the symplectic group $USp(4)$ and $Spin(6)$ is isomorphic to $SU(4)$. Thus, in these two cases the minimal spinor representations are four-dimensional.

8.4 Lorentz transformations form a group

In Chapter 2 we discussed Lorentz transformations and showed that they could be described by 2×2 matrices, which characterize linear transformations of the coordinates x and t. We also discussed the fact that successive Lorentz transformations give a Lorentz transformation, which we now recognize as the group property. In particular, in Exercise 2.19 you showed that $L(\xi_1)L(\xi_2) = L(\xi_1 + \xi_2)$ corresponds to the velocity addition rule. The case of one spatial dimension is somewhat special, however. For the more relevant case of three spatial dimensions, the possibilities are richer. For example, one can imagine multiplying a Lorentz transformation in one direction by a Lorentz transformation in another. The result turns out not to be a pure Lorentz boost, but rather a combination of a boost and a rotation. Therefore, in order to describe the group that contains Lorentz transformations, it is necessary to view it as an extension of the group of rotations. In other words the rotation group is a subgroup of the Lorentz group. This does not contradict what we found when there was only one spatial direction, since in that case there are no rotations.

Lorentz transformations can be viewed, from a mathematical point of view, as generalized rotations in a spacetime with an indefinite metric, as was already discussed in Chapter 5. A rotation about the origin of a D-dimensional Euclidean space transforms the coordinates according to the rule $\vec{x} \to \tilde{\vec{x}}$, where $\tilde{x}^l = \sum_j R^{\tilde{l}}_j x^j$ and R is an orthogonal matrix satisfying $RR^T = R^TR = I$, or equivalently $R^TIR = I$. This can be written in terms of components as

$$\sum_{\tilde{k}\tilde{l}} R^{\tilde{k}}_i \delta_{\tilde{k}\tilde{l}} R^{\tilde{l}}_j = \delta_{ij}. \tag{8.34}$$

The reason we have written the formulas this way is to emphasize that the transformation preserves the "metric" I, which corresponds to the matrix δ_{ij}. These formulas just represent the fact that the vectors $\vec{x}$ and $\tilde{\vec{x}}$ have the same length s, where the length squared is given by the usual formula, namely $s^2 = \sum_i (x^i)^2 = \sum_{ij} \delta_{ij} x^i x^j$.

Lorentz transformations have a similar rule, namely they are linear transformations of the coordinates that preserve the invariant interval

$s^2 = -(x^0)^2 + \sum_i (x^i)^2$. We can rewrite this in the convenient form

$$s^2 = \sum_{\mu\nu} \eta_{\mu\nu} x^\mu x^\nu, \tag{8.35}$$

where the matrix η is the Lorentz metric. We use lower case Greek indices to label spacetime directions, with $\mu = 0$ corresponding to the time direction and $\mu = i$ corresponding to the spatial directions. Specifically, the Lorentz metric η is a diagonal matrix, whose diagonal components are $\eta_{00} = -1$ and $\eta_{ii} = 1$ for $i = 1, 2, \ldots D$. The usual case is $D = 3$, of course.

Let us now consider a linear transformation of spacetime given by $\tilde{x}^\lambda = \sum_\nu L^{\tilde{\lambda}}_\nu x^\nu$, as before. The only difference from before is that the indices now range from 0 to D rather than from 1 to D. We require that this transformation preserve the invariant interval s^2, since this is the defining property of a Lorentz transformation. Expressed in terms of matrices this yields the condition $L^T \eta L = \eta$. This can be written in terms of coordinates as

$$\sum_{\tilde{\rho}\tilde{\lambda}} L^{\tilde{\rho}}_\mu \eta_{\tilde{\rho}\tilde{\lambda}} L^{\tilde{\lambda}}_\nu = \eta_{\mu\nu}. \tag{8.36}$$

Any $(D + 1) \times (D + 1)$ matrix that satisfies this equation corresponds to a Lorentz transformation. The converse is also true: each Lorentz transformation corresponds to a matrix A satisfying this equation. Therefore Eq. (8.36) is the defining representation of the Lorentz group. This representation is $(D + 1)$-dimensional when there are D spatial dimensions.

We can now show that the set of all real matrices L satisfying Eq. (8.36) form a group. It is quite obvious that $L = I$, which is the identity element, is a solution. Also, if L is a solution, then so is the inverse matrix $L^{-1} = \eta L^T \eta$. It is only a little more work, which we leave to the reader, to verify the closure property: namely, if L_1 and L_2 are solutions, then so is their product $L_1 L_2$. The mathematical name for this group of matrices is $O(D, 1)$, since it preserves a metric with D plus signs and one minus sign.

The Lorentz group has four components

The group $O(D, 1)$ includes components that describe spatial reflections and that reverse the direction of time. To illustrate the latter, note that $L = \eta$ is an element of the group that reverses the sign of x^0 without changing the spatial coordinates. Similarly, there is a diagonal L that changes the sign of a spatial coordinate, which describes a reflection. Altogether, the group has four components: (1) no reflection or time reversal, (2) reflection but no time reversal, (3) time reversal but no reflection, and (4) reflection and time reversal. The determinant of the matrix L is $+1$ for components (1) and (4), and -1 for components (2) and (3). The "proper Lorentz group" is defined to consist of transformations belonging to component (1) only. They form a subgroup of $O(D, 1)$. This subgroup is usually denoted $SO(D, 1)$. Note that this is a more severe restriction than requiring that the determinant is $+1$, since that would also allow transformations of the fourth type.[3] Components (1) and (2), taken together, form the subgroup of the full Lorentz group that preserves the direction of time. The Lorentz transformations with this property are sometimes called orthochronous Lorentz transformations.

In thinking about Lorentz transformations, it is sometimes helpful to try to visualize their effect on light cones. For example, time reversal clearly interchanges the forward and backward light cones, whereas spatial reflections preserve light cones.

The difference between space and spacetime rotations

Both the Lorentz group $O(D, 1)$ and the proper Lorentz group $SO(D, 1)$ contain the group of spatial rotations $SO(D)$ as a subgroup. Note that this means that spatial rotations are considered to be special cases of Lorentz transformations. In more elementary discussions one often distinguishes rotations from boosts. However, from a more general viewpoint, this is awkward, since the generic transformation is a combination of a boost and a rotation. So it is much more natural to call all of these transformations Lorentz transformations and to identify spatial rotations as those Lorentz transformations that are contained in the rotation subgroup.

A much more profound distinction between the rotation group $SO(D)$ and the proper Lorentz group $SO(D, 1)$ is that the former is a *compact* Lie group and the latter is a *noncompact* Lie group. This is an important topological distinction that has far-reaching consequences for the structure of the groups and their representations.

A group is compact if the group manifold has finite volume. To make this statement precise one ought to explain how one measures the volume of a group. Here let us just describe it intuitively. A rotation is necessarily bounded in size: after you reach an angle of 2π, you're back where you started. As a consequence any group of rotations has finite volume. Lorentz boosts, on the other hand, can be arbitrarily large. The boost factor $\gamma = (1 - v^2/c^2)^{-1/2}$ approaches infinity as $v \to c$. Therefore Lorentz groups have infinite volume. Another fact that illustrates the same point concerns the surfaces that are preserved by the respective groups. Rotation groups preserve spheres, which have finite volume, whereas Lorentz groups preserve hyperboloids, which have infinite volume.

---

[3] An alternative notation in which this does include the fourth component is also used. In that case, some additional symbol, such as a vertical arrow, is needed to represent the restriction to the first component.

Infinitesimal Lorentz transformations

Let us now construct the infinitesimal generators for Lorentz transformations. The formulas are given by straightforward generalization of those for rotations that were obtained in Section 8.3. We just have to be sure to replace unit matrices by the Lorentz metric in the appropriate places. In that section we found the matrix that describes the infinitesimal generator of a rotation in the x^i–x^j plane is given by

$$(L_{ij})_{kl} = i(\delta_{ik}\delta_{jl} - \delta_{il}\delta_{jk}). \tag{8.37}$$

We need a generalization to the Lorentz group generators that reduces to this formula when we restrict all indices to spatial directions. The natural (and correct) generalization to spacetime is

$$(L_{\mu\nu})^\lambda_\rho = i(\eta_{\mu\rho}\delta^\lambda_\nu - \eta_{\nu\rho}\delta^\lambda_\mu). \tag{8.38}$$

Here we find it convenient to write the row index ρ as a subscript and the column index λ as a superscript. This ensures that when we multiply matrices the summation combines a subscript with a superscript. This is a convenient way to ensure that the Lorentz metric is properly taken into account. Note that for this choice of $L_{\mu\nu}$ the only nonzero elements of the matrix L_{ij} are those that are given in Eq. (8.37). However, this equation does more. It also gives the infinitesimal generators for Lorentz boosts. A boost in the x^i direction is generated by the matrix L_{0i}. From Eq. (8.38) we can read off that the only nonzero elements of the matrix L_{0i} are $(L_{0i})_{i0} = (L_{0i})_{0i} = -i$.

The Lie algebra of $SO(D, 1)$ is obtained by computing the commutators of the matrices that describe the infinitesimal transformations. The result of this computation is

$$[L_{\mu\nu}, L_{\rho\lambda}] = i(\eta_{\mu\rho}L_{\nu\lambda} - \eta_{\mu\lambda}L_{\nu\rho} + \eta_{\nu\lambda}L_{\mu\rho} - \eta_{\nu\rho}L_{\mu\lambda}). \tag{8.39}$$

Again generalizing what we did previously for rotations, the infinitesimal generators of $SO(D, 1)$ can be represented by differential operators that act on the coordinates. This yields

$$L_{\mu\nu} = -i\left(\eta_{\mu\rho}x^\rho \frac{\partial}{\partial x^\nu} - \eta_{\nu\rho}x^\rho \frac{\partial}{\partial x^\mu}\right). \tag{8.40}$$

In this equation it is understood that the repeated index ρ is summed. In particular, for the generators of Lorentz boosts we have

$$L_{i0} = -i\left(x^i \frac{\partial}{\partial x^0} + x^0 \frac{\partial}{\partial x^i}\right). \tag{8.41}$$

One can verify that these differential operators satisfy commutation relations given above.

To illustrate these equations, consider an infinitesimal boost in the x^1 direction generated by L_{10}:

$$\begin{pmatrix} x'^0 \\ x'^1 \end{pmatrix} = (1 + i\theta L_{10} + O(\theta^2)) \begin{pmatrix} x^0 \\ x^1 \end{pmatrix} = \begin{pmatrix} x^0 + \theta x^1 \\ x^1 + \theta x^0 \end{pmatrix} + O(\theta^2). \qquad (8.42)$$

Exponentiating this to give a finite Lorentz boost gives

$$\begin{pmatrix} x'^0 \\ x'^1 \end{pmatrix} = \exp[i\theta L_{10}] \begin{pmatrix} x^0 \\ x^1 \end{pmatrix} = \begin{pmatrix} \cosh\theta x^0 + \sinh\theta x^1 \\ \cosh\theta x^1 + \sinh\theta x^0 \end{pmatrix}. \qquad (8.43)$$

We recognize these as the standard formulas for a Lorentz boost. The velocity $v = \beta c$ associated to the boost is given by $\cosh\theta = 1/\sqrt{1 - \beta^2}$ and $\sin h\theta = \beta/\sqrt{1 - \beta^2}$.

SL(2, C)

The proper Lorentz group for four-dimensional spacetime, $SO(3, 1)$, contains the rotation group $SO(3)$ as a subgroup. As we explained in Section 8.3, $SO(3)$ has a covering group, $SU(2)$, which contains two elements (of opposite sign) for every element of $SO(3)$. This extension of the rotation group is required to describe fermions in quantum theory. Suppose now we want to construct relativistic (that is, Lorentz-invariant) quantum theories that contain fermions. What is the appropriate symmetry group in such a case? Clearly, we require a covering group of $SO(3, 1)$ that contains $SU(2)$ as a subgroup. Fortunately, such a group exists, and it is called $SL(2, C)$.

The group $SL(2, C)$ consists of all 2×2 complex matrices M with $\det M = 1$ (unimodular matrices). This group has three complex dimensions (or six real dimensions). A generic element has the form

$$M = \begin{pmatrix} a & b \\ c & d \end{pmatrix} \qquad ad - bc = 1, \qquad a, b, c, d \in C. \qquad (8.44)$$

Since there are six independent real parameters, there are six independent infinitesimal generators. One can show that they give precisely the same Lie algebra as $SO(3, 1)$.

The relationship between $SL(2, C)$ and $SO(3, 1)$ can be described by an extension of the reasoning that we used to explain the relationship between $SU(2)$ and $SO(3)$. For this purpose, we associate the Hermitian matrix

$$X = \begin{pmatrix} x^0 + x^3 & x^1 - ix^2 \\ x^1 + ix^2 & x^0 - x^3 \end{pmatrix} \qquad (8.45)$$

to the spacetime point x^μ. If one performs a transformation, by an element M of the group $SL(2, C)$ one obtains a transformed matrix

$$X' = M^\dagger X M. \qquad (8.46)$$

Note that X' is also Hermitian, and therefore it corresponds to a spacetime point x'^μ consisting of four real numbers that are related to x^μ by a real linear transformation. Furthermore, this linear transformation is a Lorentz transformation. To see this, we note that $-\det X = s^2 = -(x^0)^2 + (x^1)^2 + (x^2)^2 + (x^3)^2$, and $\det X' = \det X$. Therefore the vectors x^μ and x'^μ have the same invariant length. In this way one associates a unique Lorentz transformation to each element of the group $SL(2, C)$. In fact, the mapping is still two-to-one, because the matrices M and $-M$ correspond to the same Lorentz transformation.

The group $SL(2, C)$ has two inequivalent two-dimensional irreducible representations.[4] One of them is the defining representation: $D^{(L)}(M) = M$ and the other is the complex conjugate representation: $D^{(R)}(M) = M^*$. As representations of $SL(2, C)$ these are inequivalent, though they become equivalent upon restriction to the $SU(2)$ subgroup.

Two-component vectors that are transformed by these representation matrices are called *Weyl spinors*. Ones that transform by $D^{(L)}$ are called "left-handed" or "positive chirality," whereas the ones that transform by $D^{(R)}$ are called "right-handed" or "negative chirality." Explicitly, if ψ_α is a left-handed Weyl spinor, under a Lorentz transformation M,

$$\psi_\alpha \to \psi'_\alpha = M_\alpha{}^\beta \psi_\beta \quad \alpha, \beta = 1, 2. \qquad (8.47)$$

The Hermitian conjugate $\psi^\dagger$ is a right-handed Weyl spinor.

8.5 The Poincaré group

A relativistic theory is required to be invariant under the group of Lorentz transformations. However, the Lorentz group is only a subgroup of the whole symmetry group. Additional symmetries, which are not part of the Lorentz group, are shifts (or "translations") in space and time. Such translations are given by $x^\mu \to x^\mu + a^\mu$, where the a^μs are $D + 1$ arbitrary real constants that characterize the translations. Note that a^0/c is the time translation. Adjoining these transformations to the Lorentz group yields a Lie group called the Poincaré group. It is a nontrivial extension in the sense that translations do not commute with Lorentz transformations. The subgroup generated by spatial translations and rotations is the

---

[4] These representations are not unitary, but that's okay, since the group is noncompact.

Euclidean group $E(D)$, which was discussed earlier. It does not require relativity, so it is also a symmetry of nonrelativistic theories.

Space and time translations

Space and time translations can be represented by differential operators. To see how this works consider a scalar function $f(x^\mu)$. Performing a translation gives $f(x^\mu + a^\mu)$. For an infinitesimal translation we can expand to first order

$$f(x^\rho + a^\rho) = f(x^\rho) + a^\mu \frac{\partial}{\partial x^\mu} f(x^\rho) + O(a^2) \tag{8.48}$$

Therefore $\partial/\partial x^\mu$ generates a translation of x^μ.

Translations form a noncompact abelian Lie group. One can exponentiate the infinitesimal transformations to obtain finite ones. Specifically,

$$f(x^\rho + a^\rho) = \exp\left(a^\mu \frac{\partial}{\partial x^\mu}\right) f(x^\rho). \tag{8.49}$$

To prove this, expand the exponential in a power series. Then one sees that this gives the standard formula for the power series expansion of a function about a shifted point. All that is required is that the function f should have sufficient analyticity to ensure the convergence of the series.

Let us rewrite the preceding formula in the form

$$f(x^\rho + a^\rho) = \exp\left(i a^\mu P_\mu\right) f(x^\rho), \tag{8.50}$$

where

$$P_\mu = -i \frac{\partial}{\partial x^\mu}. \tag{8.51}$$

Aside from a factor of $\hbar$, the spatial part of P_μ is the operator that represents momentum in quantum mechanics, as was discussed in Chapter 7. The Lie algebra for translations is simply

$$[P_\mu, P_\nu] = 0, \tag{8.52}$$

so the group of translations is Abelian.

Commutators of translations and rotations

In the preceding subsection, we have introduced the generator of translations, P_μ, and presented a representation in terms of differential operators that act on the spacetime coordinates x^μ. In Section 8.4 we did the same thing for the generators of Lorentz transformations $L_{\mu\nu}$. In each case we could show that these operators satisfied the appropriate Lie algebras, namely

$$[P_\mu, P_\nu] = 0 \tag{8.53}$$

and

$$[L_{\mu\nu}, L_{\rho\lambda}] = i(\eta_{\mu\rho}L_{\nu\lambda} - \eta_{\mu\lambda}L_{\nu\rho} + \eta_{\nu\lambda}L_{\mu\rho} - \eta_{\nu\rho}L_{\mu\lambda}). \tag{8.54}$$

To complete the characterization of the Poincaré algebra, we must also specify the commutators of the translation generators with the Lorentz generators. Using the representation in terms of differential operators, it is straightforward to compute this algebra. One finds that

$$[P_\mu, L_{\nu\rho}] = i(\eta_{\mu\nu}P_\rho - \eta_{\mu\rho}P_\nu). \tag{8.55}$$

This formula encodes the fact that P_μ transforms as a spacetime vector.

The preceding formulas are a complete description of the Poincaré algebra. Arbitrary finite Poincaré group transformations can be written as the sum of a translation and a Lorentz transformation. That is what one obtains by exponentiating a linear combination of the infinitesimal transformations.

Representations of the Poincaré group

The Poincaré group can be represented in terms of differential operators and in terms of matrices. We have already given the representation in terms of differential operators, so we now want to describe a representation in terms of matrices. We know such a representation for the Lorentz subgroup, namely in terms of $(D + 1)$-dimensional matrices in the defining representation of $SO(D, 1)$. So the problem is to extend this representation to the entire Poincaré group.

If we are to find such a matrix representation, we must be able to represent translations by matrices. To do that we must confront the fact that translations are inhomogeneous transformations. This means that the transformation $x^\mu \to x^\mu + a^\mu$ has a term that is not a multiple of x^μ. Therefore the right-hand side cannot possibly be written as an x-independent matrix acting on the vector x^μ. The trick for overcoming this obstacle is to use matrices with an extra row and column as follows:

$$\begin{pmatrix} x^\mu \\ 1 \end{pmatrix} \to \begin{pmatrix} \delta^\mu_\nu & a^\mu \\ 0 & 1 \end{pmatrix} \begin{pmatrix} x^\nu \\ 1 \end{pmatrix} = \begin{pmatrix} x^\mu + a^\mu \\ 1 \end{pmatrix}. \tag{8.56}$$

To generalize this construction to the entire Poincaré group, let us write an arbitrary Poincaré group transformation in the form $\tilde{x}^\mu = L^{\tilde{\mu}}_\nu x^\nu + a^\mu$, where $L^{\tilde{\mu}}_\nu$

is a $SO(D, 1)$ matrix. Clearly this transformation is reproduced by the matrix formula

$$\begin{pmatrix} \tilde{x}^{\mu} \\ 1 \end{pmatrix} = \begin{pmatrix} L_{\nu}^{\tilde{\mu}} & a^{\tilde{\mu}} \\ 0 & 1 \end{pmatrix} \begin{pmatrix} x^{\nu} \\ 1 \end{pmatrix} = \begin{pmatrix} L_{\nu}^{\tilde{\mu}} x^{\nu} + a^{\tilde{\mu}} \\ 1 \end{pmatrix}. \tag{8.57}$$

The Poincaré group has two classes of particle representations – massless and massive. This is important since both kinds of particles are known to exist, for example, electrons and protons are massive, but photons and gravitons are massless. Massless particles travel with the speed of light in any inertial reference frame. In the massive case the Lorentz-invariant combination of momentum and energy $\vec{P}^2 c^2 - E^2 = -M^2 c^4 \neq 0$, and one can choose a frame of reference, called the rest frame, such that $E = Mc^2$ and $\vec{P} = 0$. Note that the energy of a particle is understood here to include its rest energy as well as its kinetic energy.

Let us consider a massive particle at rest in four-dimensional spacetime. The subgroup of the Poincaré group (called the *little group*) that keeps the particle at rest consists of all rotations and translations. In quantum theory, the particle is represented by a multicomponent wave function that transforms according to a particular irreducible representation of $SU(2)$, which corresponds to the spin s of the particle.

In the massless case it is not possible to bring the particle to rest, and the analysis is different. One can restrict the choice of Lorentz frame by requiring that the particle travels in a particular direction, for example the z direction. Then we have $P_z c = E$ and $P_x = P_y = 0$. In this case the little group includes only those rotations that preserve the direction of motion, that is, rotations about the z axis. The representations of this group play the role of spin in this case. These representations are one-dimensional, since the group $SO(2)$ is Abelian. These one-dimensional representations are labeled by a quantum number called *helicity*. However, if spatial inversion (parity) is also a symmetry the group is enlarged to $O(2)$, which is nonabelian. Then a positive and a negative helicity can belong to the same irreducible representation.

Exercises

8.1   Describe the transformation of the plane given by an arbitrary element g of the group E_2 (described in Section 8.1). Let g_1 and g_2 denote two elements of the group E_2. Compute the transformation that is described by the group product $g_1 g_2$.

8.2   Work out the multiplication table for the group D_n described in Section 8.1.

8.3   Which of the following form a group under ordinary multiplication: the nonzero integers, the nonzero real numbers, the nonzero complex numbers? Why have we excluded zero?

8.4   Enumerate the elements of the group S_3. Is this group Abelian? Which elements are even and which are odd? Verify that the even elements form a subgroup.

8.5   Prove that the set of all unitary $N \times N$ matrices form a group.

8.6   Prove that the mapping $A \to \det A$ is a one-dimensional representation of the group $U(N)$.

8.7   Show explicitly that the matrix in Eq. (8.12) satisfies Eq. (8.13).

8.8   Verify that the differential operators $L_i = i\epsilon_{ijk}x^j \partial_k$ satisfy the same Lie algebra as the 3×3 matrices in Eqs. (8.12), (8.15) and (8.16).

8.9   Verify that the map from $SU(2)$ to $SO(3)$ described in Eqs. (8.27) and (8.28) is a two-to-one homomorphism.

8.10   Verify that the differential operators in Eq. (8.33) satisfy the commutation relations in Eq. (8.32).

8.11   Using the representation in Eq. (8.41), show that the commutator of two boost generators $[L_{i0}, L_{j0}]$ gives a rotation generator.

8.12   Construct the infinitesimal generators of $SL(2, C)$. Show that they give the same Lie algebra as $SO(3, 1)$.

8.13   Verify the commutation relations in Eq. (8.55).

9

Supersymmetry and superspace

In Chapter 8 we learned that Lorentz boosts and rotations are part of a group called the Lorentz group. Adding translations in space and time, one gets an even larger group, called the Poincaré group. An understanding of this group is very helpful in the construction and interpretation of relativistic theories. Since a relativistic theory has the same basic equations in any inertial frame, these equations should be invariant – or transform among themselves – under all Poincaré group transformations. This places a severe restriction on the structure of allowed theories.

It is natural to ask whether theories could be restricted further by requiring that they have additional symmetries beyond those described by the Poincaré group. The answer, of course, is that there is a variety of possibilities for other types of symmetry besides those that are contained in the Poincaré group. In considering additional symmetries, we can make the following distinction: either the added symmetries act independently from the Poincaré symmetries or they do not. In the first case the additional symmetries form a group G by themselves, and the total symmetry of the theory is a product of the Poincaré symmetries and the G symmetries. In particular, this implies that the G symmetries commute with rotations, and therefore the generators of G must be rotationally invariant. In the second case – which is the one that is realized in supersymmetrical theories – there are new symmetry generators that are not rotationally invariant. The total symmetry is a nontrivial extension of the Poincaré group, not just a product.

Besides the distinction discussed above, symmetries are characterized by various other sorts of distinctions. In particular, they can be discrete or continuous, global or local, unbroken or broken. Poincaré symmetry, for example, is an unbroken continuous global symmetry. Another important example of a fundamental symmetry is the gauge symmetry of the standard model of elementary particles, which is characterized by the Lie group $SU(3) \times SU(2) \times U(1)$. This symmetry is a broken local continuous symmetry.[1] The standard model will be described in Chapter 10.

In this chapter we wish to consider supersymmetry, which is characterized mathematically by a generalization of a Lie group called a *super Lie group*. This will be defined later in this chapter. The first examples of super Lie groups (and super Lie algebras) were found by physicists (around 1970) before the mathematical theory was worked out. Specifically, the four-dimensional *super-Poincaré group* (discussed below) was formulated in the former Soviet Union by Gol'fand and Likhtman, a work that went unnoticed for several years. At about the same time studies of string theory by Ramond and by Neveu and Schwarz led to the introduction of an infinite-dimensional super Lie algebra, which describes superconformal transformations of the two-dimensional string world sheet. (The discussion of strings and superstrings in Chapter 10 will make this sentence more intelligible.)

There are various possible versions of supersymmetry, depending on the dimension of spacetime and the number of supersymmetry charges. Collectively, they give the unique set of possibilities for nontrivial extensions of the Poincaré group symmetry of relativistic quantum theories. For this reason we believe it is appropriate to include a brief introduction to this subject in a book about special relativity.

A supersymmetry extended Poincaré group is called a super-Poincaré group. Since such a super-Lie group generalizes the group of spacetime symmetries, it is natural to seek a geometric understanding of the extension. As we will explain, a natural one is provided by a generalization of spacetime, called *superspace*, as the arena on which the super-Poincaré group acts. Superspace has additional coordinates beyond those of ordinary space and time, which differ from ordinary spacetime coordinates in a way that will be described.

In quantum mechanical theories, there are two distinct categories of particles, called bosons and fermions. An important property of supersymmetry is that the symmetry transformations relate bosons and fermions to one another. In other words, the irreducible representations of the super-Poincaré group combine bosons with fermions. Since the boson and fermion concepts are so central to an understanding of supersymmetry, the first section of this chapter will describe, in more detail than we have done so far, the basic facts about bosons and fermions.

Before plunging into a discussion of supersymmetry, it seems only fair to point out that supersymmetry is not yet established experimentally to be an ingredient in the correct description of the real world. However, there are many reasons

---

[1] The subgroup that remains unbroken is $SU(3) \times U(1)$. The $SU(3)$ factor in this subgroup is associated to the strong color force and the associated quantum field theory is quantum chromodynamics (QCD). The $U(1)$ factor is associated to the electromagnetic force and the associated quantum field theory is quantum electrodynamics (QED).

to believe that it is likely to play an essential role. These reasons, which will be discussed in Chapter 10, are partly esthetical, partly theoretical, and partly phenomenological. The answer should be provided, hopefully before 2010, by experiments that will be carried out at the Large Hadron Collider (LHC), which is under construction at CERN in Geneva, Switzerland.

If supersymmetry is relevant to nature, it must appear as a broken symmetry. The reason is very simple. If it were unbroken, the bosons and fermions belonging to the same supersymmetry multiplet would have the same mass. For example, the electron, which is a fermion, would have a bosonic partner with exactly the same mass. This is known experimentally not to be the case. Indeed, the fact that such particles have not turned up in the experiments that have already been carried out provides lower bounds for the allowed masses of each of the various supersymmetry partners of the known elementary particles. These lower bounds vary from case to case, but they are typically around 50 to 200 times the mass of the proton, that is, 50–200 GeV.

9.1 Bosons and fermions

In Chapter 7 we gave a brief introduction to some of the basic concepts of quantum mechanics. One of the most remarkable of these concepts is that of identical particles. The indistinguishability of two particles of the same type means that if you start with a definite quantum state containing two identical particles and then imagine making a new quantum state by interchanging them, this new quantum state would be indistinguishable from the quantum state that one started with. This is to be contrasted with classical physics in which one could imagine harmlessly putting a small green dot on one object and a small red dot on the other one so as always to be able to tell which is which. When dealing with quantum states, the dots are not harmless, which means that they change the objects in an important way.

This rule of indistinguishability can be illustrated by considering the wave function for a system of two identical particles, $\Psi(1, 2) = \Psi(x_1, \alpha_1, x_2, \alpha_2)$. Here x_1 and x_2 represent the spatial positions of particles 1 and 2, and α_1 and α_2 represent specific choices for any other degrees of freedom that they possess, such as their spin states. There may also be time dependence. Interchange of the two particles gives the wave function $\Psi(2, 1) = \Psi(x_2, \alpha_2, x_1, \alpha_1)$, and the principle of indistinguishability implies that this must be equivalent to the original wave function. Since the overall phase of the wave function is not a physical observable, this equivalence is satisfied if $\Psi(1, 2) = e^{i\theta}\Psi(2, 1)$. For point particles in three or more spatial dimensions there are only two possibilities for the phase factor: ± 1. (In two spatial dimensions, other phases would be possible.) Particles for which the choice $+1$ is appropriate are called *bosons*, whereas particles for which the choice -1 is appropriate are called *fermions*.[2] These are named for the physicists Bose and Fermi, who made pioneering contributions to understanding these matters.

Bosons

The rule we have just stated for the behavior of a wave function for a pair of identical bosons can be generalized to an arbitrary number. In fact, we already stated the rule in Section 8.2, where we discussed the permutation group. Recall that the permutation group S_N (also called the symmetric group) is the group of permutations of N objects, a group that has $N!$ elements. The context we wish to consider now is when the N objects are identical bosons with a wave function $\Psi(1, 2, \ldots, N)$, where we use the same condensed notation described above. Then the rule for bosons is very simple. The wave function has total symmetry. This means that it is unchanged under any of the $N!$ possible permutations of the N identical bosons. As explained in Section 8.2, this rule corresponds to choosing the trivial representation of the permutation group.

An example of a boson is the photon, the quantum of the electromagnetic field. The fact that the photon is a boson underlies the structure of the Planck formula for blackbody spectrum discussed in Section 7.1. That formula is a consequence of the total symmetry of a many-photon wave function. In this context one usually says that bosons satisfy "Bose–Einstein statistics."

Fermions

In the case of N identical fermions the wave function $\Psi(1, 2, \ldots, N)$ should have total antisymmetry. As was explained in Section 8.2, this rule corresponds to changing the sign of the wave function when the particles are permuted by an odd permutation. (Recall that a simple interchange is an odd permutation, and that all odd permutations can be expressed as a product of an odd number of inter-changes.) This rule is referred to as *Fermi–Dirac statistics*. This simple distinction of a minus sign gives fermions physical properties that are dramatically different from those of bosons.

An important example of a fermion is the electron. Because the wave function vanishes when two electrons in the same spin state approach one another, there is an effective repulsion that is over and above the obvious electrostatic repulsion.

---

[2] The other possibilities that exist for $D = 2$ are called "anyons." The special feature of two dimensions is that when one particle is transported around another one, and returned to its original position, there is a topologically meaningful winding number that can be defined. Thus arbitrary multiples of 2π can be distinguished.

This feature, known as the *Pauli exclusion principle*, is largely responsible for the chemical properties of atoms and the structure of the periodic table of the elements.

Spin and statistics

We have stated that photons are bosons and electrons are fermions. Now we want to give the general rule. The rule is a consequence of a famous theorem, called the *spin and statistics theorem* that holds (very generally) for relativistic quantum theories.

To explain the relation between spin and statistics, we need to recall some facts about spin that were presented in Section 8.3. In that section we discussed the group of rotations in D spatial dimensions $SO(D)$. In the important example of three dimensions, we discussed a closely related group, namely $SU(2)$, which is the covering group of $SO(3)$. As we explained, the irreducible representations of $SU(2)$ are labelled by a "spin" j, which can take the values $j = 0, 1/2, 1, 3/2, \ldots$. The dimension of the spin j representation is $2j + 1$. The integer spin representations (of odd dimension) are also representations of the subgroup $SO(3)$, whereas the half-integer spin representations (of even dimension) are not representations of $SO(3)$.

According to the spin and statistics theorem, a particle whose spin is an integer must be a boson, whereas a particle whose spin is a half-integer must be a fermion. (For example, the photon[3] whose spin is 1 is a boson, and the electron whose spin is $1/2$ is a fermion.)

In contemporary research there is a lot of interest in theories with additional spatial dimensions. The appropriate generalization of the spin and statistics theorem is easily stated. As mentioned in Section 8.3, the D-dimensional rotation group $SO(D)$ has a covering group called $Spin(D)$. (For example, $Spin(3)$, the covering group of $SO(3)$, is another name for $SU(2)$.) The group $Spin(D)$ has two classes of irreducible representations, called tensor representations and spinor representations. The tensor representations are also representations of $SO(D)$, and (by the spin and statistics theorem) must be used for bosons. The spinor representations of $Spin(D)$, which are not representations of $SO(D)$, must be used for fermions. One consequence of these rules, which is true in any dimension, is that a rotation by 2π radians about any axis results in a change of sign for a fermion wave function (or the wave function of an odd number of fermions).

---

[3] The fact that the photon is massless changes the group theory somewhat, but the conclusion is true as stated nonetheless.

Coleman–Mandula theorem

The possibilities for extending the Poincaré group symmetry of relativistic quantum field theories were investigated in 1967 by Coleman and Mandula (CM). The outcome of their investigation was formulated as a theorem. Given a certain set of assumptions, which seemed very reasonable at the time, they proved that any symmetry described by a Lie group must have the structure of a direct product of the Poincaré group and another group G that describes any other continuous symmetries that the theory might possess. This implies that the Lie algebra generators of G, which correspond to conserved charges, commute with the generators of the Poincaré group. Therefore they must be rotationally invariant (scalar) operators.

Fortunately, this no-go theorem did not stop physicists from discovering supersymmetry a few years later. The CM theorem was evaded by dropping one of its assumptions. The assumption that turned out to be unnecessary was that the enlarged symmetry should be described by a Lie group. A Lie group has an associated Lie algebra that is defined by commutation relations of the generators X_i. As discussed in Appendix 3, such an algebra takes the form

$$[X_i, X_j] = f_{ij}{}^k X_k, \tag{9.1}$$

where the numerical coefficients $f_{ij}{}^k$ are called structure constants.

In quantum field theory, there is a quantum field associated to each elementary particle in the theory. When the particle is a boson, the associated field can be expanded in creation and annihilation operators, as discussed in Section 7.5, and as a result the field operator ϕ and its canonically conjugate momentum π satisfy a simple commutation relation of the schematic structure $[\phi, \pi] = i\delta$ that was described in more detail in that section. This builds in Bose–Einstein statistics. As result, when one constructs conserved charges X_i out of bosonic fields one is naturally led to consider commutation relations for the charges. These turn out to have the structure of a Lie algebra.

In a quantum theory that contains fermions a new possibility opens up. When a fermion field (such as the Dirac field) is quantized one builds in Fermi–Dirac statistics by requiring *anticommutation* relations for the fermion field and its conjugate momentum. In the same condensed notation as above, these have the structure $\{\phi, \pi\} = \delta$. Here we have used an anticommutator, whose definition is $\{A, B\} = AB + BA$. This differs by a sign from the definition of the commutator $[A, B] = AB - BA$. This simple change of sign makes all the difference. Not only does it distinguish bosons from fermions but it opens up the possibility of a symmetry algebra with *fermionic generators*. The point is that if there are symmetry generators that are constructed out of an odd number of fermion fields, and any number of boson fields, then these generators will be fermionic. For a pair of fermionic generators the natural bracket that one can form is an anticommutator.

In this way one is led to consider superalgebras that contain both bosonic generators B_i and fermionic generators F_α. The general structure of a superalgebra is described in Appendix 4. It contains three types of relations: (1) commutators of two bosonic generators give bosonic generators, (2) commutators of bosonic generators with fermionic generators give fermionic generators, (3) anticommutators of two fermionic generators give bosonic generators.

The bosonic generators of a superalgebra form a subalgebra, which is an ordinary Lie algebra. The CM theorem applies to the Lie group generated by this subalgebra. This means that this Lie group must be a direct product of the Poincaré group with some other bosonic symmetry group G. However, the assumptions of the CM theorem do not allow for the possibility of fermionic generators, and the supergroup does not have this direct product structure. One way to see this is to note that the spin and statistics theorem implies that a fermionic generator must transform under rotations as an operator of half integer spin. Having nonzero spin, such an operator is not rotationally invariant, and therefore it does not commute with the rotation generators that are contained in the Poincaré algebra. It follows that the entire superalgebra is necessarily a nontrivial extension of the Poincaré algebra, not just a direct product of the Poincaré algebra with the rest.

For relativistic theories in which the Poincaré symmetry is extended by the addition of fermionic symmetry generators, the fermionic symmetries are called supersymmetries. We have argued that they must have half-integer spin. In fact, spin $1/2$ turns out to be the only case that is possible. Other spins (such as $3/2$) can be excluded by the same sort of reasoning that goes into the proof of the CM theorem, where it is used to exclude all positive integer spins. Spin $1/2$ supersymmetry generators are the unique loophole to the theorem,[4] and their presence implies that the symmetry is described by a superalgebra. Altogether, the symmetry algebra consists of a super-Poincaré algebra that contains the Poincaré generators and the supersymmetry generators and (perhaps) the bosonic generators of another Lie group G which appears as a direct product with the rest (as in the CM theorem).

9.2 Superspace

Supersymmetry is a profound mathematical possibility, because (as we have explained) it is the unique possibility for a nontrivial extension of the known symmetries of space and time. This suggests that supersymmetric theories should be

---

[4] There are also conformally invariant field theories that have a different algebraic structure. However, these theories do not describe particles, so they are outside the scope of the current discussion.

formulated in terms of some new sort of geometry that extends the usual flat space-time of special relativity. Flat spacetime is retained, but there are additional degrees of freedom, which can be described as extra dimensions of a very peculiar kind. Specifically, they are not parametrized by ordinary numbers, but rather by *Grassmann numbers*. This looks very bizarre if you have not encountered it before, and like so many things in relativity and quantum theory, takes a certain amount of getting used to.

The notion of fermions is a quantum-mechanical concept that is not part of ordinary classical physics. So, since Grassmann numbers are associated with fermions, these new dimensions are, in a sense, quantum dimensions. The entire space spanned by ordinary spacetime plus the additional Grassmann dimensions is called *superspace*.

Grassmann numbers and Grassmann algebras

Ordinary numbers (be they real or complex) commute. This means that two numbers x and y satisfy $xy = yx$. Grassmann numbers are defined to anticommute. Suppose we have N of them, called θ_α, where $\alpha = 1, 2, \ldots, N$. Then the rule of anticommutativity is that for any pair

$$\{\theta_\alpha, \theta_\beta\} = 0. \tag{9.2}$$

As a special case (putting $\alpha = \beta$), this implies that the square of any particular Grassmann number is zero. Such "numbers" can be made to seem somewhat less strange by remarking that this algebra could be represented in terms of square matrices of appropriate size made out of ordinary numbers. (This remark should make you more comfortable about the mathematical consistency of working with Grassmann numbers, but it is not very useful in practice.) The Grassmann numbers are required to commute with ordinary numbers, that is, $[\theta, x] = 0$.

A set of N Grassmann numbers, as above, generates a Grassmann algebra. This arises by combining them with ordinary (real or complex) numbers in all possible ways. To describe this more precisely, let us begin by considering monomials made by multiplying k of the Grassmann numbers. The $k \theta$s that one multiplies must all be different because the square of any one of them is zero. For example, $\theta_1 \theta_2 \theta_1 = -\theta_1 \theta_1 \theta_2 = 0$. Thus the number of distinct nonzero monomials of degree k is given by the binomial coefficient $N!/k!(N-k)!$. The largest possible value of k is N for which there is a unique monomial, namely the product of the N θs. The order of multiplication only affects an overall plus or minus sign. Altogether, summing the binomial coefficients from $k = 0$ to $k = N$, one sees that there are 2^N distinct monomials. The most general element of the associated Grassmann algebra is given by an arbitrary linear combination of these 2^N monomials. Any two elements of the Grassmann algebra can be multiplied by using the rules described above.

The monomials of Grassmann numbers can be divided into two sets called even and odd. The distinction is whether k, the number of θs, is even or odd. There are 2^{N-1} even monomials and 2^{N-1} odd ones. An element of the Grassmann algebra that is made out of the even monomials only is called even and one that is made out of the odd monomials only is called odd. With these definitions one has the following simple rules: even $\times$ even $=$ even, even $\times$ odd $=$ odd, and odd $\times$ odd $=$ even. Note that the even elements of the algebra are commuting quantities just like ordinary numbers, whereas the odd elements are anticommuting quantities just like the θs.

Odd and even functions

Given N Grassmann numbers θ_α, as above, one can introduce functions $f(\theta_\alpha)$ that depend on them. Such a function is defined completely by its power series expansion, which is necessarily a finite series:

$$f(\theta_\alpha) = f_0 + f_\alpha \theta_\alpha + \frac{1}{2} f_{\alpha\beta} \theta_\alpha \theta_\beta + \cdots + f_{12\ldots N} \theta_1 \theta_2 \cdots \theta_N. \qquad (9.3)$$

If all the coefficients in this expansion were ordinary numbers, then this would be just a representation of an arbitrary element of the Grassmann algebra. However, that turns out not to be the most convenient thing to do.

What is more useful in the application to supersymmetrical theories is to consider functions that are themselves even or odd. The two possibilities are realized as follows. Suppose we want f to be an even function. All the terms with an even number of θs are even if their coefficients are ordinary numbers. The new ingredient arises for the terms with an odd number of θs. Each coefficient of a term with an odd number of θs should be chosen to be a Grassmann number – not one of the θs but some other Grassmann number.

The way to make an odd function f is a simple variant of the preceding. The coefficients of terms with an odd number of θs should be chosen to be ordinary numbers and the coefficients of terms with an even number of θs should be chosen to be Grassmann numbers. Then every term in f is odd.

One can define differentiation with respect to Grassmann numbers. The rules are very simple, since there is a unique linear operation that one can define.[5] One simply defines the θ derivative of θ to be 1 and the θ derivative of 1 to be 0. This then extends to a Grassmann algebra with N θs in an obvious way. The only

---

[5] Unlike ordinary differentiation, differentiation with respect to Grassmann numbers is not defined as a limit. It is better regarded as an algebraic operation.

thing to be careful of is the minus signs that arise from anticommuting θs past one another. For example,

$$\frac{\partial}{\partial\theta_1}(\theta_1\theta_2) = \theta_2 \tag{9.4}$$

and

$$\frac{\partial}{\partial\theta_2}(\theta_1\theta_2) = -\theta_1. \tag{9.5}$$

Superspace

As we have already mentioned, supersymmetry can be described in a geometrical way as a symmetry that acts on a generalization of spacetime. In this generalization, the usual space and time coordinates x^μ that parametrize Minkowski space are supplemented by quantum mechanical extra dimensions that are parametrized by Grassmann numbers θ_α, $\alpha = 1, 2, \ldots, N$. The reason the Grassmann coordinates have an index is that they transform under the Lorentz group as a spinor. In other words they transform like the spin $1/2$ fields described in Chapter 7. They also are the same kind of spinors as the supersymmetry charges Q_α that appear in the supersymmetry algebra.

There are various possible types of spinors depending on the dimension of spacetime and other possible restrictions. What is true in general is that the number of components N is always a power of 2. For example, if there is one time dimension and D space dimensions, the minimal allowed value is $N = 2$ for $D = 2$, $N = 4$ for $D = 3$, $N = 16$ for $D = 9$, and $N = 32$ for $D = 10$. It turns out that $D = 10$ is the largest number of dimensions in which it is mathematically possible to construct an interacting supersymmetrical theory. (Adding in time, the spacetime dimension is 11.) These issues can become somewhat technical, because the rules for constructing spinors are somewhat special in each dimension. This makes it difficult to discuss all cases at once, unless one is somewhat schematic in the treatment. That is what we will do here. Readers who want to learn more are referred to more specialized treatises.

It is also possible to construct theories that have multiple supersymmetries. In this case we need to attach another index I to the supersymmetry charges $Q_{\alpha I}$, $I = 1, 2, \ldots, \mathcal{N}$. In this case one is describing $\mathcal{N}$-extended supersymmetry. However, there are restrictions on the possible values of $\mathcal{N}$. Basically the rule is that the total number of supersymmetry charges, which is $N\mathcal{N}$, should not exceed 32. Therefore the maximum possible value of $\mathcal{N}$ is 16 for $D = 2$, 8 for $D = 3$, 2 for $D = 9$, and 1 for $D = 10$. In the following discussion we will assume that $\mathcal{N} = 1$ and not display the index I.

As we have said, supersymmetrical theories can be described as theories in superspace. This means that instead of fields defined on spacetime, such as a scalar field $\phi(x)$ or a spinor field $\psi_\alpha(x)$, one considers fields on superspace $\Phi(x, \theta)$. There are two viewpoints that one might adopt to this. One is that superspace is a profound generalization of ordinary spacetime on which new types of symmetry transformations (namely, supersymmetry transformations) can be defined.

The alternative viewpoint is that a *superfield*, such as $\Phi(x, \theta)$, is just a convenient algebraic trick for packaging 2^N ordinary fields in a single mathematical expression. This is the case, as we explained in the preceding section, because Φ has an expansion of the form

$$\Phi(x, \theta) = \phi(x) + \psi_\alpha(x)\theta_\alpha + \frac{1}{2}\chi_{\alpha\beta}(x)\theta_\alpha\theta_\beta + \cdots + \zeta_{12\ldots N}(x)\theta_1\theta_2\ldots\theta_N.$$

(9.6)

The point we want to make is that any theory that is formulated in terms of superfields defined on superspace can also be formulated in terms of the component fields that appear in the expansion without any reference to the Grassmann coordinates θ. Therefore one could regard them as just a convenient bookkeeping device.

In this example we have combined a scalar field $\phi(x)$, a spinor field $\psi_\alpha(x)$, and a bunch of other fields together into a single entity. This packaging is often convenient for formulating supersymmetrical theories in ways that make their supersymmetry properties manifest (to the trained observer). Note that this superfield is an even element of the Grassmann algebra if ϕ is even, ψ is odd, and so forth, just as required by the spin and statistics theorem.

9.3 Supersymmetry transformations

Now that we have introduced the concept of superspace and superfields we wish to use this formalism to discuss the structure of supersymmetry transformations and the super-Poincaré algebra. Since we don't want to delve into the technicalities of spinors in all dimensions, we will write formulas that are correct in most cases and can be modified easily to cover the cases where they are not precisely right. Specifically, we will only consider spinors that are real.[6] By definition, a Dirac spinor ψ is complex and can be written in terms of a pair of real spinors in the usual way ($\psi = \psi_1 + i\psi_2$).

A translation of spacetime by an amount a^μ is given by the simple rule $x^\mu \to x^\mu + a^\mu$. This rule describes the translation subgroup of the Poincaré group. This

---

[6] The precise technical name for such spinors is *Majorana*. Spinors take Grassmann values, but one can define a Hermitian conjugation operation on them and use it to define what one calls "real." Majorana spinors exist only when the spacetime dimension is 2, 3, or 4 modulo 8.

group is abelian because a translation by an amount a^μ followed by a translation by an amount b^μ gives the same result $(x^\mu + a^\mu + b^\mu)$ as when the two translations are performed in the opposite order. Supersymmetry can be described by a similar rule in superspace, but we will see that the supergroup that arises as a generalization of the translation group is not abelian.

Supertranslations

The basic supersymmetry transformations work as follows. Let ϵ_α denote a Grassmann-valued spinor that is the parameter for a supersymmetry transformation (or *supertranslation*) "by an amount ϵ." Then the rule is that this acts as a translation of the Grassmann coordinates

$$\theta_\alpha \to \theta_\alpha + \epsilon_\alpha. \tag{9.7}$$

This is not the whole story. The spacetime coordinate x^μ transforms at the same time according to the rule

$$x^\mu \to x^\mu + i\bar{\epsilon}_\alpha \gamma^\mu_{\alpha\beta} \theta_\beta, \tag{9.8}$$

where the repeated spinor indices are summed. The matrices γ^μ are the Dirac matrices constructed in Section 7.4, which we recall satisfy the Dirac algebra

$$\{\gamma^\mu, \gamma^\nu\} = -2\eta^{\mu\nu}. \tag{9.9}$$

That section was specific to four dimensions ($D = 3$), in which case the Dirac matrices were 4×4, but there are analogous constructions for other values of D. When there are more dimensions, the size of the Dirac matrices needs to be increased appropriately so that a matrix solution of the Dirac algebra exists. Note that Eq. (9.9) implies that the square of γ^0 is the unit matrix, and that the square of any spatial component γ^i is the negative of the unit matrix. As a result of this it is always possible to construct a representation in which $\gamma^0 = -\gamma_0$ is Hermitian and $\gamma^i = \gamma_i$ is antihermitian.

The bar on top of the epsilon in formula (9.8) is a standard notation introduced by Dirac. It involves a complex conjugation as well as a multiplication by the matrix $C = \gamma_0$, which is sometimes referred to as the charge-conjugation matrix. Explicitly, for any spinor ψ we have

$$\bar{\psi}_\alpha = \psi_\beta^* C_{\beta\alpha}. \tag{9.10}$$

Note that the matrix product $C\gamma^\mu$ is Hermitian for all values of μ. In fact, in the Majorana representation of the Dirac algebra, which is assumed in this discussion, these matrices are real and symmetric. These rules ensure that the entire quantity in Eq. (9.8) transforms as a spacetime vector under Lorentz transformations.

We can write Eq. (9.8) more succinctly by suppressing the spinor indices as $x^\mu \to x^\mu + i\bar\epsilon\gamma^\mu \theta$. The factor of i ensures reality, since (for the types of spinors we are considering) $\bar\epsilon\gamma^\mu\theta$ goes into its negative under complex conjugation. To see this we can rewrite the expression as $(C\gamma^\mu)_{\alpha\beta}\epsilon_\alpha\theta_\beta$. The first matrix is Hermitian and the two spinors are each real. When one takes the Hermitian conjugate, the order of the two spinors is reversed. The fact that they anticommute is then the origin of the minus sign.

The supersymmetry transformations of superspace given above are valid for finite transformations, where the parameter ϵ is not required to be small. However, we can restrict them to the case of infinitesimal transformations in which case it is convenient to recast them as expressions for the infinitesimal change in θ_α and x^μ:

$$\delta\theta_\alpha = \epsilon_\alpha \quad \text{and} \quad \delta x^\mu = i\bar\epsilon\gamma^\mu \theta. \tag{9.11}$$

You may wonder what it means for a Grassmann number to be small. One way of thinking about this is to imagine a fixed Grassmann number which can then be multiplied by a real number which is allowed to become infinitesimal. However, it is simpler to incorporate that factor in the definition of what one calls ϵ rather than to display it separately.

The commutator of two supertranslations

We have now got the essential ingredients to understand the basic structure of the supersymmetry algebra. To do this, what is required is to compute the commutator of two infinitesimal supersymmetry transformations δ and δ' with parameters ϵ_α and ϵ'_α. Specifically, this means that $\delta\theta_\alpha = \epsilon_\alpha$, $\delta x^\mu = i\bar\epsilon\gamma^\mu\theta$ and $\delta'\theta_\alpha = \epsilon'_\alpha$, $\delta' x^\mu = i\bar\epsilon'\gamma^\mu\theta$.

Using these formulas we wish to apply the commutator operator $[\delta, \delta']$ to x^μ and θ_α and see what transformations result. The calculation goes as follows. First, $\delta\delta'\theta = \delta\epsilon' = 0$. The only point in this calculation that needs to be emphasized is that the infinitesimal Grassmann parameters ϵ and ϵ' should be treated as constants. Therefore their variations vanish. From this we conclude that $[\delta, \delta']\theta = 0$.

Let us now compute $[\delta, \delta']x^\mu$ in a similar manner:

$$\delta\delta' x^\mu = \delta(i\bar\epsilon'\gamma^\mu\theta) = i\bar\epsilon'\gamma^\mu\delta\theta = i\bar\epsilon'\gamma^\mu\epsilon. \tag{9.12}$$

Therefore

$$[\delta, \delta']x^\mu = i\bar\epsilon'\gamma^\mu\epsilon - i\bar\epsilon\gamma^\mu\epsilon' = 2i\bar\epsilon'\gamma^\mu\epsilon. \tag{9.13}$$

Thus, altogether, we have learned that $[\delta, \delta']\theta = 0$ and $[\delta, \delta']x^\mu = a^\mu$, where $a^\mu = 2i\bar\epsilon'\gamma^\mu\epsilon$. These formulas correspond exactly to a spacetime translation by an amount a^μ.

In conclusion, we have learned that the commutator of two supersymmetry transformations is precisely a spacetime translation! This illustrates what we mean when we say that supersymmetry is a nontrivial extension of the usual symmetries of spacetime. There is one somewhat subtle point, however, namely, the infinitesimal parameters a^μ that we found are not ordinary numbers. Rather, they are even elements of a Grassmann algebra.

Differential operators on superspace

Let us now consider the representation of the supersymmetry algebra by differential operators. To do this, we need to generalize the discussion in Section 8.5 to superspace. Recall that there we showed that

$$\partial_\mu = \frac{\partial}{\partial x^\mu} \tag{9.14}$$

generates spacetime translations. The crucial fact is that under the infinitesimal transformation $\delta x^\mu = a^\mu$, the variation of a function $f(x^\mu)$ is given (to first order in a) by $\delta f = a^\mu \partial_\mu f$.

We now wish to find analogous differential operators in superspace that implement the infinitesimal supersymmetry transformations in Eq. (9.11). In other words, acting on a superfield Φ they should give

$$\delta \Phi(x^\mu, \theta_\alpha) = \Phi(x^\mu - i\bar\theta\gamma^\mu\epsilon, \theta_\alpha + \epsilon_\alpha) - \Phi(x^\mu, \theta_\alpha) \tag{9.15}$$

to first order in the infinitesimal Grassmann parameter ϵ. By expanding to first order in each of the variables one sees that this is achieved by

$$\delta \Phi(x, \theta) = \epsilon_\alpha Q_\alpha \Phi(x, \theta) \tag{9.16}$$

where $\epsilon_\alpha Q_\alpha$ is the differential operator on superspace

$$\epsilon_\alpha Q_\alpha = \epsilon_\alpha \frac{\partial}{\partial\theta_\alpha} - i\bar\theta\gamma^\mu\epsilon \frac{\partial}{\partial x^\mu}. \tag{9.17}$$

The meaning of a θ derivative was explained in Section 9.2. Therefore, including a minus sign that arises from moving ϵ past θ, one deduces that

$$Q_\alpha = \frac{\partial}{\partial\theta_\alpha} + i(\bar\theta\gamma^\mu)_\alpha \frac{\partial}{\partial x^\mu}. \tag{9.18}$$

To summarize, what we have found is that (in the framework of superspace) the differential operator ∂_μ generates spacetime translations and the differential operator Q_α generates supersymmetry transformations.

Supersymmetry generators and the supersymmetry algebra

As we have explained, the commutator of two supersymmetry transformations gives a spacetime translation. Let us investigate what this implies for the algebra of the generators Q_α. In Eq. (9.16) we saw that an infinitesimal supersymmetry transformation can be represented on superfields by the differential operator $\epsilon_\alpha Q_\alpha$. Let us now examine the commutator of two such differential operators

$$[\epsilon_\alpha Q_\alpha, \epsilon'_\beta Q_\beta] = -\epsilon_\alpha \epsilon'_\beta \{Q_\alpha, Q_\beta\}, \tag{9.19}$$

where we have used the fact that an ϵ anticommutes with a Q as well as with another ϵ. Note that for this reason, we are forced to consider the anticommutator of a pair of supersymmetry generators, which explains the origin of the super-algebra structure.

The nonzero terms in the anticommutator arise from when a θ derivative in one Q differentiates the θ that is present in the other Q. There are two terms of this type corresponding to the two terms in the anticommutator. This gives

$$\{Q_\alpha, Q_\beta\} = 2i(C\gamma^\mu)_{\alpha\beta} \frac{\partial}{\partial x^\mu}. \tag{9.20}$$

Thus the anticommutator of two supersymmetry generators gives certain structure constants times a spacetime translation generator. In the quantum context one can replace the translation generator by the energy–momentum vector $P_\mu = -i\partial_\mu$, as we explained in Chapter 7. Then the supersymmetry algebra takes the form

$$\{Q_\alpha, Q_\beta\} = -2(C\gamma^\mu)_{\alpha\beta} P_\mu. \tag{9.21}$$

As a special case, imagine that a supersymmetric theory contains a particle of mass M, and suppose that this particle is at rest. In this case $P_0 = M$ and $P_i = 0$. Then using that $(C\gamma^0)_{\alpha\beta} = -\delta_{\alpha\beta}$, we find that

$$\{Q_\alpha, Q_\beta\} = 2M\delta_{\alpha\beta}. \tag{9.22}$$

Thus, in the rest frame, the square of each component of the supercharge spinor is precisely the mass, and the different components anticommute with one another. For massless particles a somewhat modified analysis is needed. It will be discussed later.

From Poincaré to super-Poincaré

The Poincaré group has generators P_μ associated to spacetime translations and an antisymmetric tensor of generators $L_{\mu\nu}$ associated to spatial rotations and Lorentz boosts. The commutation relations of these generators were given in Section 8.5, and we repeat them here

$$[P_\mu, P_\nu] = 0, \tag{9.23}$$

$$[L_{\mu\nu}, L_{\rho\lambda}] = i(\eta_{\mu\rho}L_{\nu\lambda} - \eta_{\mu\lambda}L_{\nu\rho} + \eta_{\nu\lambda}L_{\mu\rho} - \eta_{\nu\rho}L_{\mu\lambda}), \tag{9.24}$$

$$[P_\mu, L_{\nu\rho}] = i(\eta_{\mu\nu}P_\rho - \eta_{\mu\rho}P_\nu). \tag{9.25}$$

Our goal here is to give the complete superalgebra that generates the super-Poincaré group. To do this we need to adjoin additional commutation and anti-commutation relations that involve the supersymmetry generators Q_α. We found above that

$$\{Q_\alpha, Q_\beta\} = -2(C\gamma^\mu)_{\alpha\beta} P_\mu. \tag{9.26}$$

To complete the presentation of the super-Poincaré algebra it remains to give the commutation relations of the Qs with the Ps and the Ls. The first of these is trivial

$$[P_\mu, Q_\alpha] = 0. \tag{9.27}$$

This is easily verified using the representations of P_μ and Q_α in terms of differential operators.

Now let us turn to the evaluation of $[Q_\alpha, L_{\mu\nu}]$. In fact, the answer can be found from very general considerations. The commutator of any operator with the Lorentz group generators $L_{\mu\nu}$ encodes the rule for how that operator transforms under rotations and Lorentz transformations. In this case the correct answer should be that it transforms as a spacetime spinor. This gives the requirement

$$[Q_\alpha, L_{\mu\nu}] = \frac{i}{2}(\gamma^{\mu\nu})_{\alpha\beta} Q_\beta = -\frac{i}{2}(Q\gamma^{\mu\nu})_\alpha, \tag{9.28}$$

where we have defined $\gamma^{\mu\nu} = [\gamma^\mu, \gamma^\nu]/2$.

Note that in Chapter 8 we gave a formula for the Lorentz group generators that in the present notation is $L_{\mu\nu} = x_\mu P_\nu - x_\nu P_\mu$. However, if we used this formula to compute the commutator in Eq. (9.28) we would not get the right answer. The reason for this is that the Lorentz generators have an additional contribution in superspace that comes from the Grassmann coordinates. The correct formula, which (with a bit of algebra) does reproduce Eq. (9.28) is

$$L_{\mu\nu} = x_\mu P_\nu - x_\nu P_\mu - \frac{i}{2}\frac{\partial}{\partial\theta}\gamma^{\mu\nu}\theta. \tag{9.29}$$

9.4 $\mathcal{N} = 1$ supersymmetry in four dimensions

We have discussed the general structure of supersymmetry as an extension of the symmetry of Minkowski spacetime in a formalism that is applicable to various possible dimensions of spacetime. However, only three spatial dimensions have ever been observed. It is conceivable that indications of others might emerge in the future in very-high-energy experiments, which is one reason that they are considered. However, at ordinary energies that are accessible in contemporary experiments, it is clear that $D = 3$ is the right dimension of space.

As experiments progress to higher energy, there are good reasons to believe (though it is not a complete certainty) that evidence for supersymmetry will emerge before there is evidence of extra dimensions. If that is the case, a correct description would require the construction of a supersymmetric theory in four-dimensional spacetime ($D = 3$). For this reason, there has been extensive work on constructing supersymmetric extensions of the standard model of elementary particles, and studying the ways in which they could be tested in current and future high-energy experiments.

There are relatively straightforward four-dimensional extensions of the standard model with $\mathcal{N} = 1$ supersymmetry. However, there are obstacles to constructing viable four-dimensional theories with more supersymmetry than that. One aspect of this is the fundamental fact that the weak nuclear interactions involve parity violation (an inherent asymmetry under mirror reflection). The standard model has this feature built into it. It turns out that parity violation can also be incorporated in a supersymmetric extension with $\mathcal{N} = 1$ supersymmetry, but this is not possible if there are more supersymmetries than that. Therefore all phenomenological studies are based on $\mathcal{N} = 1$, which corresponds to having four real supercharges Q_α, labelled by the spinor index α.

As we mentioned earlier, supersymmetry is not an exact symmetry of nature, but it could still play a fundamental role as a broken symmetry. When this happens there is a characteristic energy scale associated with the supersymmetry breaking, such that the symmetry is effectively restored at energies above that scale. An important aspect of this is that each of known particles has a supersymmetry partner with a mass that differs by an amount that is comparable to the supersymmetry-breaking scale. Thus, at energies lower than that scale, particle collisions cannot produce the supersymmetry partner particles by conversion of energy into mass, but at higher energies this becomes possible. Indeed, the failure to produce supersymmetry partners in this way so far is the basis for the current experimental lower limits on their masses. Supersymmetry determines the strength of the interactions by relating them to those among the known particles, so the masses of the partners are the key unknowns.

A plausible scenario is that the scale at which $\mathcal{N} = 1$ supersymmetry is broken is approximately 1 TeV, which is a few times higher than the current experimental limits. At a much higher energy scale (perhaps 10^{16}–10^{18} GeV) extra dimensions open up. At the same time, more supersymmetries would have to come into play, because the higher-dimensional spinors have more components. This energy (often referred to as the unification scale) is also close to the scale where string theory and quantum gravity effects are expected to become important. If this is the correct story, then a quantum field theory with $\mathcal{N} = 1$ supersymmetry would give a correct description of microscopic physics over a very large range of energies. These issues will be explored further in Chapter 10.

Two-component and four-component notation

The description of supersymmetry given in the preceding section in terms of real (or Majorana) spinors is applicable to $\mathcal{N} = 1$ supersymmetry in four-dimensional spacetime. In this notation the spinor index takes four values and there are four real supercharges. There is an alternative (equivalent) formalism that is commonly used in this particular case, however. We will describe it briefly here, since it is essential for reading the literature on this subject. In this formalism the spinors are two-component complex objects. This is just a different way of assembling the same information. The existence of this formalism has an interesting group-theoretical basis, which was touched on in Chapter 8.

In Section 8.4 we discussed the group $SL(2, C)$, which bears the same relationship to the Lorentz group for $D = 3$ that $SU(2)$ does to the three-dimensional rotation group $SO(3)$. $SL(2, C)$ consists of all 2×2 complex matrices M with $\det M = 1$. Recall that the connected component of the Lorentz group is the subgroup of $O(3, 1)$ that does not contain time reversal or space inversion. This group has a double cover $SL(2, C) = Spin$ (3,1). The double cover is required in each case in order to admit half-integral spin representations.

The group $SL(2, C)$ has two inequivalent two-dimensional irreducible representations. One of them is the defining representation: $D^{(L)}(M) = M$ and the other is the complex conjugate representation: $D^{(R)}(M) = M^*$. Two-component spinors that transform by $D^{(L)}$ are called "left-handed"(LH) or "positive chirality," whereas ones that transform by $D^{(R)}$ are called "right-handed"(RH) or "negative chirality." Explicitly, if ψ_α is a LH Weyl spinor, then under a Lorentz transformation specified by M

$$\psi_\alpha \to \psi'_\alpha = M_\alpha{}^\beta \psi_\beta \qquad \alpha, \beta = 1, 2. \tag{9.30}$$

The Hermitian conjugate $\psi^\dagger$ is a RH Weyl spinor. To emphasize this fact the indices are written with dots. The definition of $\bar{\psi}_{\dot\alpha}$ is

$$\bar{\psi}_{\dot\alpha} \equiv (\psi_\alpha)^\dagger. \tag{9.31}$$

Note that $\dot{\alpha}$ also takes the values 1, 2. Under the *same* $SL(2, C)$ transformation

$$\bar{\psi}_{\dot{\alpha}} \rightarrow \bar{\psi}'_{\dot{\alpha}} = M_{\dot{\alpha}}^{*\ \dot{\beta}} \bar{\psi}_{\dot{\beta}}. \tag{9.32}$$

Four-vectors (like P_μ) can be reexpressed as bispinors $P_{\alpha\dot{\alpha}}$:

$$P_{\alpha\dot{\alpha}} \equiv \sigma^\mu_{\alpha\dot{\alpha}} P_\mu. \tag{9.33}$$

Group-theoretically, this makes sense because $2_L \times 2_R = 4$. The matrices $\sigma^1, \sigma^2, \sigma^3$ are the standard Pauli matrices and

$$\sigma^0 = \begin{pmatrix} -1 & 0 \\ 0 & -1 \end{pmatrix}. \tag{9.34}$$

Thus

$$P_{\alpha\dot{\alpha}} = \begin{pmatrix} -P_0 + P_3 & P_1 - iP_2 \\ P_1 + iP_2 & -P_0 - P_3 \end{pmatrix}. \tag{9.35}$$

Note, that this is *Hermitian*. The matrices are chosen so that under a Lorentz transformation

$$P \rightarrow P' = MPM^\dagger. \tag{9.36}$$

In two-component notation the supersymmetry algebra of Eq. (9.21) takes the form

$$\{Q_\alpha, \bar{Q}_{\dot{\beta}}\} = 2P_{\alpha\dot{\beta}} \tag{9.37}$$

$$\{Q_\alpha, Q_\beta\} = \{\bar{Q}_{\dot{\alpha}}, \bar{Q}_{\dot{\beta}}\} = 0. \tag{9.38}$$

It is also conventional to define *invariant tensors* $\epsilon^{\alpha\beta}, \epsilon_{\alpha\beta}, \epsilon^{\dot{\alpha}\dot{\beta}}, \epsilon_{\dot{\alpha}\dot{\beta}}$, which satisfy

$$M\epsilon M^T = \epsilon, \quad \text{etc.} \tag{9.39}$$

$\epsilon^{\alpha\beta}$ and $\epsilon^{\dot{\alpha}\dot{\beta}}$ are antisymmetric with $\epsilon^{12} = -\epsilon^{21} = 1$. With lower indices the sign is the opposite: $\epsilon_{21} = -\epsilon_{12} = 1$. Then these can be used to raise and lower indices:

$$\psi^\alpha = \epsilon^{\alpha\beta} \psi_\beta \qquad \psi_\alpha = \epsilon_{\alpha\beta} \psi^\beta, \tag{9.40}$$

since $\epsilon_{\alpha\beta}\epsilon^{\beta\gamma} = \delta_\alpha^\gamma$. Another definition of the same kind is

$$\bar{\sigma}^{\mu\dot{\alpha}\alpha} \equiv \epsilon^{\dot{\alpha}\dot{\beta}} \epsilon^{\alpha\beta} \sigma^\mu_{\beta\dot{\beta}} \qquad \begin{pmatrix} \bar{\sigma}^0 = \sigma^0 \\ \bar{\sigma}^i = -\sigma^i \end{pmatrix}. \tag{9.41}$$

Note that the order of the dotted and undotted indices has been interchanged in passing from σ^μ to $\bar{\sigma}^\mu$. This ensures that the matrix multiplications $\sigma^\mu \bar{\sigma}^\nu$ and

$\bar\sigma^\mu \sigma^\nu$ are group-theoretically sensible. Then the Dirac algebra becomes:

$$(\sigma^\mu \bar\sigma^\nu + \sigma^\nu \bar\sigma^\mu)_\alpha^{\ \beta} = -2\eta^{\mu\nu}\delta_\alpha^{\ \beta} \tag{9.42}$$

$$(\bar\sigma^\mu \sigma^\nu + \bar\sigma^\nu \sigma^\mu)^{\dot\alpha}_{\ \dot\beta} = -2\eta^{\mu\nu}\delta^{\dot\alpha}_{\ \dot\beta}. \tag{9.43}$$

In this notation the supersymmetry transformations of superspace take the form

$$\delta x^\mu = i(\theta\sigma^\mu\bar\epsilon - \epsilon\sigma^\mu\bar\theta)$$

$$\delta\theta_\alpha = \epsilon_\alpha, \qquad \delta\bar\theta_\alpha = \bar\epsilon_{\dot\alpha}. \tag{9.44}$$

To relate these formulas to the four-component notation one can take

$$\gamma^\mu = \begin{pmatrix} 0 & \sigma^\mu \\ \bar\sigma^\mu & 0 \end{pmatrix} \tag{9.45}$$

to recover the more familiar form of the Dirac algebra

$$\{\gamma^\mu, \gamma^\nu\} = -2\eta^{\mu\nu}. \tag{9.46}$$

In this basis a four-component *Dirac spinor* is obtained by combining two Weyl spinors of opposite chirality

$$\Psi_D = \begin{pmatrix} \chi_\alpha \\ \bar\psi^{\dot\alpha} \end{pmatrix}. \tag{9.47}$$

In particular, $\psi = \chi$ for a *Majorana spinor*, so that

$$\Psi_M = \begin{pmatrix} \chi_\alpha \\ \bar\chi^{\dot\alpha} \end{pmatrix}. \tag{9.48}$$

9.5 Massless representations

Let us now analyze *massless* representations of the supersymmetry algebra. Usually, these are the only ones that are used in applications. When there are masses, the massive representations can be obtained by combining massless ones in suitable ways. In the massless case we can choose the inertial frame so that the energy is E and the motion is along the positive z axis. Then the four-momentum becomes $P_\mu = (-E, 0, 0, E)$. This choice is convenient for analyzing the representations.

Setting $P_\mu = (-E, 0, 0, E)$ in Eq. (9.37), the supersymmetry algebra becomes

$$\{Q_\alpha, \bar Q_{\dot\beta}\} = 4 \begin{pmatrix} E & 0 \\ 0 & 0 \end{pmatrix}_{\alpha\dot\beta}. \tag{9.49}$$

Note that the operators Q_2 and $\bar Q_{\dot 2}$ are now central (they commute or anticommute with everything). This implies that any states that can be written in the form $Q_2|\psi\rangle$

or $\bar{Q}_{\dot{2}}|\psi\rangle$ decouple in any physical process. Thus, such states can be omitted, and we need only consider half as many fermionic generators as in the massive case. Let us then define

$$b = \frac{1}{2\sqrt{E}}Q_1, \qquad b^\dagger = \frac{1}{2\sqrt{E}}\bar{Q}_1. \tag{9.50}$$

These satisfy

$$\{b, b^\dagger\} = 1$$

$$\{b, b\} = \{b^\dagger, b^\dagger\} = 0. \tag{9.51}$$

Consider a state $|0; \lambda\rangle$ describing a particle of helicity (angular momentum component along the direction of motion) λ. The notation is meant to imply that

$$b|0; \lambda\rangle = 0, \tag{9.52}$$

and

$$J_z|0; \lambda\rangle = \lambda|0; \lambda\rangle. \tag{9.53}$$

Acting on this with the fermionic raising operator $b^\dagger$, we can make another state of the form $b^\dagger|0; \lambda\rangle$. Acting on a state with a $b^\dagger$ increases the helicity by $1/2$, because the Lorentz transformation rules (the $[Q, L]$ formula) imply that

$$[J_z, b^\dagger] = \frac{1}{2}b^\dagger. \tag{9.54}$$

Therefore, this state has helicity $\lambda + 1/2$:

$$J_z b^\dagger |0; \lambda\rangle = (\lambda + \frac{1}{2})b^\dagger|0; \lambda\rangle. \tag{9.55}$$

Altogether, these are the only two states that can be obtained by acting with the operators b and $b^\dagger$. Therefore there is a massless irreducible representation of the supersymmetry algebra that contains two states with helicities λ and $\lambda + 1/2$. One of these is a boson and the other is fermion.

Let us now list some of the most important massless supersymmetry multiplets by enumerating their helicities. As we have explained, an irreducible massless multiplet for supersymmetry in four dimensions contains two states. However, in each case we need to adjoin another supermultiplet containing the antiparticles, which have the opposite helicities. The three cases that we will consider are (1) $\lambda = 0$ (helicities 0 and $1/2$), which is called a *chiral supermultiplet*, (2) $\lambda = 1/2$ (helicities $1/2$ and 1), which together with its antiparticle conjugate is called a *vector supermultiplet*, (3) $\lambda = 3/2$ (helicities $3/2$ and 2), which together with its antiparticle conjugate is called a *gravity supermultiplet* or a *supergravity multiplet*.

Chiral supermultiplets

A chiral supermultiplet describes particles with helicities 0 and 1/2. The antiparticle conjugate multiplet, which contains helicities $-1/2$ and 0 is called an *antichiral* supermultiplet. When these multiplets acquire mass (through the Higgs mechanism, for example) a chiral supermultiplet and an antichiral supermultiplet combine to give a single massive supersymmetry multiplet. This massive multiplet contains two spin 0 fields and one spin 1/2 field. Thus, for example, in a supersymmetric theory the electron has two negatively charged spin 0 partners, which are called selectrons. Similarly, the positron has two positively charged spin 0 partners. The number of fermions is equal to the number of bosons in each case, provided we count each spin state separately.

The chiral supermultiplet is described by fields

$$A(x), \quad \psi_\alpha(x), \quad F(x), \tag{9.56}$$

a total of two bosonic field (A and F) and two fermionic fields (ψ_α). This is twice the number of physical particle states. The explanation for this is that the equations of motion of the theory relate F to the other fields, so that it doesn't give an independent degree of freedom. Similarly, they also relate the two components of ψ_α, which therefore only contributes one physical degree of freedom.

It is an interesting challenge to describe a chiral supermultiplet in terms of a superfield. Consider, for example, a scalar superfield $\Phi(x, \theta, \overline{\theta})$. Since $\theta, \overline{\theta}$ are four independent Grassmann numbers, a power series expansion of Φ in θ and $\overline{\theta}$ contains $2^4 = 16$ distinct functions of x, half of which are Grassmann even (bosonic fields) and half of which are Grassmann odd (fermionic fields). This is 4 times the number of fields that we want! The solution to this dilemma is to constrain Φ in a way that eliminates 3/4 of its component fields and is consistent with supersymmetry.

The basic observation is that if Φ only depended on θ_α and not on $\overline{\theta}_{\dot\alpha}$, then it would be a function of two Grassmann variables. Then its expansion in powers of θ would give two bosonic fields and two fermionic fields as desired. This is not quite right, however, because such a truncation is inconsistent with the requirements of supersymmetry. Fortunately, there is a slight variant of this rule that is consistent with supersymmetry.

If one defines a shifted spacetime coordinate

$$y^\mu = x^\mu + i\theta\sigma^\mu\overline{\theta}, \tag{9.57}$$

Then $\Phi(y, \theta)$ is a supersymmetric field with the right number of degrees of freedom. The rationale for this construction is that $\Phi(y, \theta)$ satisfies $\overline{D}_{\dot\alpha}\Phi = 0$, where

$$\overline{D}_{\dot\alpha} = \frac{\partial}{\partial\overline{\theta}^{\dot\alpha}} - i\theta^\alpha\sigma^\mu_{\alpha\dot\alpha}\frac{\partial}{\partial x^\mu}. \tag{9.58}$$

The operator $\overline{D}$ is a *supercovariant derivative*, which means that it anticommutes with Q and $\overline{Q}$, the differential operators that generate the supersymmetry transformations in Eq. (9.44). Such a superfield is called a *chiral superfield*, and its Hermitian conjugate is an antichiral superfield. There is a highly developed technology for constructing and analyzing supersymmetric theories with chiral (and antichiral) superfields. This subject originated in 1973 with pioneering work by Wess and Zumino.

Vector supermultiplets

A vector supermultiplet contains component fields

$$v_\mu(x), \quad \lambda_\alpha(x), \quad \overline{\lambda}_{\dot\alpha}(x), \quad D(x). \tag{9.59}$$

The vector field $v_\mu(x)$, which is exactly like a Maxwell field, only describes helicities ± 1, because of the principle of gauge invariance, which is the equivalence of $v_\mu(x)$ with $v_\mu(x) + \partial_\mu f(x)$, for an arbitrary function $f(x)$. This is a manifest property of the field strength tensor $\partial_\mu v_\nu(x) - \partial_\nu v_\mu(x)$, for example. The spinors λ and $\overline{\lambda}$ describe helicities $\pm 1/2$, and $D(x)$ is an auxiliary scalar field analogous to the F field of the chiral supermultiplet. Altogether four bosonic fields and four fermionic fields are describing two bosonic and two fermionic helicity states.

A real superfield $V(x, \theta, \overline{\theta})$ contains eight bosonic and eight fermionic fields, which is twice the desired number. The trick for eliminating the extra ones is different from that used for the chiral superfield. Rather than introduce a constraint, one introduces a superspace generalization of the gauge invariance described above. The appropriate rule is the gauge equivalence of V and $V + \Lambda + \Lambda^\dagger$, where Λ is an arbitrary chiral superfield. It turns out that this leaves exactly the desired field content. Then V is called a *vector superfield*.

Vector fields are of special importance in quantum field theory because they are in correspondence with the generators of the Lie algebra that characterizes the local symmetry of the theory. The formalism discussed above is appropriate to the Abelian case, when the group is $U(1)$. This is what one needs for a supersymmetric extension of QED, for example. When the group is nonabelian, such as $SU(2)$ or $SU(3)$, the gauge fields are self-interacting, and this leads to somewhat more complicated formulas.

For a vector supermultiplet to become massive it needs to join together with a chiral supermultiplet and an antichiral supermultiplet. The resulting massive vector supermultiplet contains a massive spin 1 particle, a massive Dirac spin 1/2 particle, and a massive spin 0 particle. Altogether, counting each possible spin polarization, this is a total of four bosons and four fermions.

The gravity supermultiplet

In Einstein's general relativity, gravity is described in terms of the geometry of spacetime described by a metric tensor $g_{\mu\nu}(x)$, which is a symmetric tensor field. (More details are given in Chapter 10.) The flat spacetime metric of special relativity, $\eta_{\mu\nu}$, corresponds to the special case in which gravity is "turned off," so to speak. When one attempts to treat general relativity as a quantum theory, one discovers that after quantization the metric tensor becomes a quantum field for a massless spin 2 particle, called the *graviton*. This particle is the quantum of gravity, which transmits the gravitational force, in much the same way that the spin 1 photon is the quantum of the electromagnetic field that transmits electromagnetic forces.

If one formulates a supersymmetric theory that contains general relativity, the graviton acquires a spin 3/2 supersymmetric partner, called the *gravitino*. Since they are massless, the graviton can have helicities ± 2 and the gravitino can have helicities $\pm 3/2$. Supersymmetric theories that contain gravity are called *supergravity theories*. The first such theory was introduced in 1976 by Freedman, Ferrara, and van Nieuwenhuizen. The superspace formulation of supergravity has been worked out. The basic idea is that, just as gravity is described by the geometry of spacetime, so supergravity should be described by the geometry of superspace. The details are beyond the scope of this book.

Supergravity has an important significance that ought to be pointed out. We have mentioned how vector fields have gauge invariance properties that enable them to describe theories in which a Lie group symmetry is a local symmetry. (Recall that this means that the symmetry transformations are allowed to vary in spacetime. In other words, the parameters of the Lie group are allowed to be functions of the spacetime coordinates.) The gravitino field has an analogous role for supersymmetry. In other words, it is a type of gauge field that allows supersymmetry to become local symmetry. As a consequence of this, supersymmetry in nongravitational theories is a global symmetry, whereas in gravitational theories it is a local symmetry. The fact that supersymmetry needs to be broken, means that the gravitino should acquire mass (while the graviton remains massless) by a supersymmetric analog of the Higgs mechanism.

Like general relativity, supergravity theories are beautiful classical field theories, but they are not consistent quantum field theories, due to the problem of nonrenormalizable ultraviolet divergences. In Chapter 10 we will discuss how this problem can be overcome by embedding them in supersymmetric string theories.

Exercises

9.1 Suppose that one is given four distinct Grassmann numbers. How many monomials of each order can one form? Which ones are even and which ones are odd?

9.2 Consider two functions $f(\theta)$ and $g(\theta)$ with expansions of the type described in Eq. (9.3) for $N = 2$. Relate the expansion coefficients of the product $h = fg$ to those of f and g.

9.3 Verify that $C\gamma^\mu$ is Hermitian for all values of μ, as asserted in Section 9.3.

9.4 Verify Eq. (9.20).

9.5 Verify that Q_α and $L_{\mu\nu}$, defined in Eqs. (9.18) and (9.29), satisfy Eq. (9.28).

9.6 Derive Eq. (9.54).

9.7 Verify the equivalence of Eqs. (9.37) and (9.26).

9.8 Construct the superspace differential operators that generate the infinitesimal transformations in Eq. (9.44) and verify that they anticommute with $\overline{D}_{\dot{\alpha}}$ in Eq. (9.58).

9.9 Verify that $\overline{D}_{\dot{\alpha}}\Phi(y, \theta) = 0$.

9.10 Derive the three additional Jacobi identities involving terms of the structure B^2F, BF^2, and F^3, explained at the end of Appendix 4.

10

Looking onward

In the previous chapters we have occasionally included remarks about advanced topics. For example, we mentioned that the relativistic description of gravity requires an extension of special relativity, called general relativity. We also referred on occasion to the standard model of elementary particles and some of its properties. We described basic facts about supersymmetry, but we did not pursue how it might fit into a realistic theory. References were also made to string theory, which is the leading approach for constructing a fully unified quantum theory of particles and forces. In this chapter we will say a little more about each of these topics. The purpose is to convey a general impression and to whet the reader's appetite for exploring them further. The chapter concludes by discussing some important unsolved problems.

10.1 Relativity and gravity

When Einstein formulated special relativity he understood that a relativistic generalization of Newton's theory of gravity was not a simple matter. Rather, it requires new concepts and a new mathematical framework. Einstein dedicated himself to this project during the subsequent decade, which culminated in 1916 with a new theory of gravity, which he called *general relativity*. As will be discussed below, general relativity makes testable predictions for new phenomena that are not accounted for in Newton's theory. Some were verified right away and others in more recent times. At present, general relativity is tested to good precision (better than 1 percent in some cases) and no discrepancies have been observed. However, general relativity is only a classical (that is, not quantum) theory, though quantum effects are beyond what can be observed at the present time. It is understood on theoretical grounds that when quantum effects come into play there are new issues to confront. These are addressed by string theory, which is discussed in Sections 4 and 5.

Basic concepts of general relativity

General relativity is a beautiful theory based on profound and elegant principles.[1] The principles are called the *equivalence principle* and *general coordinate invariance*. These principles lead to a description of gravity in terms of the geometry of spacetime. The relevant branch of mathematics is called *differential geometry*. Using the techniques of *Riemannian geometry* one describes spacetime as a four-dimensional (or d-dimensional, if one wants to be more general) manifold.

In Riemannian geometry one starts with a smooth manifold and then introduces on it a *metric tensor* that defines distances. Specifically, in terms of spacetime coordinates x^μ that parametrize the spacetime manifold, the metric takes the form $g_{\mu\nu}(x)$. The invariant distance ds between a point with coordinates x^μ and a point with coordinates $x^\mu + dx^\mu$ is given by

$$ds^2 = g_{\mu\nu}(x)dx^\mu dx^\nu. \tag{10.1}$$

The metric tensor is symmetric ($g_{\mu\nu}(x) = g_{\nu\mu}(x)$). A special case is Minkowski spacetime, for which one can choose $g_{\mu\nu}(x) = \eta_{\mu\nu}$. As this special case illustrates, the matrix $g_{\mu\nu}(x)$ has 3 (or D) positive eigenvalues associated to the spatial directions and one negative eigenvalue associated to the time direction.[2]

General coordinate invariance means that the physics does not depend on the choice of coordinate system but only on the intrinsic geometry of the spacetime manifold. In a new coordinate system $\tilde{x}^\mu$ there is a new metric tensor $\tilde{g}_{\mu\nu}(\tilde{x})$, but the invariant distance between two points is the same. Thus

$$ds^2 = g_{\mu\nu}(x)dx^\mu dx^\nu = \tilde{g}_{\mu\nu}(\tilde{x})d\tilde{x}^\mu d\tilde{x}^\nu. \tag{10.2}$$

Given formulas $\tilde{x}^\mu(x)$ relating the two coordinate systems, this implies that

$$g_{\mu\nu}(x) = \frac{\partial \tilde{x}^\rho}{\partial x^\mu}\frac{\partial \tilde{x}^\lambda}{\partial x^\nu}\tilde{g}_{\rho\lambda}(\tilde{x}), \tag{10.3}$$

which is the transformation law for the components of a second-rank tensor.

The equivalence principle is the intuitive notion that in an inertial frame there is no gravitational force. The standard example of this is a freely falling elevator. The idea is that acceleration is equivalent to gravitation and for suitable acceleration gravity is effectively cancelled. Mathematically, this translates into the statement that given a particular point p in spacetime, there exists a coordinate system x^μ such that p corresponds to the origin and $g_{\mu\nu}(0) = \eta_{\mu\nu}$, the Lorentz metric. Thus, an observer at p will find that the rules of special relativity are applicable in her/his

---

[1] Such statements are inherently subjective. However, this sentiment is shared very widely in this case.
[2] Strictly speaking, the mathematical definition of a Riemannian manifold requires a positive definite metric (spatial directions only), and a manifold with a metric of indefinite signature is called pseudo-Riemannian.

immediate vicinity. Mathematically, one says that the tangent space to the space-time manifold at the point p is Minkowski spacetime. The extent to which the manifold and its tangent space differ in the vicinity of p depends on the curvature of the spacetime manifold at p.

In Riemannian geometry one can construct a fourth-rank curvature tensor with components $R^\mu{}_{\nu\rho\lambda}(x)$ out of the metric tensor and its derivatives. (We will not present the explicit formulas.) This tensor controls the way in which a vector gets rotated when it is parallel-transported around a small loop in the immediate vicinity of x. In the case of Minkowski spacetime, the derivatives of $\eta_{\mu\nu}$ are zero, and therefore the curvature tensor vanishes. If the curvature tensor vanishes in one coordinate system, it vanishes in any other coordinate system. So the vanishing of the curvature tensor constitutes a coordinate-invariant definition of flat Minkowski spacetime.[3] Indeed, when it vanishes, there exists a coordinate system x for which $g_{\mu\nu}(x) = \eta_{\mu\nu}$.

By contracting a pair of indices one obtains the Ricci tensor $R_{\nu\lambda} = R^\mu{}_{\nu\mu\lambda}(x)$. To form the scalar curvature, one needs to contract the remaining two indices. To do this one introduces the inverse of the metric tensor, which is written with upper indices. By the definition of the inverse, one has $g^{\mu\nu}(x)g_{\nu\rho}(x) = \delta^\mu_\rho$. Using this inverse metric, we can define the scalar curvature $R = g^{\mu\nu}R_{\mu\nu}$.

Like other classical field theories, general relativity can be formulated in terms of an action principle. In other words there exists a Lagrangian density $\mathcal{L}$, with its integral over spacetime defining an action S. The classical equations of motion are the Euler–Lagrange equations for which this action is stationary.

In general relativity it is important to define the volume element of spacetime carefully. In flat space one simply writes $d^4x = dx^0 dx^1 dx^2 dx^3$. However, this is not a good choice for curved spacetime, because it is not coordinate-invariant. Under a change of coordinates one picks up the Jacobian $d^4\tilde{x} = J d^4x$, where J is the determinant of the matrix $\partial_\mu \tilde{x}^\nu$. Taking the determinant of both sides of Eq. (10.3) one learns that $\det g = J^2 \det \tilde{g}$. Defining $g = \det g_{\mu\nu}$, we deduce that the combination $\sqrt{-g}\, d^4x$ is a coordinate invariant volume element. The minus sign is inserted inside the square root because the determinant g is negative. (Recall that it has one negative eigenvalue.) The density function $\sqrt{-g}$ reduces to unity for Minkowski spacetime.

We can now describe the Einstein–Hilbert action. It is the general coordinate invariant expression

$$S_{\text{EH}} = -\frac{1}{16\pi G}\int R\sqrt{-g}\, d^4x, \tag{10.4}$$

[3] This is true locally. Some topological freedom remains. For example, if some of the spatial directions are periodic they describe a flat torus.

where G is Newton's constant. This elegant formula gives the classical theory of gravity in the absence of any other fields. The associated Euler–Lagrange equations for which it is stationary are $G_{\mu\nu} = 0,^4$ where

$$G_{\mu\nu} = R_{\mu\nu} - \frac{1}{2}g_{\mu\nu}R \qquad (10.5)$$

is called the Einstein tensor. When gravity interacts with matter, additional coordinate invariant terms describing the matter fields should be added to the action. The Einstein field equations then take the form

$$G_{\mu\nu} = -8\pi G T_{\mu\nu}, \qquad (10.6)$$

where $T_{\mu\nu}$ is the energy–momentum tensor of the matter obtained by varying the matter part of the Lagrangian with respect to the metric tensor.

The classical experimental tests

In Newtonian gravity it is exactly true that a bound two-body system (such as a binary star) moves in elliptical orbits about the center of mass. This is no longer true in general relativity. However, the general relativity effects are usually very small. In the case of Mercury, the planet for which the effect is largest, one must first account for perturbations due to other planets, the finite size of the Sun and so forth. Once this is done one finds that the observed orbit can be described as an ellipse that is slowly rotating (or precessing). The observed precession rate is about 43 arc seconds per century. This discrepancy with Newtonian theory was known in 1916. Einstein computed the prediction of general relativity for the precession of the perihelion of Mercury and found that it accounted for the discrepancy within the errors. (They are currently somewhat less than 1 percent.)

A second test of general relativity that Einstein proposed concerned the bending of light by a large mass concentration. Specifically, he proposed that if a star could be observed near the limb of the Sun during a solar eclipse, its apparent position would be shifted by about 1.75 seconds of arc from what one would otherwise expect. In the famous Eddington expedition of 1919 this prediction was confirmed with an experimental uncertainty of about 10 percent. Even though Einstein was already famous in the world of science, this success made him a celebrity on the world stage.

There are a number of additional predictions of general relativity that have been verified in more recent times. These include high-precision tests of the equivalence principle, observation of gravitational redshifts, and observation of time

---

[4] $G_{\mu\nu} = 0$ is equivalent to $R_{\mu\nu} - 0$ (for $d \neq 2$).

delay effects in radar ranging of the Moon and planets. Another prediction made by Einstein that has been observed is the formation of multiple images of a distant quasar when light passes though an intervening galactic cluster, which acts as a "gravitational lens." There is also indirect evidence for gravitational radiation, which will be discussed below.

Black holes

One of the most famous solutions of the Einstein field equations ($G_{\mu\nu} = 0$) describes the spacetime geometry due to a large point mass source M. The solution was found by Schwarzschild in 1916, and therefore it is called the Schwarzschild metric. This solution, which has spherical symmetry, can be written in the form

$$ds^2 = -(1 - 2MG/r)\,dt^2 + (1 - 2MG/r)^{-1}\,dr^2 + r^2(d\theta^2 + \sin^2\theta d\phi^2). \tag{10.7}$$

This metric describes a black hole of Schwarzschild radius $r_S = 2MG$. The Schwarzschild radius of the Sun, for example, is about 3 km. However, in the case of the Sun, this solution is only a good description of the spacetime geometry for r larger than the radius of the Sun, which is about 700 000 km.

As written, the metric looks singular at $r = r_S$. However, this is an artefact of the coordinate system, and it is possible to prove that the solution is entirely smooth in the neighborhood of this spherical region, which is called the *horizon* of the black hole.[5] The horizon is special, however. For $r < r_S$ the coefficient of dt^2 is positive and the coefficient of dr^2 is negative. This means that inside the horizon t is a spatial coordinate and r is a time coordinate! Note that the Sun does not have a horizon and is therefore not a black hole. For a horizon to form, its mass would all have to be concentrated within its Schwarzschild radius.

There is compelling observational evidence for a variety of black holes throughout the Universe. Some, with masses of a few solar masses, form as remnants of supernova explosions of very massive stars. (Supernova explosions of less massive stars leave neutron star remnants.) The centers of galaxies typically contain super-heavy black holes. For example, our Milky Way galaxy has a black hole at its center with a mass of approximately 3 million solar masses. It can be studied in some detail by observing stars that are orbiting around it. Some other galaxies contain black holes that are up to a 100 times more massive than that. Quasars and active galactic nuclei (AGNs) are probably powered by massive central black holes. There may be other categories of black holes that still remain to be identified.

---

[5] There really is a singularity at $r = 0$, where the scalar curvature diverges. Such singularities are believed to represent a breakdown of general relativity. The true physics in the vicinity of the singularity is expected to be modified by a better theory, such as string theory.

Various generalizations of the Schwarzschild metric have been constructed over the years. For example, the black hole can have angular momentum, which distorts the geometry. Also, it can carry electric (or magnetic) charge, which results in electric (or magnetic) fields and also affects the geometry. All of these possibilities have been generalized in recent years to $d \neq 4$ as well as to black holes in spacetimes that are curved far away from the black hole. There are also generalizations to mass sources that instead of being points are lines, planes, etc. All of these generalizations are quite nontrivial because of the nonlinearity of the Einstein field equations. There are also supersymmetric solutions in supergravity theories. In some cases these are actually *easier* to study because of special mathematical tricks that are available in supersymmetric theories.

Gravitational radiation

Another fact about general relativity that Einstein understood early on is that it predicts the existence of gravitational radiation. An elementary approach to studying them is to start with the Einstein field equations $G_{\mu\nu} = 0$, substitute $g_{\mu\nu} = \eta_{\mu\nu} + h_{\mu\nu}$, and expand to first order in h to get a linear differential equation that describes small departures from Minkowski spacetime. The equation obtained in this way is essentially the wave equation, suitably generalized for a second-rank tensor. The tensor structure implies that the polarization of the waves has a quadrapole-type structure. Thus, if the wave is propagating in the z direction, the metric describes a space that is contracting and expanding in the x direction. At the same time it is expanding and contracting in the y direction, but with the opposite phase. In other words, contraction in x is simultaneous with expansion in y and vice versa. Solutions to the full nonlinear equations are also known, and they exhibit the same qualitative features.

Gravitational waves have been proven to exist in brilliant work by Taylor and Hulse, which was rewarded with a Nobel prize. In 1974 they discovered the relativistic binary pulsar system B1913 + 16. One of the orbiting objects is a millisecond pulsar that flashes in an extremely stable way. By observing the pulsar for a long time, the masses of the two objects and all the orbital parameters can be deduced to very high precision. Because of the high eccentricity of the orbit and the high orbital velocities (the period is about 8 h), the two objects are expected to emit a significant amount of gravitational radiation. That they actually do can be inferred by observing the effect on the decay of the orbit. From observations of this pulsar over many years Taylor and his collaborators have measured the decay rate of the orbital period. After correcting for galactic acceleration, they find the result $2.4086 \pm 0.0052 \times 10^{-12}$ s/s. This can be compared with the prediction of general relativity due to the emission of gravitational radiation, which is $2.40247 \pm 0.00002 \times 10^{-12}$ s/s. The agreement is considerably better than 1 percent, which is very impressive.

The Taylor–Hulse result is an indirect observation of gravitational radiation. Several major efforts are currently underway to detect it directly. For example, the laser interferometric gravitational observatory (LIGO) project consists of two large detectors, one located in Hanford, Washington and the other in Livingston, Louisiana. Each consists of two orthogonal arms that are about 4 km long. The idea is to detect one arm getting longer while the other is getting shorter (and vice versa). Since there are many possible sources of fake signals, one requires simultaneous detection at the two different sites (and eventually at detectors elsewhere in the world). The expected signals from various astrophysical sources are all extremely tiny, so extraordinary precision is required. As of this writing, no gravitational waves have been detected. The current sensitivity is about 1 or 2 orders of magnitude less than is required to see the expected signals. However, the sensitivity continues to increase as improvements are made to the detectors. So this endeavor should eventually be successful. In the long run, one expects the observation of gravitational waves to open up an entirely new window on the Universe.

The Universe

General relativity is also used to describe the entire Universe. The basic simplifying assumption that goes into the analysis is that the Universe is nearly homogeneous and isotropic on large scales. Certainly there are large structures such as galaxies and clusters of galaxies, but the largest of these only extends about 1 percent of the way across the visible Universe. On larger scales, homogeneity and isotropy are excellent approximations. With these simplifying assumptions one can find a spacetime geometry that describes the entire Universe.

A good description of the geometry of the Universe has to incorporate a couple of observational facts that were not known when general relativity was introduced. The first is that the Universe is expanding. On average, distant galaxies are receding from us with a speed proportional to their distance. This is Hubble's law. It is a consequence of a uniform expansion of the entire Universe (like a balloon being inflated) and is not due to the Earth being in a privileged location. One way to describe this expansion is to attribute a scale factor (or size) $R(t)$ to the Universe. Then the expansion rate is characterized by the Hubble parameter $H = \dot{R}/R$. At the present time its value is approximately 70 m/s kpc^{-1} with an observational uncertainty of about 10 percent.[6]

---

[6] A parsec is about 3 light years. H is approximately equal to the inverse of the age of the Universe, which is 13.7 ± 0.2 billion years. This is not exact because $R(t)$ is not linear.

Extrapolating backwards in time one infers that the entire Universe was in-finitely dense about 13.7 billion years ago. Thus the picture one has is that the Universe began with a bang (a *big bang* to be precise) and has been expanding ever since. The subsequent expansion of the Universe (at least after a very early spurt of growth) is described by a spacetime geometry, called a Friedman–Robertson–Walker (FRW) geometry, that incorporates the assumptions of homogeneity and isotropy. It depends on the Hubble parameter, and two additional parameters called Ω and Λ. The parameter Ω is a measure of the average energy density of the Universe and Λ is the so-called cosmological constant. Let us now discuss each of these.

A fundamental question concerning the expansion of the Universe is whether it will continue forever or else eventually slow down and turn around to give a con-traction that ultimately leads to a *big crunch*. The physical cause of a slowing down of the expansion is the gravitational attraction of all the matter and energy in the Universe. There is a critical density ρ_c, proportional to H^2/G, that characterizes the dividing line between eternal expansion and eventual contraction. By definition $\Omega = \rho/\rho_c$, where ρ is the actual density. There are theoretical reasons[7] to believe that Ω should be exactly equal to 1. The current observations are consistent with that and have an uncertainty of a few percent.

The cosmological constant Λ was introduced by Einstein in 1917, a year after he introduced general relativity, in an attempt to construct a static Universe. (The fact that it is expanding was not yet known.) This constant has the interpretation of the energy density of the vacuum. In principle Λ could be either positive, negative, or zero. This constant modifies the Einstein equation (10.6) to

$$G_{\mu\nu} = -8\pi G T_{\mu\nu} + \Lambda g_{\mu\nu}. \tag{10.8}$$

Recent observations strongly suggest that Λ is positive with a value such that vac-uum energy accounts for about 70 percent of the total energy density ρ.[8] This fraction is the current value, since it actually changes with time. In the past it was smaller and in the future it will approach 100 percent. It is a mystery why we happen to live in an epoch when the vacuum energy density is comparable to that due to matter. The cosmological constant acts like a repulsive force and this has a dramatic consequence. Namely, in the epoch when it is dominant (including the present time) the expansion of the Universe is actually accelerating! There is good evidence for this acceleration in observations of certain distant supernovas. This conclusion is also supported by other studies involving the cosmic microwave background radiation and the large-scale structure of galaxy distributions.

---

[7] We are referring to the theory of inflation, which describes a period of exponential growth of the very early Universe. It was alluded to in the previous paragraph.

[8] Alternative explanations that might account for the observations are also being explored.

10.2 The standard model of elementary particle physics

There is a specific quantum field theory that is spectacularly successful in describing a diverse range of phenomena over an enormously broad range of energies. This theory, developed with contributions from many physicists in the late 1960s and early 1970s, has come to be known as the *standard model of elementary particle physics* or simply as the standard model. Despite its successes, we will argue that it cannot be the last word in fundamental physics, but only a significant intermediate step.

The standard model is a renormalizable quantum field theory of the Yang–Mills type. This means that it is a four-dimensional theory containing quantum fields of spin 0, 1/2, and 1, and a certain limited set of interaction terms. Renormalizability ensures that unambiguous calculations of quantum corrections can be carried out despite the infinities that typically arise. Basically, the infinities can be absorbed into redefinitions of the fundamental parameters (charges and masses) and wave function normalizations.

The modern philosophy is that renormalizable quantum field theory is not fundamental. Rather, the underlying theory, whatever it may be, has a very-high-energy unification scale, perhaps in the range $10^{16}-10^{19}$ GeV. At energies far below this scale only the very light degrees of freedom and the renormalizable couplings are *relevant*. This means that any corrections are suppressed by a power of the typical energy of a reaction divided by the fundamental unification scale. These tiny corrections are only observable in very special circumstances and can be neglected for most purposes.

Theories of the Yang–Mills type are characterized by a Lie group. Any compact Lie group is mathematically possible. The correct choice is determined by experiment and turns out to be $SU(3) \times SU(2) \times U(1)$. The $SU(3)$ group, called the *color* group, describes the strong nuclear interactions. The $SU(2)$ and $U(1)$ groups together describe the electromagnetic and weak nuclear forces. These are referred to together as the electroweak force. The $SU(2)$ factor is called the *weak isospin* group and the $U(1)$ factor is called the *weak hypercharge* group.

Yang–Mills fields

There is a spin 1 Yang–Mills field associated to each of the generators of the gauge group. Thus there are eight gluon fields in the adjoint representation of color $SU(3)$, three W fields in the adjoint of weak isospin $SU(2)$, and one B field for the weak hypercharge $U(1)$. There is an adjustable coupling constant associated with each of these groups. The nonabelian gauge fields are charged, and so they couple to one another by standard rules of Yang–Mills theory. These fields couple to the other "matter" fields in the theory in ways that are entirely determined by group theory. All that one needs to know is the representations of the gauge groups that the matter fields belong to.

Quarks and leptons

The only fermions in the standard model are spin 1/2 quarks and leptons and their antiparticles, which are called antiquarks and antileptons. The left-handed and right-handed quarks and leptons belong to different representations of the electroweak group. (It is this fact that is ultimately responsible for parity violation.) We will list all the left-handed fermions, giving their representations. Their right-handed antiparticles belong to the complex conjugate representations.

The quarks

The left-handed quarks form three doublets (that is, two-dimensional representations) of weak isospin

$$\begin{pmatrix} u \\ d' \end{pmatrix}_L, \quad \begin{pmatrix} c \\ s' \end{pmatrix}_L, \quad \begin{pmatrix} t \\ b' \end{pmatrix}_L, \tag{10.9}$$

each of which is also a color triplet. The six "flavors" of quarks are called *up, down, charm, strange, top, bottom*. The quark fields d', s', and b' that appear here are related to the mass eigenstates d, s, and b by a unitary transformation (called the CKM matrix). The u, c, and t quarks each have electric charge $+2/3$ and the d, s, and b quarks each have electric charge $-1/3$.

The antiquarks

The left-handed antiquarks form six singlets of weak isospin

$$(\bar{u})_L \quad (\bar{c})_L \quad (\bar{t})_L$$
$$(\bar{d})_L \quad (\bar{s})_L \quad (\bar{b})_L \tag{10.10}$$

each of which is also a color triplet.

The leptons

The left-handed leptons form three doublets of weak isospin

$$\begin{pmatrix} \nu_e \\ e \end{pmatrix}_L, \quad \begin{pmatrix} \nu_\mu \\ \mu \end{pmatrix}_L, \quad \begin{pmatrix} \nu_\tau \\ \tau \end{pmatrix}_L, \tag{10.11}$$

each of which is a color singlet. The three types of neutrinos are all electrically neutral, whereas the electron, muon, and tau each have electric charge -1.

The antileptons

The left-handed antileptons form three singlets of weak isospin

$$(\bar{e})_L \quad (\bar{\mu})_L \quad (\bar{\tau})_L. \tag{10.12}$$

These are color singlets with electric charge $+1$.

The Higgs mechanism

If the theory only contained the fields listed above all the symmetries would be unbroken and all the associated particles would be massless. In fact, only nine of the twelve spin 1 gauge bosons are massless (the eight gluons and the photon) whereas the $W^{\pm}$ has a mass of about 80 GeV and the Z has a mass of about 90 GeV. Also the quarks and charged leptons have mass, whereas the neutrinos are massless (according to the standard model). All of these features are achieved in the standard model by the addition of certain spin 0 fields, called Higgs fields, with suitable self couplings and couplings to the quarks and leptons.

The basic idea of the Higgs mechanism is that the symmetry is spontaneously broken. What this means is that the symmetry is an exact feature of the underlying theory and its equations of motion. The symmetry is broken by virtue of the fact that the solution of the theory that corresponds to the quantum ground state (or vacuum) is asymmetric. In other words, the self-interactions of the Higgs fields are such that the solution of lowest energy does not share the symmetry of the underlying theory.

The symmetry-breaking only affects the electroweak group. The $SU(2) \times U(1)$ symmetry is broken to a $U(1)$ subgroup that is associated with electromagnetism. One subtlety is that the unbroken electromagnetic $U(1)$ symmetry is a linear combination of the original $U(1)$ and a $U(1)$ subgroup of the $SU(2)$. The specific linear combination that occurs is characterized by an electroweak mixing angle θ_W, which has been measured to high precision. The fact that many different methods of determining θ_W experimentally all give consistent results constitutes powerful confirmation of the standard model.

Why there must be more

Despite the impressive success of the standard model, there are compelling reasons for believing that it cannot be the whole story, only an excellent approximation to a better theory for a broad range of energies. The reason that is most certain is the fact that gravity is not part of the standard model. Incorporating gravity in a manner consistent with quantum theory is a highly nontrivial matter. The generally accepted solution is provided by string theory, as discussed later in this chapter.

If the lack of gravity were the only shortcoming of the standard model, its breakdown could be permanently beyond our ability to probe. The reason is that gravity only reaches a significant strength (between elementary particles) at extremely high energies approaching the Planck scale (about 10^{19} GeV). In fact, there are good reasons to believe that corrections to the standard model become significant at energies of about 1 TeV, where gravity can still be ignored and the formalism of quantum field theory should still be adequate.

A second experimental piece of evidence for physics beyond the standard model comes from studies of neutrinos. Studies of solar neutrinos have shown that electron neutrinos formed in fusion reactions in the core of the Sun can turn into muon and tau neutrinos by the time they reach us. Also, muon neutrinos formed in the upper atmosphere (as a result of collisions of cosmic rays) can turn into tau neutrinos in passing through the atmosphere and the Earth. These effects require nonzero neutrino masses, something that is not part of the standard model. Only differences of masses squared are determined in this way. However, if one assumes that there isn't near degeneracy of neutrino masses, then the experimental results imply that the heaviest neutrino should have a mass of about 10^{-2} eV. This is some 8 orders of magnitude less than the electron mass, the lightest of the other fermions.

Other arguments for physics beyond the standard model are of a more theoretical/esthetical character. One is the gauge hierarchy problem, which is discussed below in connection with supersymmetry. Another is the fact that the standard model has approximately 20 arbitrary parameters that are adjusted to give the correct quark and lepton masses, CKM matrix, symmetry-breaking Higgs potential, and so forth. One hopes that in a deeper theory some (or possibly all) of these might be computable from first principles. One would also like to understand the choice of the gauge groups and the representations used for the quarks, leptons, and Higgs fields in terms of deeper principles.

10.3 Supersymmetry

There are several lines of reasoning that suggest that supersymmetry is an essential feature of physics beyond the standard model. Two of the most compelling are that it solves the gauge hierarchy problem and that it greatly improves the high-energy unification of the forces at high energies. Each of these is discussed below.

There are also esthetic arguments. As discussed in Chapter 9, supersymmetry is the unique mathematical possibility for a nontrivial extension of the standard Poincaré group symmetries of spacetime. It would be very disappointing if nature failed to utilize such a wonderful opportunity. A somewhat more serious argument is the fact that supersymmetry is required in the construction of a consistent string theory (discussed below), which in turn seems to be required to reconcile gravity with quantum theory.

All these arguments are circumstantial, however. We won't know the status of supersymmetry for sure until there is direct experimental evidence for supersymmetry particles with the predicted properties. Such a discovery would be a landmark achievement, which would set the agenda of high-energy physics for the ensuing decades.

The gauge hierarchy problem

There are two related questions having to do with the large ratio between the electroweak scale (10^2 GeV) and the unification scale ($10^{16} - 10^{18}$ GeV) at which new physics (perhaps string theory) cuts off the ultraviolet divergences that appear in the standard model.

- Where does the incredibly small ratio between the electroweak and the unification scales come from? Pure numbers that result from computations that don't depend on very small or very large numbers are normally expected to be within a few orders of magnitude of unity.
- How do we ensure that once this ratio is established in the classical approximation, it is not destroyed by quantum corrections?

The solution of the first problem is achieved in scenarios where the natural quantity to compute is the logarithm of the ratio. This requires additional ingredients beyond supersymmetry. The second problem, called the *gauge hierarchy problem* (GHP), is solved by supersymmetry.

Technically, the main point of the GHP is that the masses of scalar fields – namely, the Higgs fields – are not protected by any symmetry of the standard model from acquiring masses of the order of magnitude of the high energy unification scale as a result of radiative corrections, thereby destroying any large mass ratio that is input at the classical level. This is to be contrasted with the situation for the gauge fields and the fermions (quarks and leptons). They would be massless, as a result of gauge symmetry, if the gauge symmetry were unbroken. The masses that they do acquire are controlled by the values of the Higgs field, which are naturally of the same order of magnitude as the Higgs mass. Thus, protecting the hierarchy against radiative corrections boils down to protecting the Higgs mass. The basic idea in using supersymmetry to solve the problem is to tie the scale of electroweak symmetry-breaking to the scale of supersymmetry-breaking. The way supersymmetry solves the GHP is by building-in cancellations between bosonic and fermionic contributions to radiative corrections to the Higgs mass so that they are only logarithmically divergent rather than quadratically divergent.

The minimal supersymmetric standard model

The MSSM is the extension of the standard model (SM). Even though the idea that nature has chosen to utilize supersymmetry may be fairly compelling, there is certainly no compelling argument that the supersymmetric extension of the SM should be the minimal one possible. Still, in the absence of a specific reason to consider something more complicated, the minimal extension seems like a reasonable place to start. One important class of extensions of the MSSM that we will discuss below are supersymmetric grand unified theories, often referred to by the rather unappetizing name of SUSY GUTS.

The MSSM is an $\mathcal{N} = 1$ theory with $SU(3) \times SU(2) \times U(1)$ gauge symmetry. It contains the following supermultiplets:

- Gauge fields belonging to vector supermultiplets. Each gauge field has a spin $1/2$ *gaugino* superpartner.
- Quarks and leptons belonging to chiral supermultiplets. Each quark or lepton has two spin 0 partners called *squarks* and *sleptons*.
- Higgs fields that belong to chiral supermultiplets and have spin $1/2$ partners called *higgsinos*.

No pair of known particles has the correct properties to make them candidates to be supersymmetry partners of one another. Rather, a new superpartner must be postulated for each known particle. The fact that they have not yet been discovered is presumably due to the fact that the superpartners are too heavy to have been produced in sufficient numbers to be detected in the experiments that have been carried out so far. These negative results provide lower limits for the various masses.

A very important fact about supersymmetry extensions of the SM is that (at least) two Higgs doublets, H_1 and H_2, are required. This is twice as many as in the SM, where there is just one Higgs doublet H, which can give mass to all charged quarks and leptons.[9] The Higgs doublets contain pairs of scalar fields with the indicated electric charges

$$H_1 = \begin{pmatrix} H_1^+ \\ H_1^0 \end{pmatrix} \qquad H_2 = \begin{pmatrix} H_2^0 \\ H_2^- \end{pmatrix}. \tag{10.13}$$

---

[9] In the standard model $H_\alpha = \epsilon_{\alpha\beta} H^{\beta*}$. In the SUSY extension such a relation is not possible, because the complex conjugate of a chiral superfield is an antichiral superfield.

Including the antichiral conjugates, altogether there are four charged Higgs fields $(H_1^{\pm}, H_2^{\pm})$ and four neutral ones $(H_1^0, \overline{H}_1^0, H_2^0, \overline{H}_2^0)$. When the Higgs mechanism is implemented, three of these provide the longitudinal polarizations of massive $W^{\pm}$ and Z^0 bosons, leaving five physical scalars denoted $H^{\pm}$, h, H, A. The last three are neutral.

The Higgs content of the MSSM is to be contrasted with that of the SM, where the physical spectrum contains only one neutral scalar h. However, finding this Higgs spectrum experimentally would not establish supersymmetry by itself, because it is possible to extend the SM to contain two Higgs doublets, like those of the MSSM, without making the theory supersymmetric. One really needs to find superpartners.

Charginos

There are two charged spin $1/2$ (Dirac) particles that are supersymmetry partners of the W bosons and the charged Higgs bosons. These fields can undergo quantum mechanical mixing. The resulting particles are called *charginos* and denoted $\tilde{\chi}_i^{\pm}$, $i = 1, 2$.

Neutralinos

The four neutral spin $1/2$ gauginos and higgsinos are the supersymmetry partners $\tilde{W}^3$, $\tilde{B}$, $\tilde{H}_1^0$, $\tilde{H}_2^0$. They can mix to give four *neutralinos* $\tilde{\chi}_i^0$, $i = 1, 2, 3, 4$. These particles are quite difficult to observe experimentally. If one wants to extract lower bounds on their allowed masses from the data it is necessary to make further assumptions. Using plausible model-dependent assumptions, the 2000 particle data tables list lower bounds of $32.5\,\text{GeV}, 55.9\,\text{GeV}$, and $106.6\,\text{GeV}$ for the first three neutralino masses.

R parity and the lightest supersymmetric particle

Two important quantum numbers in the SM are baryon number B and lepton number L; B is $1/3$ for the quarks, $-1/3$ for the antiquarks, and zero for all other fields.[10] Similarly, L is 1 for the leptons, -1 for the antileptons, and zero for all other fields. Out of these and the spin s one can construct a quantity called R parity as follows

$$P_R = (-1)^{3B-L+2s}. \tag{10.14}$$

It is quite plausible that this is an exact discrete multiplicative symmetry, even if B and L are not exact.[11]

---

[10] With this definition, the proton and neutron have $B = 1$.

[11] A multiplicative symmetry has the property that the product of all the values for the particles in the initial state of a reaction is the same as the corresponding product for the particles in the final state.

The important fact about R parity is that all the particles in the SM have $P_R = 1$. Moreover, in a supersymmetric extension, all their supersymmetric partners (sparticles) have $P_R = -1$. This is obvious because B and L are the same, but $2s$ is shifted up or down by 1. If R parity is an exact symmetry, as is generally assumed, then in collider experiments, where the initial state is R-even (that is, $P_R = 1$), R-odd sparticles must be pair-produced. Furthermore, the lightest supersymmetry particle (LSP) would be absolutely stable. In all likelihood, every other sparticle would eventually decay into a state that contains an odd number of LSPs plus ordinary R-even particles.

In a large class of viable models, the LSP is the lightest neutralino. Its mass should be comparable to the electroweak scale – about 100 GeV give or take a factor of 2 – and its interactions should have weak interaction strength. An LSP neutralino would be important for cosmology.[12] As the Universe expands, they drop out of thermal equilibrium and decouple when the annihilation rate becomes comparable to the expansion rate. This allows one to compute the residual density as a function of the neutralino mass. One finds that the contribution to the total mass of the Universe is proportional to the mass of the neutralino, approaching unity for $M \sim 300$ GeV. The precise value is model-dependent. Among other things it depends on exactly what mixture of gauginos and higgsinos it is made of. The desired fraction attributed to dark matter is about 0.3, so the expected mass is around 100 GeV.

It is remarkable that this estimate of the LSP mass agrees with the mass estimates based on completely unrelated lines of reasoning: one is based on using supersymmetry to solve the gauge hierarchy problem and the other is based on unification of the gauge couplings at a high scale (discussed below). Encouraged by these facts, the search continues, both in accelerator experiments and in non-accelerator dark matter searches.

Grand unification

The idea of "grand unification" is that at sufficiently high energies the three gauge forces should form a unified structure, and at the same time the distinction between quarks and leptons should disappear. These distinctions would then only be properties of low-energy physics below the scale at which the higher symmetry is broken. Even though it is not possible to carry out experiments at such high energies, there is a surprising amount of evidence in support of such a point of view. We will summarize the evidence and argue that it works better with supersymmetry than without it.

---

[12] The LSP belongs to the generic class of dark matter candidates called weakly interacting massive particles (WIMPs).

The SM gauge group can be embedded in various simple Lie groups as shown[13]

$$SU(3) \times SU(2) \times U(1) \subset SU(5) \subset SO(10) \subset E_6 \subset \ldots \tag{10.15}$$

This suggests various possibilities for a larger symmetry structure that could be somewhat more predictive.

One advantage of embedding the SM group in a simple or semi-simple group is that it accounts for charge quantization. When there is a $U(1)$ factor in the gauge group, as in the case of the SM, it is not clear whether the associated group manifold should be an infinite line or a circle. The distinction is important because in the noncompact case the allowed representations – and hence the allowed values of electric charge – are labeled by a continuous index, whereas in the compact case, the allowed charges are discrete (integer multiples of a basic charge). When one embeds the $U(1)$ in a compact group one knows that the $U(1)$ is compact and that charge is quantized.

In the $SU(5)$ case the 15 quarks and leptons in each of the three standard model families fit nicely into a pair of $SU(5)$ multiplets: $\bar{5} + 10$, without postulating any new fermionic fields. The Yang–Mills gauge fields should consist of the adjoint 24 of $SU(5)$. The decomposition of this representation into representations of the SM gauge group indicates that 12 new vector supermultiplets transforming as $(3, 2) + (\bar{3}, 2)$ are required. Turning to the Higgs doublets, the simplest possibility is to extend (H_1) to a 5 and (H_2) to a $\bar{5}$. Thus H_1 contains a new color triplet, since $5 = (3, 1) + (1, 2)$, and H_2 contains a new color antitriplet, since $\bar{5} = (\bar{3}, 1) + (1, 2)$. The 12 new spin 1 gauge field particles and the new triplet Higgs particles must have very heavy masses in order not to conflict with experimental bounds.

The fact that there is a single gauge coupling constant for a simple GUT group G (such as $SU(5)$ or $SO(10)$) means that the three coupling constants of the SM are related by the symmetry, when evaluated at the unification scale M_X. In particular, it turns out that the electroweak mixing angle at M_X is given by $\sin^2 \theta_W = 3/8$.

Renormalization theory implies that the three SM gauge group couplings g_i are functions of the energy scale μ at which they are evaluated and that their "running" is determined by the "beta functions" $\beta_i(g_1, g_2, g_3)$, which can be computed in perturbation theory. The relation is given by the renormalization group equations

$$\mu \frac{\partial}{\partial \mu} g_i(\mu) = \beta_i(g) \tag{10.16}$$

for $\mu < M_X$. Numerically, it turns out that the first-order (or one-loop) quantum effects give a very good approximation. The two-loop and top quark contributions

---

[13] There are also other possibilities for unification groups that are not simple. For example, there is an attractive scheme based on $SU(4) \times SU(2) \times SU(2)$.

Table 10.1. *Contributions to the one-loop couplings in Eq. (10.18) for an arbitrary number of families and Higgs doublets.*

	SM	MSSM
b_1	$4n_f/3 + n_h/10$	$2n_f + 3n_h/10$
b_2	$4n_f/3 + n_h/6 - 22/3$	$2n_f + n_h/2 - 6$
b_3	$4n_f/3 - 11$	$2n_f - 9$

Table 10.2. *Contributions to the one-loop couplings in Eq. (10.18). The SM case is evaluated for three families and one Higgs doublet, whereas the MSSM case is for three families and two Higgs doublets.*

	SM	MSSM
b_1	$41/10$	$33/5$
b_2	$-19/6$	1
b_3	-7	-3

gives only a few percent correction to the one-loop results and do not affect the qualitative picture that emerges. They are important, however, for making full use of precision experimental data. Defining $\alpha_i = g_i^2/4\pi$, unification means that

$$\alpha_i(M_X) = \alpha_u \qquad \text{for} \quad i = 1, 2, 3, \tag{10.17}$$

where α_u is the grand unification coupling constant.

At one loop, Eqs. (10.16) have a simple analytical solution. The three equations decouple, and one finds

$$\frac{1}{\alpha_i(\mu)} = \frac{1}{\alpha_u} + \frac{b_i}{2\pi} \log\left(\frac{M_X}{\mu}\right). \tag{10.18}$$

Each b_i is a pure number, determined by the particle quantum numbers, given in Table 10.1. The symbols n_f and n_h represent the number of families of quarks and leptons and the number of Higgs doublets. Putting in the standard values for the number of families and the number of Higgs doublets[14] gives the results shown in Table 10.2.

---

[14] This means $n_f = 3$ in both cases, but $n_h = 1$ for the SM and $n_h = 2$ for the MSSM.

In both cases – SM and MSSM – we can try to solve for α_u and M_X in terms of the measured $\alpha_i(m_Z)$.[15] This involves solving three equations for two unknowns. So there is a relation that the measured α_is should satisfy

$$\frac{\alpha_1^{-1} - \alpha_2^{-1}}{\alpha_2^{-1} - \alpha_3^{-1}} = \frac{b_1 - b_2}{b_2 - b_3}. \tag{10.19}$$

Note that each α is a function of energy, but this combination is predicted to be constant. Substituting the experimental numbers

$$\frac{\alpha_1^{-1} - \alpha_2^{-1}}{\alpha_2^{-1} - \alpha_3^{-1}} = 1.38 \pm 0.02. \tag{10.20}$$

Using the numbers in the Table 10.2 gives the results

$$\text{SM:} \quad \frac{b_1 - b_2}{b_2 - b_3} = 1.90$$

$$\text{MSSM:} \quad \frac{b_1 - b_2}{b_2 - b_3} = 1.40. \tag{10.21}$$

Comparing to the experimental value, it is clear that the MSSM prediction gives much better agreement than the SM one.

More refined analyses have been carried out, but the basic conclusion remains the same: unification works much better for the MSSM than for the SM. Accepting this, one finds for the MSSM that unification is achieved at a single point, M_X, provided that the supersymmetry breaking scale M_S is less than about 5 TeV. So the bottom line is that supersymmetric unification works very well for

$$M_X = (2 \pm 1) \times 10^{16}\,\text{GeV}$$

$$\alpha_u^{-1} = 25$$

$$M_S \leq 5\,\text{TeV}. \tag{10.22}$$

This is the most intriguing "experimental success" of supersymmetry to date. To be fair, it should be stressed that the case for supersymmetric unification is only circumstantial, and it is still conceivable that the SM plus some other new physics also works, or that there is no unification at a high scale.

One consequence of grand unification is that baryon number B is no longer exactly conserved. As a consequence, the proton is expected to be unstable. There are many possible decay modes. In specific GUT models one can compute the decay rate for each of the various possible decay modes. However, the calculations are

---

[15] The numbers are best measured for $\mu = m_Z$ (in studies of Z decay). The results are $\sin^2\theta_W = 0.232$, $\alpha_{em}^{-1}(m_Z) = 128$, $\alpha_1^{-1}(m_Z) = 59.0$, $\alpha_2^{-1}(m_Z) = 29.7$. The errors are dominated by $\alpha_3^{-1}(m_Z)$, which is about 8.5 ± 0.3.

subject to various uncertainties that only make them reliable within 1 or 2 orders of magnitude. With that uncertainty, nonsupersymmetric $SU(5)$ theory predicts a lifetime of about 10^{28} years, whereas supersymmetric $SU(5)$ theory predicts a lifetime of about 10^{33} years. Other unification groups, such as $SO(10)$, give approximately the same predictions. Experimental observations of a very large tank of ultrapure water instrumented with phototubes and located deep underground (to shield out cosmic rays) by the super-Kamiokande detector have determined a lower bound of approximately 10^{33} years for the lifetime. This result rules out the nonsupersymmetric models. It also indicates that further searching might discover the decays expected for supersymmetric models.

10.4 The relativistic string

In the late 1960s, prior to the discovery of quantum chromodynamics (QCD), a lot of effort was devoted to trying to construct a theory of the strong nuclear force in the framework of an approach called S matrix theory. A notable success was achieved by Veneziano and others who discovered formulas that gave a consistent set of scattering amplitudes for a theory with an infinite spectrum of particles including ones with arbitrarily high spin. Initially this theory was called *dual resonance theory*. In 1970 it was discovered independently by Nambu, Susskind, and Nielsen that these formulas describe the quantum theory of an extended object with one spatial dimension, which was called a string, and therefore the subject eventually was renamed *string theory*. The infinite spectrum was identified as the various different normal modes of vibration of the string.

Though it had many qualitatively correct features, string theory was not fully successful in describing the strong nuclear force. So when QCD came along in the early 1970s, most people stopped working on string theory. One of its problems was that the string spectrum necessarily includes a massless spin 2 particle and there is nothing like that in the spectrum of strongly interacting particles. In 1974 Scherk and Schwarz suggested identifying this particle as the graviton and reinterpreting string theory as a unified quantum theory of gravity and the other forces, rather than just a theory of the strong nuclear force. It took another decade for this proposal to receive widespread acceptance, but now it is the mainstream viewpoint.

In conventional quantum field theory the elementary particles are mathematical points, whereas in perturbative string theory the fundamental objects are one-dimensional loops (of zero thickness). Strings have a characteristic length scale, which can be estimated by dimensional analysis. Since string theory is a relativistic quantum theory that includes gravity it must involve the fundamental constants c (the speed of light), $\hbar$ (Planck's constant divided by 2π), and G (Newton's gravitational constant). From these one can form a length, known as the Planck length

$$\ell_p = \left(\frac{\hbar G}{c^3}\right)^{3/2} = 1.6 \times 10^{-33} \, \text{cm}. \tag{10.23}$$

Similarly, the Planck mass is

$$m_p = \left(\frac{\hbar c}{G}\right)^{1/2} = 1.2 \times 10^{19} \, \text{GeV}/c^2. \tag{10.24}$$

Experiments at energies far below the Planck energy cannot resolve distances as short as the Planck length. Thus, at such energies, strings can be accurately approximated by point particles. From the viewpoint of string theory, this explains why quantum field theory has been so successful.

As a string evolves in time it sweeps out a two-dimensional surface in spacetime, which is called the *world sheet* of the string. This is the string counterpart of the world line for a point particle.

In quantum field theory, analyzed in perturbation theory, contributions to amplitudes are associated to Feynman diagrams, which depict possible configurations of world lines. In particular, interactions correspond to junctions of world lines. Similarly, perturbative string theory involves string world sheets of various topologies. A particularly significant fact is that these world sheets are smooth. The existence of interactions is a consequence of world-sheet topology rather than a local singularity on the world sheet. This difference from point-particle theories has two important implications. First, in string theory the structure of interactions is uniquely determined by the free theory. There are no arbitrary interactions to be chosen. Second, the occurrence of ultraviolet divergences in point-particle quantum field theories can be traced to the fact that interactions are associated to world-line junctions at specific spacetime points. Because the string world sheet is smooth, without any singular behavior at short distances, string theory has no ultraviolet divergences.

World-line description of a point particle

Before describing the basic mathematics of string theory, it is useful to review the analogous description of point particles. A point particle sweeps out a trajectory (or world line) in spacetime. This can be described by functions $x^\mu(\tau)$ that describe how the world line, parametrized by τ, is embedded in the spacetime, whose coordinates are denoted x^μ. For simplicity, let us assume that the spacetime is flat Minkowski spacetime with the Lorentz-invariant line element given by

$$ds^2 = \eta_{\mu\nu} dx^\mu dx^\nu. \tag{10.25}$$

In units $\hbar = c = 1$, the action for a particle of mass m is given by

$$S = -m \int \sqrt{-ds^2}.$$

(10.26)

In terms of the embedding functions, $x^\mu(\tau)$, the action can be rewritten in the form

$$S = -m \int d\tau \sqrt{-\eta_{\mu\nu}\dot{x}^\mu\dot{x}^\nu},$$

(10.27)

where dots represent τ derivatives.

An important property of this action is invariance under local reparametrizations of the world line $\tau \to \tau(\tilde{\tau})$. The reparametrization invariance is a one-dimensional analog of the four-dimensional general coordinate invariance of general relativity. The reason for this invariance is the fact that S is proportional to the invariant length of the world line, which exists independently of any particular choice of parametrization.

The reparametrization invariance of S allows us to choose $x^0 = \tau$. For this choice (renaming the parameter t) the action becomes

$$S = -m \int \sqrt{1 - v^2}\, dt,$$

(10.28)

where

$$\vec{v} = \frac{d\vec{x}}{dt}.$$

(10.29)

Requiring this action to be stationary under an arbitrary variation of $\vec{x}(t)$ gives the Euler–Lagrange equations

$$\frac{d\vec{p}}{dt} = 0,$$

(10.30)

where

$$\vec{p} = \frac{\delta L}{\delta \vec{v}} = \frac{m\vec{v}}{\sqrt{1 - v^2}},$$

(10.31)

which is standard relativistic kinematics.

World-sheet description of a relativistic string

We can now generalize the analysis of the massive point particle to a string of tension T. Before doing that, let us introduce the general framework by briefly discussing an even more general problem, namely a p-brane of tension T_p. A p-brane is an object that has p spatial dimensions. (Thus a point particle has $p = 0$

and a string has $p = 1$.) The action in this case involves the reparametrization-invariant $(p + 1)$-dimensional volume and is given by

$$S_p = -T_p \int d\mu_{p+1},$$ (10.32)

where the invariant volume element is

$$d\mu_{p+1} = \sqrt{-\det(\eta_{\mu\nu}\partial_\alpha x^\mu \partial_\beta x^\nu)} d^{p+1}\sigma.$$ (10.33)

Here the embedding of the p-brane into d-dimensional spacetime is given by functions $x^\mu(\sigma^\alpha)$. The index $\alpha = 0, \ldots, p$ labels the $(p + 1)$ coordinates σ^α of the p-brane world volume and the index $\mu = 0, \ldots, d - 1$ labels the d coordinates x^μ of the d-dimensional spacetime. We have defined

$$\partial_\alpha x^\mu = \frac{\partial x^\mu}{\partial \sigma^\alpha}.$$ (10.34)

The determinant operation acts on the $(p + 1) \times (p + 1)$ matrix whose rows and columns are labeled by α and β. The tension T_p is interpreted as the mass per unit volume of the p-brane. For a 0-brane, it is just the mass. The action S_p is reparametrization-invariant. In other words, substituting $\sigma^\alpha = \sigma^\alpha(\tilde\sigma)$, it takes the same form when expressed in terms of the coordinates $\tilde\sigma^\alpha$. Again, the reason is that the invariant volume has a meaning that does not depend on any particular choice of parametrization.

Let us now specialize to the string, $p = 1$. Evaluating the determinant gives

$$S[x] = -T \int d\sigma d\tau \sqrt{(\dot x \cdot x')^2 - \dot x^2 x'^2},$$ (10.35)

where we have defined $\sigma^0 = \tau, \sigma^1 = \sigma$, and

$$\dot x^\mu = \frac{\partial x^\mu}{\partial \tau}, \qquad x'^\mu = \frac{\partial x^\mu}{\partial \sigma}.$$ (10.36)

This action, called the Nambu–Goto action, is equivalent to the action

$$S[x, h] = -\frac{T}{2} \int d^2\sigma \sqrt{-h} h^{\alpha\beta} \eta_{\mu\nu} \partial_\alpha x^\mu \partial_\beta x^\nu,$$ (10.37)

where $h_{\alpha\beta}(\sigma, \tau)$ is the world-sheet metric, $h = \det h_{\alpha\beta}$, and $h^{\alpha\beta}$ is the inverse of $h_{\alpha\beta}$. The Euler–Lagrange equations obtained by varying $h^{\alpha\beta}$ are

$$T_{\alpha\beta} = \partial_\alpha x \cdot \partial_\beta x - \frac{1}{2} h_{\alpha\beta} h^{\gamma\delta} \partial_\gamma x \cdot \partial_\delta x = 0.$$ (10.38)

The equation $T_{\alpha\beta} = 0$ can be used to eliminate the world-sheet metric from the action, and when this is done one recovers the Nambu–Goto action. (To show this take the determinant of both sides of the equation $\partial_\alpha x \cdot \partial_\beta x = \frac{1}{2} h_{\alpha\beta} h^{\gamma\delta} \partial_\gamma x \cdot \partial_\delta x$.)

In addition to reparametrization invariance, the action $S[x, h]$ has another local symmetry, called conformal invariance (or Weyl invariance). Specifically, it is invariant under the replacement

$$h_{\alpha\beta} \rightarrow \Lambda(\sigma, \tau) h_{\alpha\beta}$$
$$x^{\mu} \rightarrow x^{\mu}. \tag{10.39}$$

for an arbitrary function $\Lambda(\sigma, \tau)$. This local symmetry is special to the $p = 1$ case (strings).

The two reparametrization invariance symmetries of $S[x, h]$ allow us to represent the three functions $h_{\alpha\beta}$ (this is a symmetric 2×2 matrix) in terms of just one function. A convenient choice is the "conformally flat gauge"

$$h_{\alpha\beta} = \eta_{\alpha\beta} e^{\phi(\sigma,\tau)}. \tag{10.40}$$

Here $\eta_{\alpha\beta}$ denotes the two-dimensional Lorentzian metric of a flat world sheet. However, because of the factor e^{ϕ}, $h_{\alpha\beta}$ is only "conformally flat." Classically, substitution of this gauge choice into $S[x, h]$ yields the gauge-fixed action

$$S = \frac{T}{2} \int d^2\sigma \, \eta^{\alpha\beta} \partial_{\alpha} x \cdot \partial_{\beta} x. \tag{10.41}$$

Quantum mechanically, the story is more subtle. When this is done correctly, one finds that in general ϕ does not decouple from the answer. Only for the special case $d = 26$ does the quantum analysis reproduce the formula we have given based on classical reasoning. Otherwise, as was explained by Polyakov, there are correction terms whose presence can be traced to a quantum-mechanical breakdown of the conformal invariance.

Mathematically, (10.41) is the same as a theory of d free scalar fields in two dimensions. The equations of motion are simply free two-dimensional wave equations:

$$\ddot{x}^{\mu} - x''^{\mu} = 0. \tag{10.42}$$

This is not the whole story, however, because we must also take account of the constraints $T_{\alpha\beta} = 0$.

To go further, one needs to choose boundary conditions. There are three important types. For a closed string one should impose periodicity in the spatial parameter σ. Choosing its range to be π (as is conventional)

$$x^{\mu}(\sigma, \tau) = x^{\mu}(\sigma + \pi, \tau). \tag{10.43}$$

For an open string (which has two ends), each end can be required to satisfy either Neumann or Dirichlet boundary conditions (for each value of μ).

$$\text{Neumann:} \quad \frac{\partial x^\mu}{\partial \sigma} = 0 \quad \text{at } \sigma = 0 \text{ or } \pi \tag{10.44}$$

$$\text{Dirichlet:} \quad \frac{\partial x^\mu}{\partial \tau} = 0 \quad \text{at } \sigma = 0 \text{ or } \pi. \tag{10.45}$$

The Dirichlet condition can be integrated, and then it specifies a spacetime location on which the string ends. As explained by Polchinski and collaborators, the only way this makes sense is if the open string ends on a physical object – called a D-brane. (D stands for Dirichlet.) If all the open-string boundary conditions are Neumann, then the ends of the string can be anywhere in the spacetime. The modern interpretation is that this means that there are spacetime-filling D-branes present.

Quantization

Starting with the gauge-fixed action in Eq. (10.41), the canonical momentum of the string is

$$p^\mu(\sigma, \tau) = \frac{\delta S}{\delta \dot{x}^\mu} = T\dot{x}^\mu. \tag{10.46}$$

Canonical quantization (this is just free two-dimensional field theory for scalar fields) gives

$$[p^\mu(\sigma, \tau), x^\nu(\sigma', \tau)] = -i\hbar\eta^{\mu\nu}\delta(\sigma - \sigma'). \tag{10.47}$$

The functions x^μ and p^μ can be expanded in Fourier modes. Substituting the expansions into Eq. (10.47) one learns that (up to normalization factors) the coefficients satisfy the same algebra as quantum-mechanical harmonic oscillators. There is just one problem: because $\eta^{00} = -1$, the time components are proportional to oscillators with the wrong sign ($[a, a^\dagger] = -1$). This is potentially very bad, because such oscillators create states of negative norm, which could lead to an inconsistent quantum theory (with negative probabilities, etc.). Fortunately, in 26 dimensions the $T_{\alpha\beta} = 0$ constraints eliminate the negative-norm states from the physical spectrum.

At this point one can examine the string spectrum. For $d = 26$, all states of negative norm are eliminated, but there is another problem. The string ground state turns out to describe a particle with negative mass squared! Such a particle is called a *tachyon*. Naively, a tachyon would imply faster-than-light propagation, but this certainly is unphysical. The appearance of a tachyon means that 26-dimensional Minkowski spacetime is an unstable solution of the bosonic string theory. There may be another stable solution that it flows to, but it is not yet known what that might be. Another point of view is that one should not worry too much about the tachyon, because the bosonic string theory is only a mathematical warm-up exercise before turning to superstrings, which do not have tachyons in their spectra.

The first excited states of the open strings are massless vector particles. These correspond to Yang–Mills fields, and therefore one learns that Yang–Mills theory lives on D-branes. Given the physical importance of Yang–Mills theory in the SM, this fact makes the study of D-branes a topic of great interest. The first excited state of the closed string also has zero mass and includes a graviton. It is this remarkable fact that forms the basis of the claim that string theory requires gravity.

Perturbation theory

Until 1995 it was only understood how to formulate string theories in terms of perturbation expansions. Perturbation theory is useful in a quantum theory that has a small dimensionless coupling constant, such as quantum electrodynamics, since it allows one to compute physical quantities as power series expansions in the small parameter. In quantum electrodynamics (QED) the small parameter is the fine-structure constant $\alpha \sim 1/137$. Since this is quite small, perturbation theory works very well for QED.

For a physical quantity $T(\alpha)$, one computes (using Feynman diagrams)

$$T(\alpha) = T_0 + \alpha T_1 + \alpha^2 T_2 + \dots . \tag{10.48}$$

It is the case generically in quantum field theory that expansions of this type are divergent. More specifically, they are asymptotic expansions with zero radius convergence. Nonetheless, they can be numerically useful if the expansion parameter is small. The problem is that there are non-perturbative contributions that have the structure

$$T_{\rm NP} \sim e^{-({\rm const.}/\alpha)}, \tag{10.49}$$

which are completely missed by the perturbation expansion. In a theory such as QCD, there are certain types of problems for which perturbation theory is useful and other ones for which it is not. For problems of the latter type, such as computing the hadron spectrum, nonperturbative methods of computation, such as lattice gauge theory, are required.

In the case of string theory the dimensionless string coupling constant is determined dynamically by the value of a scalar field called the *dilaton*. There is no particular reason that this number should be small. So it is unlikely that a realistic vacuum could be analyzed accurately using perturbation theory. More importantly, these theories have many qualitative properties that are inherently nonperturbative. So one needs nonperturbative methods to understand them.

10.5 Superstrings

The bosonic string theory described in the previous section illustrates many of the essential features of relativistic string theories, but it has a few serious shortcomings. One is the absence of fermions in its spectrum and another is the presence of a tachyon ground state. The requirement of 26 dimensions is also somewhat disturbing.

A second string theory that does contain fermions was constructed in 1971. The free fermionic string was constructed first by Ramond and a couple of months later the associated bosons and the various interactions were constructed by Neveu and Schwarz. A year later it was realized that the cancellation of the anomaly that would break conformal invariance in this theory requires that the spacetime dimension should be 10. The way it was originally formulated, this theory seemed to require tachyons, but it was eventually realized by Gliozzi, Scherk, and Olive that they can (and should) be eliminated. When this is done correctly, the theories have 10-dimensional spacetime supersymmetry.

Supersymmetric strings

The addition of fermions led quite naturally to supersymmetry and hence superstrings. There are two alternative formalisms that are used to study superstrings. In the original one, called the Ramond–Neveu–Schwarz (RNS) formalism, the supersymmetry of the two-dimensional world-sheet theory plays a central role. The second approach, developed about a decade later, is called the Green–Schwarz (GS) formalism. It emphasizes supersymmetry in the 10-dimensional spacetime. Which one is more useful depends on the particular problem being studied.

In the normal mode analysis of closed strings, there are two distinct classes of excitations associated with "left-moving" and "right-moving" modes. Mathematically, these are described by functions of the world-sheet coordinate combinations $\tau + \sigma$ and $\tau - \sigma$, respectively. It turns out that the properties of the left-mover and right-movers can be chosen independently. As the names suggest, they describe excitations that travel around the string in one direction or the other.

When one uses the supersymmetric string formalism for both the left-movers and the right-movers the supersymmetries associated with the left-movers and the right-movers can have either opposite handedness or the same handedness. These two possibilities give the type IIA and type IIB superstring theories, respectively. A third possibility is type I superstring theory, which has a symmetry under interchange of left-movers and right-movers as a result of which the strings are unoriented.

A more surprising possibility is to use the formalism of the 26-dimensional bosonic string for the left-movers and the formalism of the 10-dimensional supersymmetric string for the right-movers. The string theories constructed in this way are called *heterotic*. The mismatch in spacetime dimensions may sound strange, but it is actually okay. The extra 16 left-moving dimensions must describe a torus with very special properties to give a consistent theory. There are precisely two distinct tori that have the required properties.

Five superstring theories

The first superstring revolution began in 1984 when Green and Schwarz discovered that quantum-mechanical consistency of a 10-dimensional theory with $\mathcal{N} = 1$ supersymmetry requires a local Yang–Mills gauge symmetry based on one of two possible Lie groups: $SO(32)$ or $E_8 \times E_8$. Only for these two choices do certain quantum-mechanical *anomalies* cancel. The fact that these groups were singled out caused a lot of excitement, because in ordinary quantum field theory there is no mathematical principle that makes one group better than any other. The fact that only these groups are possible suggested that string theory has a very constrained structure, and therefore it might be very predictive.

Five distinct superstring theories, each in 10 dimensions, were identified. Three of them, the type I theory and the two heterotic theories, have $\mathcal{N} = 1$ supersymmetry in the 10-dimensional sense. The minimal spinor in 10 dimensions has 16 real components, so these theories have 16 conserved supercharges. The type I superstring theory has the gauge group $SO(32)$, whereas the heterotic theories realize both $SO(32)$ (the HO theory) and $E_8 \times E_8$ (the HE theory). The other two theories, type IIA and type IIB, have $\mathcal{N} = 2$ supersymmetry (32 supercharges).

In each of these five superstring theories there are consistent perturbation expansions of physical quantities. In four of the five cases (heterotic and type II) the fundamental strings are oriented and unbreakable. As a result, these theories have particularly simple perturbation expansions. Specifically, there is a unique Feynman diagram at each order of the expansion. The Feynman diagrams depict string world sheets, and therefore they are two-dimensional surfaces. For these four theories the unique L-loop diagram is a genus-L Riemann surface, which can be visualized as a sphere with L handles. External (incoming or outgoing) particles are represented by N points (or "punctures") on the Riemann surface. A given diagram represents a well-defined integral of dimension $6L + 2N - 6$. This integral has no ultraviolet divergences.

Type I superstrings are unoriented and breakable. As a result, the perturbation expansion is more complicated for this theory, and there are various world-sheet Feynman diagrams at each order. The separate diagrams have divergences that cancel when they are combined correctly.

Compactification of extra dimensions

All five superstring theories require that spacetime should have 10 dimensions, which is six more than are observed. The reason this is not a fatal problem is that these theories contain general relativity, and therefore the geometry of spacetime is determined dynamically. In other words, the spacetime geometry must be part of a complete solution of the equations of the theory. This is a severe constraint, but it still leaves many possibilities. Among these possibilities there are ones in which the 10 dimensions consist of a product of four-dimensional Minkowski spacetime with a compact six-dimensional manifold K. If K has a typical size a, then by general principles of quantum mechanics, it would be unobservable for energies below $E_a = \hbar c/a$. The most natural guess is that this compactification scale should be comparable to the unification scale or the string scale.

The possibilities for K are quite limited, especially if one requires that there be some supersymmetry below the compactification scale. A class of manifolds K that has been studied a great deal are called Calabi–Yau spaces.[16] They have properties that ensure that $1/4$ of the original supersymmetry is unbroken at low energies. In particular, starting with the HE theory compactified on a suitably chosen Calabi–Yau space one can come quite close to making contact with a realistic SUSY GUT theory. In the late 1980s such scenarios received a great deal of attention. More recently, it has been recognized that there are a variety of other ways that superstrings could give rise to a realistic model. Some of them are based on type II superstrings.

10.6 Recent developments in superstring theory

The discovery that superstring theory can consistently unify gravity with the other forces in a quantum framework was an important development. However, the realization that there are five different superstring theories was somewhat puzzling. Certainly, there is only one physical Universe that we can ever hope to observe, so it would be most satisfying if there were only one possible theory. In the late 1980s

---

[16] A Calabi–Yau space is a special type of six-dimensional space, which can described using three complex coordinates. It has various mathematical properties which can be summarized (using mathematics jargon) as a Kähler manifold of $SU(3)$ holonomy.

it was realized that when extra dimensions are compact there is a property known as T duality that relates the type IIA and type IIB theories to one another, and also the HE and HO theories to one another. T duality can be understood within the framework of perturbation theory.

Further progress required understanding nonperturbative phenomena, something that was achieved in the 1990s. Nonperturbative S dualities and the opening up of an eleventh dimension led to new identifications. Once all of these correspondences are taken into account, one ends up with the best possible conclusion. There really is a unique underlying theory, which has no arbitrary adjustable dimensionless parameters. These dualities and related issues are described in this section.

T duality

String theory exhibits many strange and surprising properties. One that was discovered in the late 1980s is called T duality.[17] In many cases, T duality implies that two different geometries for the extra dimensions, K and $\tilde{K}$, are physically equivalent! In the simplest example, a circle of radius R is equivalent to a circle of radius ℓ^2/R, where ℓ is the fundamental string length scale.[18]

Let us sketch an argument that should make this duality plausible. When there is a circular extra dimension, the momentum along that direction is quantized: $p = n/R$, where n is an integer. Using the relativistic energy formula $E^2 = M^2 + \sum_i (p_i)^2$, one sees that the momentum along the circular dimension can be interpreted as contributing an amount $(n/R)^2$ to the mass squared as measured by an observer in the noncompact dimensions. This is true whether one is considering point particles, strings, or any other kinds of objects. Particle states with $n \neq 0$ are usually referred to as *Kaluza–Klein excitations*.

In the special case of closed strings, there is a second kind of excitation that can also contribute to the mass squared. Namely, the string can be wound around the circle, so that it is caught up on the topology of the space. The contribution to the mass squared is the square of the tension $(T = 1/(2\pi \ell^2))$ times the length of wrapped string, which is $2\pi Rm$, if it wraps m times. Multiplying, the contribution to the mass squared is $(Rm/\ell^2)^2$. These are referred to as *winding mode excitations*.

Now we can make the key observation: under T duality the role of Kaluza–Klein excitations and winding-mode excitations are interchanged. Note that the

---

[17] The letter T has no particular significance. It was the symbol used by some authors for one of the low energy fields.

[18] Another frequently used symbol is $\alpha' = \ell^2$, which is called the Regge slope parameter. It is related to the string tension by $T = (2\pi\alpha')^{-1}$.

two contributions to the mass squared are exchanged if one interchanges m and n and at the same time sends $R \to \ell^2/R$.

T duality typically relates two different theories. Two particularly important examples are

$$\text{IIA} \leftrightarrow \text{IIB} \quad \text{and} \quad \text{HE} \leftrightarrow \text{HO}.$$

Therefore type IIA and type IIB (also HE and HO) should be regarded as a single theory. More precisely, they represent opposite ends of a continuum of geometries as one varies the radius of a circular dimension. This radius is not a parameter of the underlying theory. Rather, it arises as the value of a scalar field, and therefore it is determined dynamically.

There are also fancier examples of T-duality equivalences. For example, there is an equivalence of type IIA superstring theory compactified on a Calabi–Yau space and type IIB compactified on the "mirror" Calabi–Yau space. This mirror pairing of topologically distinct Calabi–Yau spaces is a striking discovery made by physicists that has subsequently been explored by mathematicians.

T duality might play a role in cosmology, since it suggests a possible way for a big crunch to turn into a big bang. The heuristic idea is that a contracting space when it becomes smaller than the string scale can be reinterpreted as an expanding space that is larger than the string scale, without the need for any exotic forces to halt the contraction. Unfortunately, we do not yet have the tools to analyze such time-dependent scenarios reliably.

S duality

Another kind of duality – S duality – was discovered as part of the "second superstring revolution" in the mid 1990s in work of Sen, Hull, Townsend, Witten, and others. S duality relates the string coupling constant g to $1/g$ in the same way that T duality relates R to $1/R$. The two basic examples are

$$\text{I} \leftrightarrow \text{HO} \quad \text{and} \quad \text{IIB} \leftrightarrow \text{IIB}.$$

Thus we learn how these three theories behave when $g \gg 1$. For example, strongly coupled type I theory is equivalent to the weakly coupled $SO(32)$ heterotic theory.

The transformation g to $1/g$ (or, more precisely, the corresponding transformation of the dilaton field whose vacuum value is g) is a symmetry of the type IIB theory. In fact, this is an element of an infinite discrete symmetry group $SL(2, Z)$.[19]

---

[19] This is the group of 2×2 matrices whose matrix elements are integers and whose determinant is 1.

M theory

In S duality we are told how three of the five original superstring theories behave at strong coupling. This raises the question: What happens to the other two superstring theories – IIA and HE – when g is large? The answer, which came as quite a surprise, is that they grow an eleventh dimension of size $g\ell$. This new dimension is a circle in the IIA case and a line interval in the HE case.

When the eleventh dimension is large, one is outside the regime of perturbative string theory, and new techniques are required. This calls for a new type of quantum theory, for which Witten has proposed the name M theory.[20]

One can try to construct a realistic four-dimensional theory by starting in eleven dimensions and choosing a suitable seven-manifold for the extra dimensions. One way to get $\mathcal{N} = 1$ supersymmetry in four dimensions is to require that the seven-manifold have G_2 holonomy. The study of G_2 manifolds is more difficult and less well understood than that of Calabi–Yau manifolds. It is possible that some models constructed this way will turn out to be dual to ones constructed by Calabi–Yau compactification of the HE theory. That would be interesting, because the M theory picture should allow one to understand phenomena that are nonperturbative in the heterotic picture.

D-branes

Superstring theory gives various p-branes, in addition to the fundamental strings. However, all p-branes, other than the fundamental string, become infinitely heavy as $g \to 0$, and therefore they do not appear in perturbation theory. On the other hand, at strong coupling this distinction no longer applies, and they are just as important as the fundamental strings.

As was mentioned earlier, an important class of p-branes, called D-branes, has the defining property that fundamental strings can end on them. This implies that quantum field theories of the Yang–Mills type, like the SM, reside on D-branes. An interesting possibility is that we experience four dimensions because we are confined to live on three-dimensional D-branes (D3-branes), which are embedded in a spacetime with six additional spatial directions. Model-building along these lines is being explored.

Black hole entropy

The gravitational field of the D-branes causes warpage of the spacetime geometry and creates horizons, like those associated to black holes. In fact, studies of

---

[20] He suggests that M should represent "mysterious" or "magical." Others have suggested that M could stand for "membrane" or "mother."

D-branes have led to a much deeper understanding of the thermodynamic properties of black holes in terms of string theory microphysics, a fact that is one of the
most notable successes of string theory so far.

In special cases, starting with an example in five dimensions that was analyzed
by Strominger and Vafa, one can count the quantum microstates associated with
D-brane excitations and compare the result with the Bekenstein–Hawking entropy
formula. Although many examples have been studied and no discrepancies have
been found, this correspondence has not yet been derived in full generality. The
problem is that one needs to extrapolate from the weakly coupled D-brane picture to the strongly coupled black hole one, and mathematical control of this ex-
trapolation is only straightforward when there is a generous measure of unbroken
supersymmetry.

AdS/CFT duality

In a remarkable development in 1997, Maldacena conjectured that the quantum
field theory that lives on a collection of D3-branes (in the IIB theory) is actually
equivalent to type IIB string theory in the geometry that the gravitational field of
the D3-branes creates. He also proposed several other analogous M theory dualities. These dualities are sometimes referred to as AdS/CFT, because the D-brane
geometry is a product of an anti de Sitter space (AdS) and a sphere, and the field
theory has conformal invariance. A conformally invariant field theory is referred
to as a CFT. [21] This astonishing proposal has been extended and generalized in a
couple of thousand subsequent papers.

While we can't hope to convince you here that these dualities are sensible, we
can point out that the first check is that the symmetries match. An important ingredient in this matching is the fact that the symmetry group of Anti de Sitter space
in $n + 1$ dimensions is $SO(n, 2)$, the same as that of the conformal group in n
dimensions.

10.7 Problems and prospects

In this final section, we discuss some of the important issues that still need to be
resolved, if we are to achieve the lofty goals of fundamental physics. These goals
are to develop a complete theoretical description of fundamental microphysics and
to understand the origin, evolution, and fate of the Universe. As will be evident, the

---

[21] Anti de Sitter space is a maximally symmetric solution of Einstein's equation with a negative cosmological
constant. It has constant negative scalar curvature. The conformal group is an extension of the Poincaré group
that includes transformations that rescale distances.

issues discussed here represent formidable challenges, and the solution of any one of them would be an important achievement. We have come a long way already, however, so there is no reason to become pessimistic at this point.

Find a complete formulation of the theory

Although there are techniques for identifying large classes of superstring vacua, there is not yet a succinct and compelling formulation of the underlying theory of which they are vacua. Many things that we take for granted, such as the existence of a spacetime manifold, should probably be emergent properties of specific vacua rather than identifiable features of the underlying theory. If this is correct, then we need something that is quite unlike any previous theory.

Understand the cosmological constant

In a theory that contains gravity, such as string theory, the cosmological constant, which characterizes the energy density of the vacuum, is a computable quantity. As discussed in Section 10.1.5, this energy (sometimes called *dark energy*) has recently been measured to reasonable accuracy, and the result is surprising: it accounts for about 70 percent of the the total mass/energy in the Universe.

The observed value of the cosmological constant is important for cosmology, but it is extremely tiny when expressed in Planck units (about 10^{-120}). Therefore a Lorentz-invariant Minkowski spacetime, which has a vanishing vacuum energy, is surely an excellent approximation to the real world for particle physics purposes. So a reasonable first goal is give a theoretical understanding of why Λ should be zero. We can achieve an exact cancellation between the contributions of bosons and fermions when there is unbroken supersymmetry, but there does not seem to be a good reason for such a cancellation when supersymmetry is broken. Many imaginative proposals have been made to solve this problem, but none of them has gained a wide following. Once one understands how to derive zero when supersymmetry is broken, the next step will be to try to account for the tiny nonzero value that is actually observed.

Find all quantum vacua of the theory

Knowing the fundamental theory would be great, but it is not the whole story. One also needs to understand what are the possible solutions (or quantum vacua) of the theory and which one is correct. This is a tall order. Many classes of consistent supersymmetric vacua, often with a large number of parameters (called moduli), have been found. The analysis becomes more difficult as the amount of unbroken supersymmetry decreases. Vacua without supersymmetry are a real problem. In addition to the issue of the cosmological constant, one must also address the issue of quantum stability. Stable nonsupersymmetric classical solutions are often destabilized by quantum corrections.

Presumably, if one had a complete list of allowed quantum vacua, one of them would be an excellent approximation to the microscopic world of particle physics. Obviously, It would be great to know the right solution, but we would also like to understand why it is the right solution. Is it picked out by some special mathematical property, or is it just an accident of our particular corner of the Universe? The way this plays out will be important in determining the extent to which the observed world of particle physics can be deduced from first principles.

Understand black holes and spacetime singularities

Long ago, in 1976, Hawking suggested that when matter falls into black holes and eventually comes back out as thermal radiation (called Hawking radiation), quantum coherence is lost. In other words, an initially pure quantum state can evolve into a mixed state, in violation of the basic tenets of quantum mechanics. Most string theorists are convinced that this argument is not correct, but they are finding it difficult to explain exactly how string theory evades it.

Singularities in the geometry of spacetime are a common feature of non-trivial solutions to general relativity. In the case of black holes they are shielded behind a horizon. However, they can also occur unshielded by a horizon, in which case one speaks of a *naked singularity*. Not only are singularities places where general relativity breaks down, but even worse they undermine the Cauchy problem – the ability to deduce the future from initial data.

The situation in string theory is surely better. Strings respond to spacetime differently from point particles. Certain classes of spacelike singularities, which would not be sensible in general relativity, are known to be entirely harmless in string theory. However, there are other important types of singularities that are not spacelike, for which current string theory technology is unable to say what happens. Perhaps some of them are acceptable and others are forbidden, but it remains to be explained which is which and how this works.

Understand time-varying solutions

Within the past few years people have started to analyze carefully time-dependent solutions to string theory. This is important for cosmological applications. The first goal is to construct examples that can be analyzed in detail, and that do not lead to pathologies. If we had a complete list of consistent time-dependent solutions, then we would face the same sort of question that we posed earlier in the particle physics context. What is the principle by which a particular one is selected? How much of the observed large-scale structure of the Universe can be deduced from first principles? Was there a pre-big-bang era, and how did the Universe begin?

Develop mathematical tools and concepts

String theory is up against the frontiers of most major branches of mathematics. Given the experience to date, there is little doubt that future developments in string theory will utilize many mathematical tools and concepts that do not currently exist. The need for cutting-edge mathematics is promoting a very healthy relationship between large segments of the string theory and mathematics communities. Not only are fundamental forces being unified, but so are many disciplines.

Exercises

10.1 Equation (10.3) gives the transformation rule for the metric tensor under a change of coordinates $\tilde{x}^\mu(x)$. Consider the change of coordinates $\tilde{x}^\mu = x^\mu + \xi^\mu(x)$, where the functions $\xi^\mu(x)$ are infinitesimal. Derive the change in the metric tensor $\delta g_{\mu\nu} = \tilde{g}_{\mu\nu}(\tilde{x}) - g_{\mu\nu}(x)$ to first order in ξ.

10.2 Use the result of the previous exercise to derive the infinitesimal variation formula $\delta\sqrt{-g} = \partial_\mu(\xi^\mu\sqrt{-g})$.

10.3 Verify that the inverse of the Hubble parameter $(70\,\text{m/s kpc}^{-1})$ is comparable to the age of the Universe (13.7 billion years).

10.4 Consider a family of quarks and leptons. The weak hypercharge Y is defined as twice the average electric charge in a weak isospin multiplet. What are the values of Y for the left-handed quarks, antiquarks, leptons, and antileptons?

10.5 Use your results from the previous exercise to show that the sum of the Y^3 values for the 15 left-handed fermions in a family is zero. This property is important for the quantum-mechanical consistency of the theory. In particular, if either the quarks or the leptons were omitted, the theory would not be consistent.

10.6 Show that Eq. (10.18) solves Eq. (10.16) for $\beta_i(g) = (b_i/16\pi^2)g_i^3$.

10.7 Find the generalization of Eqs. (10.21) for arbitrary values of n_f and n_h. What conclusions can you draw?

10.8   How many liters of water are required for there to be one proton decay per year, if the proton lifetime is 10^{34} years?

10.9   Show that Eq. (10.35) follows from Eqs. (10.37) and (10.38).

10.10   What are the dimensions of the Lie groups $SO(32)$ and $E_8 \times E_8$? What are their ranks?

10.11   Explain why quantum mechanics requires the momentum along a circle of radius R to be of the form n/R, where n is an integer.

Appendix 1

Where do equations of motion come from?

Long after Newton's equations of motion were well known in the physics and mathematics communities, there were a series of formal developments that helped to give a deeper understanding of the significance of these equations. These developments also provided a mathematical framework for describing and understanding new theories as they arose. This framework is useful not only for systems of point particles, but also for classical field theories, such as the Maxwell theory, and even for quantum field theories and string theories.

In this appendix, we survey some of the basic concepts and tools that have been developed. When all is said and done, the main message is that the central object for describing a theory, be it a classical or a quantum theory, is the *action S*. In classical theories the equations of motion are obtained by making the action stationary. In quantum theories one computes a probability amplitude by integrating the phase $\exp(iS/\hbar)$ h over all possible trajectories.

A1.1 Classical mechanics

Each dynamical system can be characterized by a set of independent variables that specify its degrees of freedom. For example, in the case of a theory of a collection of point particles, a possible choice is the positions of the particles. Let us refer to these coordinates as $q^i(t)$. The superscript i is mean to label both the directions of space and the particles. So if there are N particles in three dimensions, it takes $3N$ values.

One may take a more abstract point of view, and simply say that the system is described by coordinates $q^i(t)$ without committing oneself to what they represent physically. Part of the freedom corresponds to the possibility of a change of coordinates to new ones that are functions of the old ones. There are also more subtle alternatives.

These variables are functions of time. Clearly this is awkward from the point of view of relativity, which has symmetry transformations that relate space and time. A formalism in which time is given such a preferred status will not make Lorentz invariance manifest, though it might have it in a hidden form. However, this is not really so bad. After all, in a relativistic theory one still wants to be able to compute the time evolution in a specific inertial frame. Moreover, if one wants, it is possible to replace the time coordinate by a more abstract parameter, so as to make the Lorentz invariance manifest. This is illustrated by the descriptions of a relativistic point particle and a relativistic string given in Chapter 10. Here we choose t to be the time coordinate in a specific inertial frame.

Given the coordinates $q^i(t)$, the dynamics is specified by a Lagrangian $L(q, \dot{q})$, which is a function of the coordinates and their first derivatives. (The indices will be dropped in most subsequent equations. They can be restored easily.) Suppose, that at some initial time t_1 the coordinates have specified values $q(t_1) = q_1$, and at a later final time t_2 they have specified values $q(t_2) = q_2$. Then one can ask to find the trajectory $q(t)$ that the system will follow from q_1 to q_2. The dynamical systems that we will be interested in typically have equations of motion given by second-order differential equations. Thus to specify a solution uniquely, ones requires two pieces of information. These are often chosen to be the initial coordinates and velocities. Here we have chosen to specify the initial and final coordinates instead.

The solution to the problem we have posed is the following: the trajectory that the system chooses is the one that makes the action

$$S[q(t)] = \int_{t_1}^{t_2} L(q(t), \dot{q}(t)) dt \tag{A1.1}$$

extremal. The extremum is usually a minimum, but in some situations it might be a maximum or a saddlepoint instead. Since S depends on the function $q(t)$ for all times t between t_1 and t_2, it is called a *functional*, which we indicate by the use of square brackets. Note that the label t is a dummy symbol.

Requiring that a functional be extremal is called a *variational principle*, and the mathematical tools for solving such problems belong to the subject called the *calculus of variations*. The idea is that given a trajectory $q(t)$ one can investigate neighboring trajectories $q(t) + \delta q(t)$ and compare the value of S for each of them. An extremal trajectory $q(t)$ is one for which S is extremal compared to the neighboring ones. This means that the first *variational derivative* vanishes. More simply put, the change δS vanishes to first order in $\delta q(t)$ for an extremal trajectory $q(t)$.

Let us now compute the variation δS using

$$\delta L(q(t), \dot{q}(t)) = \frac{\partial L}{\partial q(t)} \delta q(t) + \frac{\partial L}{\partial \dot{q}(t)} \delta \dot{q}(t). \tag{A1.2}$$

Substituting and integrating the second term by parts gives

$$\delta S[q(t)] = \int_{t_1}^{t_2} \left(\frac{\partial L}{\partial q(t)} - \frac{d}{dt} \frac{\partial L}{\partial \dot{q}(t)} \right) \delta q(t). \tag{A1.3}$$

Note that the boundary terms that arise in the integration by parts have been dropped. The reason for this is that they contain factors of $\delta q(t_1)$ and $\delta q(t_2)$. Both of these vanish, because the coordinates have specified values at these times. This vanishing of δS is required to hold for arbitrary (infinitesimal) functions $\delta q(t)$ that vanish at the endpoints. The only way to achieve this is to require that

$$\frac{\partial L}{\partial q(t)} - \frac{d}{dt} \frac{\partial L}{\partial \dot{q}(t)} = 0. \tag{A1.4}$$

This is the celebrated Euler–Lagrange equation. To complete the problem for a given L, one needs to solve this second-order differential equation with the specified boundary conditions.

There is a closely related Hamiltonian formalism, which is usually used in non-relativistic quantum mechanics. It is related to the Lagrangian formalism, which we have just described, by a *Legendre transformation*. This works as follows: introduce the Hamiltonian H by writing

$$H(q(t), p(t)) = p(t)\dot{q}(t) - L(q(t), \dot{q}(t)), \tag{A1.5}$$

where the canonically conjugate momenta $p_i(t)$ are defined by

$$p_i(t) = \frac{\partial L}{\partial \dot{q}^i(t)}. \tag{A1.6}$$

In principle, this can be solved algebraically for $\dot{q}$ to eliminate it from the right-hand side of Eq. (A1.5). Then H only depends on the qs and ps as indicated. Next we rewrite the action in terms of H as a functional of the qs and ps

$$S[q(t), p(t)] = \int_{t_1}^{t_2} (p(t)\dot{q}(t) - H(q(t), p(t))) \, dt. \tag{A1.7}$$

Requiring this to be stationary with respect to variations of $p(t)$ and $q(t)$ gives Hamilton's equations

$$\frac{\partial H}{\partial p(t)} = \dot{q}(t) \tag{A1.8}$$

$$\frac{\partial H}{\partial q(t)} = -\dot{p}(t). \tag{A1.9}$$

Note that an integration by parts was required to derive the second of these equations. Again, the boundary terms could be dropped because $q(t)$ is fixed at the endpoints. This pair of first-order equations is precisely equivalent to the second-order Euler–Lagrange equations.

Let us illustrate this formalism with the simple example of a nonrelativistic particle of mass m moving in a potential $V(q)$. In this case

$$L = \frac{1}{2}m\dot{q}^2 - V(q). \tag{A1.10}$$

The canonical momentum is $p(t) = m\dot{q}(t)$, and thus the Hamiltonian is

$$H = \frac{1}{2m}p^2 + V(q). \tag{A1.11}$$

The Euler–Lagrange equation or Hamilton's equations gives the usual Newtonian equations of motion, namely

$$m\ddot{q}(t) = -\frac{\partial V(q(t))}{\partial q(t)}. \tag{A1.12}$$

Another fundamental construct of classical mechanics is called the *Poisson bracket*. Suppose we are given a system in Hamiltonian formalism with coordinates q^i and canonically conjugate momenta p_i. Then given two arbitrary functions $A(q, p)$ and $B(q, p)$, one defines the Poisson bracket as follows:

$$\{A(q, p), B(q, p)\}_{\text{PB}} = \sum_i \left(\frac{\partial A}{\partial q^i} \frac{\partial B}{\partial p_i} - \frac{\partial A}{\partial p_i} \frac{\partial B}{\partial q^i} \right). \tag{A1.13}$$

If, as a special case, one considers $A = q^i$ and $B = p_j$, one finds that

$$\{q^i, p_j\}_{\text{PB}} = \delta^i_j \tag{A1.14}$$

and

$$\{q^i, q^j\}_{\text{PB}} = \{p_i, p_j\}_{\text{PB}} = 0. \tag{A1.15}$$

The space spanned by the coordinates q and the momenta p is called *phase space*. In modern treatments of classical mechanics one often focuses on the *symplectic geometry* of phase space. In doing so a central object is the differential form $\sum_i dq^i \wedge dp_i$.

A1.2 Classical field theory

The description of a classical field theory works in essentially the same way as point-particle systems. The crucial difference is that the coordinates $q^i(t)$ are replaced by fields $\phi^i(x)$. Here x represents a point in $(D + 1)$-dimensional space-time. Once again we will suppress the index i, which is easily restored. The Lagrangian is usually local in x. This means that it can be expressed as an integral over space of a density function $\mathcal{L}$, called the *Lagrangian density*, that is, $L = \int \mathcal{L} d^D x$. Moreover, $\mathcal{L}$ is allowed to be a function of ϕ and all of its first derivatives. We indicate this by writing $\mathcal{L}(\phi, \partial_\mu \phi)$. Here ∂_μ represents all $(D + 1)$ of the spacetime derivatives. Putting these facts together, we now have the general structure

$$S[\phi(x)] = \int dx \mathcal{L}(\phi, \partial_\mu \phi). \qquad (A1.16)$$

Here dx represents the Lorentz-invariant volume element, which in a specific inertial frame takes the form $d\tau \, d^D x$.

The Euler–Lagrange equations for this system can be obtained by exactly the same techniques as before, and one finds

$$\frac{\partial \mathcal{L}}{\partial \phi(x)} - \frac{\partial}{\partial x^\mu} \frac{\partial \mathcal{L}}{\partial [\partial_\mu \phi(x)]} = 0. \qquad (A1.17)$$

For example, if

$$\mathcal{L} = -\frac{1}{2} (\eta^{\mu\nu} \partial_\mu \phi \partial_\nu \phi + m^2 \phi^2), \qquad (A1.18)$$

one obtains the Klein–Gordon equation

$$\eta^{\mu\nu} \partial_\mu \partial_\nu \phi - m^2 \phi = 0. \qquad (A1.19)$$

Note that these equations have manifest Lorentz invariance.

Symmetries and conservation laws

Symmetries of physical theories give rise to conservation laws. For example, space translation symmetry implies conservation of momentum and time translation symmetry implies conservation of energy. Similarly, rotational symmetry implies conservation of angular momentum. These examples all involve spacetime symmetries, but the possibilities are not limited to such symmetries. The general rule for the connection between symmetries and conservation laws was worked out by Emmy Noether and is known as Noether's theorem. We will illustrate here how it works for a theory with a collection of scalar fields $\phi_i(x)$, though it can be generalized to include other types of fields.

Suppose that a specific set of infinitesimal variations of the fields

$$\delta \phi_i(x) = \epsilon \Delta_i(x) \qquad (A1.20)$$

is a symmetry of a theory, the action of which is of the form in Eq. (A1.16). These variations $\Delta_i(x)$ may be arbitrary finite functions of the fields themselves, linear or nonlinear, and the spacetime coordinates, whereas ϵ is an infinitesimal constant.

The fact that this describes a symmetry means that the resulting infinitesimal variation of the Lagrangian density

$$\delta\mathcal{L} = \frac{\partial\mathcal{L}}{\partial\phi_i(x)}\delta\phi_i(x) + \frac{\partial\mathcal{L}}{\partial[\partial_\mu\phi_i(x)]}\delta\partial_\mu\phi_i(x) \tag{A1.21}$$

must give a vanishing contribution to δS. To achieve this it is not necessary that $\delta\mathcal{L}$ vanish, though in many cases it actually does. Rather, it is sufficient that it be a total derivative of the form

$$\delta\mathcal{L} = \epsilon\partial_\mu X^\mu, \tag{A1.22}$$

for some suitable functions $X^\mu(x)$. We are assuming, of course, that the asymptotic properties of the theory are such that the integral of this expression vanishes.

The next step is to impose the equations of motion in (A1.17). This enables us to recast Eq. (A1.21) in the form

$$\delta\mathcal{L} = \partial_\mu\left(\frac{\partial\mathcal{L}}{\partial[\partial_\mu\phi_i(x)]}\delta\phi_i(x)\right). \tag{A1.23}$$

Comparing this equation with Eq. (A1.22), we learn that the Noether current

$$J^\mu(x) = \frac{\partial\mathcal{L}}{\partial[\partial_\mu\phi_i(x)]}\Delta_i(x) - X^\mu(x) \tag{A1.24}$$

satisfies the current conservation equation

$$\partial_\mu J^\mu = 0. \tag{A1.25}$$

Note that in a relativistic theory J^μ transforms as a spacetime vector under Lorentz transformations, so that the conservation equation is a Lorentz-invariant equation.

To define a charge at a particular time requires choosing a particular inertial frame. In this frame the time component of the current J^0 is interpreted as the charge density, and the total charge is given by

$$Q(\tau) = \int J^0(x)\, d^D x. \tag{A1.26}$$

Let us now consider the time derivative of this charge

$$\dot{Q}(\tau) = \int \frac{\partial J^0(x)}{\partial\tau}\, d^D x = -\int \frac{\partial J^i(x)}{\partial x^i}\, d^D x = 0, \tag{A1.27}$$

which is the statement of charge conservation. In the last step we have assumed that the spatial integral of spatial derivatives gives zero, which is usually valid. If it isn't, this would mean that charge was flowing out of the region under consideration.

A1.3 Quantization

The formal methods described in the preceding sections to describe classical systems are also very useful for quantum systems. However, as we will see, there are new issues that arise. In introductory treatments of nonrelativistic quantum mechanics, the Hamiltonian formalism is generally emphasized, and we will describe it briefly here. However, there is a beautiful alternative that uses the action, called the Feynman path integral, which we will also discuss briefly. Lest the reader get the mistaken impression that the application of the recipes we have presented is always straightforward, we will conclude by mentioning some of the complications that can arise.

Operator formalism

In the Hamiltonian formulation of quantum mechanics, observables are represented by (usually Hermitian) operators that act on Hilbert space. As discussed in Chapter 7, one way of doing this is by representing the Hilbert space as a function space and the operators as differential operators that can act on the functions.

If one is given a classical theory in Hamiltonian form, there is a simple prescription for the construction of the commutators of the various operators of the corresponding quantum theory. The rule is to replace Poisson brackets by commutators according to the rule

$$\{A(q, p), B(q, p)\}_{\text{PB}} \rightarrow \frac{1}{i\hbar}[A(q, p), B(q, p)]. \tag{A1.28}$$

In particular, this implies that

$$[q^i, p_j] = i\hbar\delta^i_j \tag{A1.29}$$

and

$$[q^i, q^j] = [p_i, p_j] = 0. \tag{A1.30}$$

There are a couple of clarifying points to be made. First, the overall normalization of the action S is inconsequential for classical physics, since rescaling by a constant factor does not affect the classical equations of motion. However, such a rescaling also rescales the conjugate momenta, and therefore is important for the quantum commutation relations in Eq. (A1.29). That equation only has a precise meaning for a specified normalization choice. The one that is made has the property that the canonical momenta are the ordinary momenta, when they can be defined. For this choice the action has dimensions ML^2T^{-1}, which are also the dimensions of Planck's constant. A second subtlety concerns the meaning of the operators in Eq. (A1.28). Classically, $A(p, q)$ is unambiguous, but quantum mechanically it depends on the ordering conventions that are used to express it in terms of the noncommuting coordinates and momenta. In specific problems there is generally a definite choice that has the desired physical properties.

If the Hilbert space is represented by functions of the coordinates q^i, then to satisfy Eq. (A1.29) the canonically conjugate momenta should be represented by $p_i = -i\hbar\partial/\partial q^i$. An equally valid alternative is to represent the Hilbert space by functions of the momenta. In this case the coordinates take the form $q^i = i\hbar\partial/\partial p_i$. These two alternatives are related by a Fourier transform.

Feynman path integral

An alternative way of deriving a quantum theory from a classical theory was proposed by Feynman. It is equivalent to the preceding in simple cases, but in more complicated ones it can have advantages. It also leads to a fascinating new understanding of the relation between classical physics and quantum physics. In quantum theory, the goal is to compute the probability amplitude A for a given physical process. The probability (or probability density, depending on the situation) is then given by $|A|^2$.

Let us return to the original problem we discussed, propagation of a particle (or system of particles) with specified positions at an initial time t_1 and a final time t_2. The classical problem was to find the trajectory that makes the action extremal. The quantum problem is to find the probability amplitude. Feynman's prescription is the following: for each path $q(t)$ from q_1 at time t_1 to q_2 at time t_2 compute the classical action $S[q(t)]$, and associate the phase factor $\exp(iS[q(t)]/\hbar)$. Then the desired amplitude $A(q_1, t_1; q_2, t_2)$ is given by summing these phases over all possible trajectories. We can express this formally as follows

$$A(q_1, t_1; q_2, t_2) = \int_{q(t_1)=q_1}^{q(t_2)=q_2} Dq(t)\exp\left(iS[q(t)]/\hbar\right). \tag{A1.31}$$

An integral in which one sums over all functions (satisfying specified rules) is called a *path integral* or a *functional integral*. It needs to be defined properly. The method suggested by Feynman was to approximate the trajectory by N points at equally spaced intermediate times and to connect the successive points by straight line segments. Next, one integrates over the coordinates of each of the N points. This gives an approximate answer. Finally, to get the exact answer, one lets $N \to \infty$, introducing a multiplicative normalization factor in the measure so as to achieve a finite limit.

The Feynman prescription means that all possible trajectories contribute to the physical amplitude, which is quite profound. It is relatively easy to acquire a qualitative understanding of the classical limit, $\hbar \to 0$, in this picture. As this limit is approached, the phase varies rapidly as the trajectory is varied and there is destructive interference. The only case in which the phases reinforce to give a significant contribution is that of trajectories in the immediate vicinity of a trajectory for which the action is stationary, that is, the classical trajectory. Thus, the picture one has is that, even in the classical case, all trajectories contribute to the amplitude, but the contributions of all those that are not in the immediate vicinity of the classical trajectory give canceling contributions.

The problem of constraints

When the coordinates and conjugate momenta are not independent, but satisfy relations, one has what is called a *constrained Hamiltonian system*. The prescriptions described above need to be modified before they can be applied to such systems.

Rather that discussing the matter in general terms, let us look at an important example. In classical Maxwell theory, the fields are the vector potential $A_\mu(x)$, which has a component for each direction in spacetime. Out of this one constructs the field strength tensor

$$F_{\mu\nu} = \partial_\mu A_\nu - \partial_\nu A_\mu. \tag{A1.32}$$

Separating space and time indices, the components F_{i0} are the electric fields and the components F_{ij} are the magnetic fields. The Lagrangian density $\mathcal{L}$ for Maxwell theory is proportional to the square $F^{\mu\nu}F_{\mu\nu}$, where indices are raised with the Lorentz metric. This expression has manifest Lorentz invariance, which proves the Lorentz invariance of Maxwell theory.

The subtlety arises when one considers quantization. The conjugate momenta are obtained by differentiating the Lagrangian density with respect to $\partial_0 A_\mu$. However, $\mathcal{L}$ does not depend on $\partial_0 A_0$, and therefore the corresponding conjugate momentum is zero. This is a simple example of a Hamiltonian constraint. Clearly, it would not make sense to impose canonical commutation relations for an operator that is zero, so a better prescription is called for. This issue has been studied exhaustively over the years and various alternative methods have been developed to deal with it. Rather than go down that road, which is too far afield for this book, we would like to point out the origin of the problem. It can be traced back to the fact that the action has a *local gauge invariance*. Specifically, $F_{\mu\nu}$, and hence S, is invariant under the gauge transformation $A_\mu(x) \to A_\mu(x) + \partial_\mu \Lambda(x)$, where Λ is an arbitrary function of spacetime.

Local symmetries of this type are quite widespread and lead to Hamiltonian constraints. They also occur in Yang–Mills theories, which has a somewhat more complicated gauge invariance. General relativity (discussed in Chapter 10) has general coordinate invariance, which is also a local symmetry. String theory actions (also discussed in Chapter 10) have local reparametrization invariance and local Weyl invariance. Supersymmetric string actions in the Green–Schwarz formalism have an additional local fermionic symmetry. This one has proved to be particularly challenging to deal with, especially if one wishes to maintain manifest Lorentz invariance.

Appendix 2

Basic group theory

Suppose that a subset H of the elements of a group G also satisfies the axioms for a group using the same multiplication rule as in the definition of G. In this case H is called a *subgroup* of G. Every group has two trivial subgroups. One is the entire group G itself, and the other is the one element set $\{e\}$ consisting of only the identity element. Any other subgroups are called *proper*.

Whenever one has a group G with a subgroup H, one can define certain sets called *cosets*. Specifically, consider the set of group elements formed by multiplying each element of H from the left by a specific group element $a \in G$. This set has the same number of elements as H, but it is not identical to H unless a happens to belong to H. It is conventional to denote this coset by aH. More specifically, since we have used left multiplication, this is a *left coset*. Right cosets Ha are defined in an analogous manner. Note that a coset contains the identity element e only if the subgroup H contains the element a^{-1}. Also, since H satisfies the axioms of a group, this is equivalent to the statement that H contains a. Thus we learn that the coset is either equal to H or it does not contain the identity element. Thus all cosets, other than H itself, are not subgroups of G. It is easy to show that any pair of left cosets are either identical or disjoint (no common elements). There is a similar statement for right cosets, of course.

A special type of subgroup is called an *invariant subgroup*. (Another name that is sometimes used is *normal subgroup*.) The concept is a bit subtle, but it's worth making some effort to understand, since it plays a basic role in group theory. Let H be a subgroup of a group G. Then, by definition, H is an invariant subgroup if and only if its left cosets and its right cosets are identical ($aH = Ha$) for every element $a \in G$. The reason this is subtle is that the equality is an equality of sets, and it is not required to be true element by element. In other words, if $h \in H$, we do not require that $ah = ha$. Rather, all that is required is that H contain some element h' such that $ah = h'a$.

There is another way of dividing the elements of a group into a collection of subsets, called *conjugacy classes*, that is quite interesting. For any group G, every element of the group is in one and only one of the conjugacy classes. So the only issue that needs to be addressed is a criterion for deciding whether or not two group elements, a and b, are in the same conjugacy class. By definition, the rule is that they are in the same class if and only if there exists some element g of the group such that $b = g^{-1}ag$ (or, equivalently, $ag = gb$). This rule is an example of an equivalence relation. A couple of immediate facts are the following: first, the identity element is always in a class by itself. This is obvious because $g^{-1}eg = e$ for every group element $g \in G$. Similarly, every element of an Abelian group is in a class by itself. Clearly, if $ag = gb$ and $a \neq b$, then the group is nonabelian.

The reader may have noticed that the definitions of conjugacy classes and of invariant subgroups involve similar considerations. The precise relation is the following: a subgroup is an invariant subgroup if and only if it is a union of conjugacy classes. The point is that it must not contain some elements of conjugacy class without containing all of them. That is possible for a subgroup, but when it happens the subgroup is not an invariant subgroup.

Let D be a representation of a group G. Then we can define a subset H of G to consist of all elements $a \in G$ such that $D(a) = I$. This subset is called the *kernel* of the representation. If D is faithful, H is the trivial subgroup consisting of the identity element only. Otherwise, it contains additional elements. A basic theorem states that H is an invariant subgroup of G. The proof involves showing that for all $a \in G$ and for all $h \in H$, $D(aha^{-1}) = I$. From this it follows that $h' = aha^{-1} \in H$, and hence the left and right cosets coincide.

To every representation D of a group G, one can associate a quantity called the *character* χ_D of the representation. It is defined for element $a \in G$ by taking a trace of the representation matrix as follows:

$$\chi_D(a) = \operatorname{tr} D(a) = \sum_{i=1}^{n} D_{ii}(a). \tag{A2.1}$$

An important fact about traces of matrices is the cyclicity property $\operatorname{tr}(AB) = \operatorname{tr}(BA)$. This enables us to show that the characters for two group elements a and b are the same if they belong to the same conjugacy class. Recall that being in the same class means that there is a group element g such that $b = g^{-1}ag$. Using basic properties of representations, this implies that $D(b) = [D(g)]^{-1}D(a)D(g)$. Now taking the trace of both sides of this equation and using the cyclic property of the trace, one deduces that $\chi_D(b) = \chi_D(a)$. Because of this property, the character of a representation is effectively a function that is defined on the conjugacy classes.

Since the character takes the same value for each element of a conjugacy class, we can associate that value to the entire class.

There are a couple of basic facts about finite groups that are worth pointing out. The first is that the number of inequivalent irreducible representations is equal to the number of conjugacy classes. The second is that the order of the group (that is, the number of elements that it contains) is equal to the sum of the squares of the dimensions of the inequivalent irreducible representations. For example, the symmetric group S_3, which has six elements, has three conjugacy classes. The dimensions of its three inequivalent irreducible representations are 1, 1, and 2.

Appendix 3

Lie groups and Lie algebras

Lie groups are continuous groups in which the group elements are labeled by continuous coordinates. Hence the groups are infinite. For example, rotations in three dimensions can be labeled by three coordinates. In this case one says the group is three-dimensional. The entire group forms a smooth space called the group manifold.

One can introduce a measure on the group manifold, which is invariant under either left or right multiplication by a fixed group element, and use it to define integration over the group manifold in a way that is compatible with the group structure. Denoting the measure by dg, one can define the volume of the group to be $\int_G dg$. If the volume is finite, the group is said to be *compact* and if it is infinite the group is said to be *noncompact*.

A simple example of a compact Lie group is the group $U(1)$, which consists of numbers (1×1 matrices) of the form $e^{i\alpha}$, where α is real. The group manifold is a circle in this case, and the invariant measure is (up to normalization) given by $d\alpha$. Clearly, this is a compact group. Another Lie group, which could be denoted $GL(1, R)^+$, consists of numbers of the form e^x, where x is real. In this case the group manifold is the real line, the invariant measure is dx, and the group is noncompact. This group can be regarded as the simply connected covering group of $U(1)$ by making the identification of α with x modulo 2π. This can be visualized as an infinite helix associated to the circle.

The structure of a Lie group in the neighborhood of the identity element is characterized by the Lie algebra. The Lie algebra is a vector space whose dimension d is the same as the dimension of the group manifold. It can be regarded as the space of tangent vectors to the group manifold at the identity. Let X_i, $i = 1, \ldots, d$, be a basis of this vector space. Then the group property translates into the statement of closure of this vector space under the operation of commutation (Lie bracket).

Explicitly, this means that one has

$$[X_i, X_j] = f_{ij}{}^k X_k, \tag{A3.1}$$

where summation on k is understood. The constants $f_{ij}{}^k$ are called the *structure constants* of the Lie algebra. Their numerical values depend on the specific choice of basis. For one-dimensional Lie groups, such as those in the preceding paragraph, the Lie algebra is simply $[X, X] = 0$. The Lie algebra becomes nontrivial when the corresponding Lie group is nonabelian.

The structure constants satisfy the symmetry constraints

$$f_{ij}{}^k = -f_{ji}{}^k. \tag{A3.2}$$

They also satisfy a nonlinear constraint known as the the Jacobi identity. It is a consequence of the identity

$$[[X_i, X_j], X_k] + [[X_j, X_k], X_i] + [[X_k, X_i], X_j] = 0, \tag{A3.3}$$

which is verified easily by writing out the commutators and observing that all the terms cancel. Substituting Eq. (A3.1) into this identity gives

$$f_{ij}{}^m [X_m, X_k] + \text{cyclic perms} = f_{ij}{}^m f_{mk}{}^n X_n + \text{cyclic perms} = 0, \tag{A3.4}$$

where the repeated indices m and n are summed. By "+ cyclic perms" we mean adding two terms in which the indices i, j, k are cycled as in Eq. (A3.3). Since the X_n are linearly independent, this implies that

$$f_{ij}{}^m f_{km}{}^n + f_{jk}{}^m f_{im}{}^n + f_{ki}{}^m f_{jm}{}^n = 0. \tag{A3.5}$$

If one learns the structure constants of a Lie algebra by commuting explicitly defined matrices or operators, then it is guaranteed that the Jacobi identity will be satisfied. It is only important to check the Jacobi identity if one introduces a Lie algebra abstractly without an explicit representation.

The Lie algebra contains a subalgebra called the *Cartan subalgebra*. It is a (nonunique) maximal subalgebra with the property that all of its elements commute with one another. The number of linearly independent elements (that is, the dimension of the corresponding vector space) is called the *rank* of the Lie algebra. The simplest examples of nonabelian Lie groups are $SU(2)$ and $SO(3)$, which have the same Lie algebra. This Lie algebra has rank $= 1$ and dimension $= 3$.

Another useful definition is the following: a Lie group is said to be *simple* if and only if it has no nontrivial invariant subgroups. (The subgroups consisting of only the identity element or the entire group are trivial invariant subgroups.) As explained in the previous appendix, a subgroup H of a group G is called invariant if and only if $gHg^{-1} = H$ for all g in G. The Abelian one-dimensional groups $U(1)$ and $GL(1, R)^+$ described above are not simple. A group that is a tensor product of simple Lie groups is called *semisimple*.

Cartan gave a complete classification of simple Lie algebras. (As in the case of $SU(2)$ and $SO(3)$, several different Lie groups can correspond to a single Lie algebra.) Cartan's classification consists of four infinite families of "classical" Lie algebras and five exceptional Lie algebras. In Cartan's notation the classical Lie algebras are A_n with $n \geq 1$, B_n with $n \geq 2$, C_n with $n \geq 3$, and D_n with $n \geq 4$. The exceptional Lie algebras are denoted G_2, F_4, E_6, E_7, and E_8. These have dimensions 14, 52, 78, 133, and 248, respectively. In each case the numerical subscript is the rank of the Lie algebra. A_n corresponds to the matrix group $SU(n+1)$ and has dimension $n(n+2)$. B_n corresponds to the matrix group $SO(2n+1)$ and has dimension $n(2n+1)$. C_n corresponds to the symplectic group $Sp(n)$ (also denoted $USp(2n)$ by many physicists) and has dimension $n(2n+1)$. D_n corresponds to the matrix group $SO(2n)$ and has dimension $n(2n-1)$. The reason for restricting the ranges of the index n in the way indicated is to avoid counting the same Lie algebra more than once. This is required because of the relations $A_1 = B_1 = C_1$, $B_2 = C_2$, and $A_3 = D_3$. Also, $D_1 = U(1)$ and $D_2 = A_1 \times A_1$ are not simple.

Appendix 4
The structure of super Lie algebras

Supersymmetric theories have symmetries that are generated by fermionic charges, in addition to the bosonic ones. Accounting for them naturally leads to a generalization of a Lie algebra that includes fermionic charges with *anticommutation relations* in addition to the (usual) bosonic charges with commutation relations. The resulting algebra, which is called a *super Lie algebra*, has the general structure

$$[B_i, B_j] = f_{ij}{}^k B_k \tag{A4.1}$$

$$[B_i, F_\alpha] = g_{i\alpha}{}^\beta F_\beta \tag{A4.2}$$

$$\{F_\alpha, F_\beta\} = h_{\alpha\beta}{}^i B_i. \tag{A4.3}$$

Here, the B_i represent bosonic charges and the F_α represent fermionic charges. Note that the first line is the commutation relations of an ordinary Lie algebra, which is the bosonic subalgebra of the super Lie algebra. The structure constants satisfy the obvious symmetry constraints

$$f_{ij}{}^k = -f_{ji}{}^k, \qquad h_{\alpha\beta}{}^i = h_{\beta\alpha}{}^i. \tag{A4.4}$$

The CM theorem is a correct statement about the $[B, B]$ algebra (by itself). It excludes the existence of conserved charges transforming under Lorentz transformations as operators of positive integer spin. The key to supersymmetry, however, is that conserved charges that transform as spin $1/2$ operators are possible in an interacting local quantum field theory. It turns out that charges with spin $3/2, 5/2, \ldots$, can be ruled out by an extension of the CM theorem.

Recall how the Jacobi identity is derived for an ordinary Lie algebra:

$$[[B_i, B_j], B_k] + \text{cyclic perms} = 0$$
$$= f_{ij}{}^m [B_m, B_k] + \text{cyclic perms}$$
$$= f_{ij}{}^m f_{mk}{}^n B_n + \text{cyclic perms}$$
$$\Rightarrow f_{[ij}{}^m f_{k]m}{}^n = 0. \tag{A4.5}$$

The square brackets represent antisymmetrization of the enclosed indices. In the case of a super Lie algebra, there are three similar identities for the general structures B^2F, BF^2, and F^3, which give relations involving the various structure constants f, g, h. Their derivation is left to the reader.

References

Cornwell, J. F. (1984). *Group Theory in Physics*, I. London: Academic Press.

Das, A. (1993). *The Special Theory of Relativity*. New York: Springer-Verlag.

DeWitt, B. (1984). *Supermanifolds*. Cambridge: Cambridge University Press.

Einstein, A. (1995) *Relativity: The Special and the General Theory*, reprint. Crown Publishers.

Faber, R. L. (1983). *Differential Geometry and Relativity Theory*. New York: Marcel Dekker, Inc.

French, A. P. (1968) *Special Relativity*. W. W. Norton and Co.

Georgi, H. (1982). *Lie Algebras in Particle Physics*. Menlo Park: Benjamin/Cummings.

Goldstein, H. (1980). *Classical Mechanics*, 2nd edn. Reading, Mass.: Addison Wesley Publishing Co.

Green, M. B., Schwarz, J. H. & Witten, E. (1987). *Superstring Theory* in 2 vols. Cambridge: Cambridge University Press.

Halliday, D. & Resnick, R. (1978). *Physics*, 3rd edn. New York: John Wiley and Sons.

Hawking, S. W. & Ellis, G. F. R. (1973). *The Large Scale Structure of Space-Time*. Cambridge: Cambridge University Press.

Jackson, J. D. (1975). *Classical Electrodynamics*, 2nd edn. New York: John Wiley & Sons.

Jones, H. F. (1998). *Groups, Representations, and Physics*, 2nd edn. Bristol: Institute of Physics.

Landau, L. D. & Lifschitz, E. M. (1997). *The Classical Theory of Fields*, 4th edn. Butterworth-Heinemann.

Lightman, A. P., Press, W. H., Price, R. H. & Teukolsky, S. A. (1975). *Problem Book in Relativity and Gravitation*. Princeton: Princeton University Press.

Lovelock, D. & Rund, H. (1989). *Tensors, Differential Forms, and Variational Principles*. Dover Publications.

Mermin, N. D. (1968). *Space and Time in Special Relativity*. McGraw Hill.

Misner, C. W., Thorne, K. S. & Wheeler, J. A. (1973). *Gravitation*. W. H. Freeman & Co.

Nakahara, M. (1990). *Geometry, Topology, and Physics*. Bristol: Adam Hilger.

Perkowitz, S. (1996). *Empire of Light*. Washington: Joseph Henry Press.

Polchinski, J. (1998). *String Theory* in 2 vols. Cambridge: Cambridge University Press.

Rindler, W. (1991). *Introduction to Special Relativity*. Oxford: Oxford University Press.

(2001). *Relativity: Special, General, and Cosmological*. Oxford: Oxford University Press.

Sartori, L. (1996). *Understanding Relativity*. Berkeley: University of California Press.

Schutz, B. F. (1985). *A First Course in General Relativity*. Cambridge University Press.
  (1980). *Geometrical Methods of Mathematical Physics*. Cambridge: Cambridge
   University Press.
Taylor, E. F. & Wheeler, J. A. (1992). *Spacetime Physics: Introduction to Special
   Relativity*. W. H. Freeman & Co.
Tung, W. K. (1985). *Group Theory in Physics*. Philadelphia: World Scientific Publishing.
Wald, R. M. (1984). *General Relativity*. Chicago: University of Chicago Press.
Weinberg, S. (1972). *Gravitation and Cosmology: Principles and Applications of the
   General Theory of Relativity*. New York: John Wiley & Sons.
Wertheim, M. (1999). *The Pearly Gates of Cyberspace*. New York: W. W. Norton & Co.
Wess, J. & Bagger, J. (1992). *Supersymmetry and Supergravity*, 2nd edn. Princeton:
   Princeton University Press.
Zajonc, A. (1993). *Catching the Light*. New York: Oxford University Press.

Index

Printed in the United States
by Baker & Taylor Publisher Services

Printed in the United States
by Baker & Taylor Publisher Services